Handbook of Insecticides

Handbook of Insecticides

Jacob Richardson
Claire Bisho

Kruger Brentt
Publishers
2026

Kruger Brentt Publishers UK. LTD.

Company Number 9728962

Regd. Office: 68 St Margarets Road, Edgware, Middlesex HA8 9UU

ISBN: 978-1-78715-665-4

For information on all our publications visit our website at http://krugerbrentt.com/

Preface

Handbook of Insecticides is a comprehensive guide designed to provide students, researchers, and practitioners in agriculture, horticulture, and pest management with essential knowledge on the effective and safe use of insecticides. Insect pests pose a significant threat to crop productivity and quality, and the proper management of these pests is critical for ensuring sustainable agricultural practices and food security. This book offers both theoretical foundations and practical guidance for understanding, selecting, and applying insecticides responsibly.

The text begins by introducing the biology and ecology of key insect pests, their life cycles, and the damage they cause to various crops. It then explains the principles of insecticide action, classification, and modes of application, helping readers understand how different products work and how to optimize their use for maximum efficacy.

A major focus of the handbook is on practical considerations, including dosage, formulation, mixing, application techniques, storage, and safety precautions. The book also addresses resistance management, integrated pest management (IPM) strategies, and regulatory guidelines to ensure compliance with national and international standards.

Case studies and real-world examples illustrate successful insecticide practices in diverse cropping systems and environments. Special emphasis is placed on sustainable approaches, including reduced chemical dependency, environmental stewardship, and the integration of biological and eco-friendly pest control methods.

We hope that Handbook of Insecticides will serve as a valuable resource for agricultural students, extension workers, farmers, and industry professionals. By combining scientific knowledge with practical application, this book aims to support informed decision-making, effective pest management, and responsible agricultural practices.

Jacob Richardson

Claire Bisho

Contents

CHAPTER 1

Introduction to Insecticide

1. INTRODUCTION: AN OVERVIEW

Insecticides are organic or inorganic chemical substances, or mixtures thereof, designed to prevent, destroy, repel, or mitigate the effects of insects, including both crawling and flying species, whether occurring naturally or synthesized (e.g., pyrethroids, organic perfumed oils, and pyrethrins). There exist multiple types of pesticides. Most synthetic insecticides are inherently dangerous to health under the conditions of their use. Insecticides vary from very toxic to those unlikely to pose any acute risk. The majority are repellents and insect growth regulators utilized in agriculture, public health, horticulture, food preservation, or other chemical agents serving analogous functions.

Insecticides have significantly enhanced food output and controlled disease vectors. Nonetheless, these economically advantageous aspects may include drawbacks when evaluated against key environmental and human health factors. Numerous freshly manufactured pesticides have unknown health and environmental repercussions.

Insecticides have been utilized in several forms, including hydrocarbon oils (tar oils), arsenical compounds, organochlorines, organophosphorus compounds, carbonates, dinitrophenols, organic thiocyanates, sulfur, sodium fluoride, pyrethroids, rotenone, and nicotine, in both solid and liquid preparations. Notably, the majority of these have been retracted due to the harmful impact of the drugs. The examination of these formulations, together with their by-products and residues, has historically facilitated an impartial re-evaluation and reassessment of these chemicals based on a benefit-risk analysis, leading to their withdrawal from use when deemed hazardous to human health and the environment. The quality and sophistication of these assays have advanced, allowing for the detection of minute concentrations of these insecticides or their residues with a high degree of specificity, precision, and accuracy..

1.1 Inorganic and organ metal insecticides

The successive organometallic and organometalloid insecticides are delineated in relation to the equivalent inorganic chemicals. The extremely hazardous and persistent chemicals, such as trichloro-bis-chlorophenyl ethane (DDT) and bis-chlorophenyl acetic acid (DDA), are produced inadvertently. The organic combination typically alters the absorption and distribution of a hazardous metal, so modifying the emphasis of its effects, although the fundamental method of action remains unchanged.. The toxic effects of insecticides depend on the elements that characterize it as inorganic or organmetal insecticides and on the specific properties of one form of the element or one of its components or merely on an inordinately high dosage. Certain very poisonous elements, like iron, selenium, arsenic, and fluorine, are vital for normal development. The organometals and organometalloids are defined in relation to their respective inorganic substances. The organic combination typically alters the absorption and distribution of a hazardous metal, thereby modifying the emphasis of its effects, while the fundamental method of action remains same.

The differentiation between synthetic substances and those derived from natural sources is somewhat artificial. In practice, similar compounds are categorized based on whether the initially recognized and utilized component in the group was of synthetic or natural origin. For instance, pyrethrum and subsequently the naturally occurring pyrethrins were recognized for several years prior to the synthesis of the first synthetic parathyroid; hence, pyrethroids are regarded as derivatives of natural compounds, despite their absence in nature and the improbability of their occurrence.

1.2 Pyrethrum and related compounds

The insecticidal properties of pyrethrum flowers (chrysanthemum cinerarae-folum) have been recognized as insect powder since the middle of last century (McLaughlin 1973). In addition to their insect-killing activity, an attractive feature of the natural pyrethrins (pyrethrum) as insecticides was their lack of persistence in the environment and their rapid action whereby flying insects very quickly become incapacited and unable to fly. Prior the development of DDT, pyrethrum was a major insecticides for both domestic and agricultural use, despite its poor light stability. Development of synthetic pyrethroid with increased light stability and insecticidal activity allows it to be used as foliar insecticide while the natural pyrethrins are now used mainly as domestic insecticides.(Elliot, 1979).

1.3 Mode of Action

Pyrethrum and the synthetic pyrethroids are sodium channel toxins which, because of their remarkable potency and selection have found application in general pharmacology as well as toxicology (Lazdunski et al., 1985). Pyrethroids have a very high affinity for membrane sodium channels with dissociation constants of the order of 4x10-8 M (Sodeland,1985), and produce subtle changes in their function.

By contrast, inexcitable cells are little affected by pyrethroids. The pyrethroids are thus referred to as open channel blockers.

1.4 Metabolism

The relative resistance of mammals to the pyrethriods is almost wholly attributable to their ability to hydrolyze the pyrethroids rapidly to their inactive acid and alcohol components, since direct injection into the mammalian CNS leads to susceptibility similar to that seen in insects (Lawrence and Casida, 1982). Metabolic disposal of the pyrethroids is very rapid (Gray et al., 1980), which means that toxicity is high by intravenous route, moderate by slower oral absorption, and often immeasurably moderate by slower oral absorption.

1.5 Poisoning syndromes

The pyrethroids are essentially functional toxins, producing their harmful effects largely secondarily, as a consequence of neuronal hyperexitability (Parker et al.1985). Despite this dependence on a relatively well-defined mode of action, the pyrethroids are capable of generating a bewildering variety of effects in mammals and insects, which although showing some analogies with those produced by other sodium channel toxins (Gray, 1985; Lazdunski et al., 1985) and with DDT (Narahashi, 1986), have many unique characteristics (Ray, 1982b). Thus, toxicity of pyrethroids is divided into two groups Table 1. Type 1 pyrethroids produce the simplest poisoning syndrome and produce sodium tail currents with relatively short time constants (Wright et al., 1988). Poisoning closely resembles that produced by DDT involving a progressive development of fine whole-body tremor, exaggerated startle response, uncoordinated twitching of the dorsal muscles, hyperexcitability, and death (Ray, 1982b). The tremor is associated with a large increase in metabolic rate and leads to hyperthermia which, with metabolic exhaustion, is the usual cause of death. Respiration and blood pressure are well sustained, but plasma noradrenalin, lactate, and adrenaline are greatly increased (Cremer and Servile 1982). Type 1 effects are generated largely by action on the central nervous system, as shown by the good correlation between brain levels of cismethrin and tremor (White et al., 1976). In addition to these central effects, there is evidence for repetitive firing in sensory nerves (Staatz-Benson and Hosko, 1986).

Table 1. I Acute toxicity of pyrethroids (Wright et al., 1988; Forshow and Ray, 1990).

Type I	Intermediate	Type II
Allethrin	Cyhenothrin	Cyfluthrin
Barthrin	Fenproponate	Cyhalothrin
Bioalethrin	Flucythrinate	Cypermethrin

Type I	Intermediate	Type II
Cismethrin		Deltamethrin
Fenfluthrin		Fenvalerate
Trans-fluorocyphenothrin		Cis-fluorocyphenothrin
Kadethrin		
Permethrin		
Phenothrin		
Pyrethrin I		
Pyrethrin II		
Resmethrin		
Tetramethrin		

The type 11 pyrethroid produces a more complex poisoning syndrome and act on a wider range of tissues. They give sodium tail currents with relative longterm constants (Wright, et al., 1988). At lower doses more suble repetitive behavior is seen (Brodie and Aldridge, 1982). As with type I pyrethroids, the primary action is on the central nervous system, since symptoms correlate well with brain concentrations (Rickard and Brodie, 1985). As might be expected, both classes of parathyroid produce large increases in brain glucose utilization (Cremer et al. 1983). A final factor distinguishing type 11 pyrethroids is their ability to depress resting chloride conductance, thereby amplifying any sodium or calcium effects (Forshaw and Ray, 1990).

Intermediate signs representing a combination of type I and type 11 are produced by some pyrethroids. These appear to represent a true combination of the type I and 11 classes (Wright et al., 1983) and thus represent a transitional group. Evidence in support of this is given by measurement of the time constants of the sodium after potential produced by the pyrethroids. Since pyrethroids appear to be essentially functional toxins, they produce few if any specific neuropathological effects.

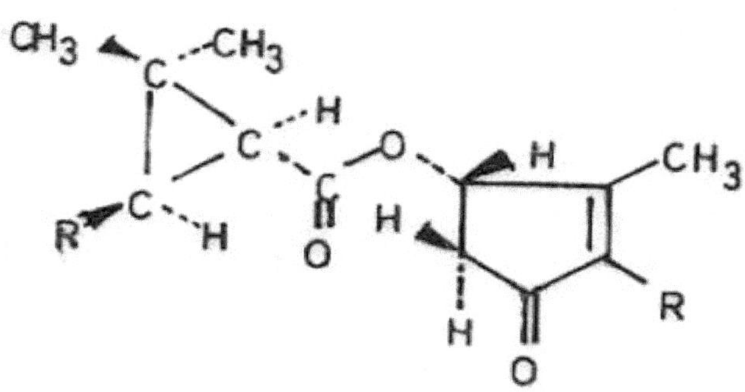

Fig. 1. Pyrethrins of the form.

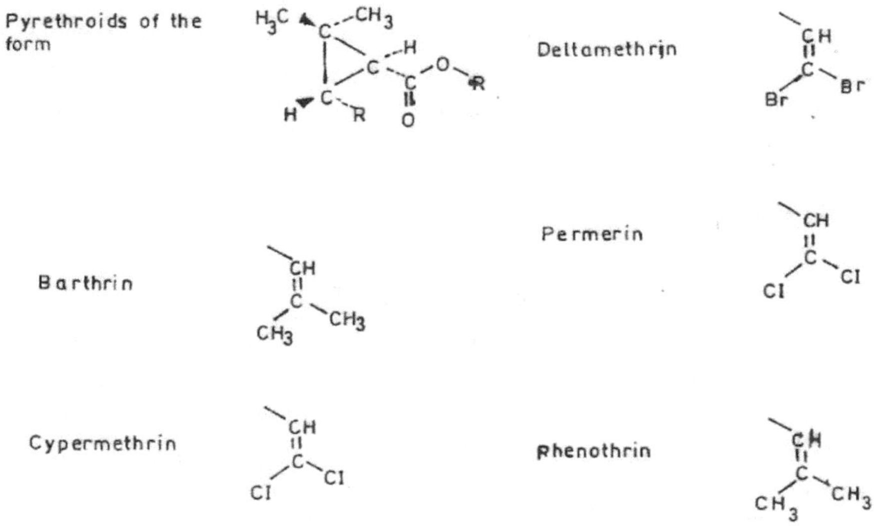

Fig. 2. Pyrethroids of the form.

1.6 Identity, properties and uses

The six known insecticidal active compounds in pyrethrum are esters of two acids and three alcohols. Insect powder made from "Dalmatian insect flower" (*Chrysanthemum cinerariaefolium*) is called pyrethrum powder or simply pyrethrum. The powder itself was formerly used as an insecticide, but now it is usually extracted. The six active ingredients are:

- Pyrethrin I – pyrethrolone ester of chrysanthemic acid
- Pyrethrin II- pyrethrolone ester of pyrethric acid
- Cinerin I- cinerolone ester of chrysanthemic acid
- Cinerin II- cinerolone ester of pyrethric acid
- Jasmolin I- jasmolone ester of chrysanthemic acid
- Jasmolin II- jasmolone ester of pyrethric acid

The six known insecticidally active compounds in pryrethrum are esters of two acids and three alcohols. Specifically, pyrethrins l is the pyrethrolone ester of chrysanthermic acid, - pyrethrin II is the pyrethrolone ester of pyrethria acid, cinerin l is the cinerolone ester of chrysanthemic acid, cinerin II is the cinerolone ester of pyrethric acid, jasmolin I is the jasmolone ester of chrysanthemic acid and jasmolin II is the jasmolone ester of pyrethric acid. There is much evidence indicating that the biological activity of these molecules depends on their configuration (Elliot, 1969, 1971).

The six active ingredients are known collectively as pyrethrins; those based on chrysanthemic acid are called pyrethrin I, and those based on pyrethric acid are called pyrethrins II. Pyrethrins, generally combined with a synergist, are used in sprays and aerosols against a wide range of flying and crawling insects. Usually about 0.5% active pyrethrum principles are formulated. They are equally effective for control of head lice and flea in dogs and cats.

1.7 Raid as insecticide

The insecticide 'Raid' belongs to a group of chemically stable pyrethrin, has widespread use in control of insects. Chemical stability, insecticide and organic phosphorus hydrocarbon have been shown to accumulate rapidly in tissues causing death and have profound effect on growth (Nebeker et al., 1994). Insecticide raid shows no observable effects on mortality and growth at lower test concentrations in rats. At higher concentration of 430 and 961 μg/g, survival decreased as concentration increases. In addition, mean total body weight of animals fed insecticides raid with concentrations of 430 and 961 μg/g were significantly decreased ($P<0.05$) than the controls. Conclusively, the higher the concentration of the insecticide Raid, the more hazardous it has on cell death (Achudume et al., 2008) (Table II). Bioaccumulation factor of insecticide Raid was observed in lipids, up to three times that of the feed at the first concentration and gradually decreases as the concentrations increase (Table III), whereas accumulation factor in the muscle (0.7), brain (0.5), and liver (0.3) was about the indicated number times that of the feed. At higher concentration of 961 μg/g, bioaccumulation factor decreased in the lipid to 1.2 and 0.6 in the muscle, 0.03 in the brain, and 0.08 in the liver. Using the mean of insecticide in feed, the tissues accumulate the insecticide in the following ascending order: brain < liver < muscle < lipid. Similarly, Table III indicates the estimated detectable levels of toxicity in rat tissues exposed to the insecticide Raid. The brain shows mild decrease in toxicity of the enzymes glucose-6-phosphatase and lactic acid dehydrogenase, whereas significant decreases were noticeable in the muscle and liver (Achudume et al. 2008). Long-term exposure of insecticide had been reported to result in systemic toxicity such that may impair the function of the nervous system and increase the risk of acute leukemia in children (Menegaux et al., 2006). Also, pesticides including organ phosphorus insecticides used against crawling and flying insects in homes have the potential of being carcinogens (Peter and Cherion, 2000). The adverse effect of insecticide Raid was demonstrated in a study by increase in alkaline phosphates activity in both plasma and liver which is a known measure of hepatic toxicity, and confirms "Raid" as a hepatotoxicant. The significant increase in alkaline phosphates activity (Table IV) may be due to hepatocellular necrosis which causes increase in permeability of cell membrane resulting in the release of this enzyme into the blood stream. The insecticide Raid significantly decreased reduced glutathione levels especially in the liver and this has implications for the ability of the animal to withstand oxidative stress. Studies have shown that GSH deficiency in cells is associated with markedly decreased survival (Kohlmeier et al., 1997), thus, chemically stable lipid-soluble, organophosphorus insecticides are hazarddous to health through mechanisms including depletion of GSH (Menegaux et al., 2006).

Table 1. II mortality and growth of wistar rats exposed to different concentrations of "Raid".

Means SD concentrations of insecticide "Raid" in feed (Mg/g)	Mortality	Means: SD body weight (g)
0.00	Nil	135=5.4
25.0=2.4	Nil	135=21.7
54.0=9.5	Nil	132=2.9
108.2=12.5	Nil	129=3.2
216.2=14.6	Nil	128=19.8
430.0=20.2	1	118=20.5
961.2=70.5	2	116=5.3

Table 1. III Tissue total raid concentrations and bioaccumulation factors (BAF) in wistar rats.

Raid Concentration in Wistar Rats (μg/g)[a] and Bioaccumulation Factor (BAF) Mean±SD Insecticide "Raid" in Feed (μg/g)	Lipid	Muscle	Brain	Liver
00.0	-	-	-	-
25.0±2.4	72.5± (2.9)	17.5(0.7)	12.5(0.5)	7.5(0.3)
54.0±9.2	86.4(1.6)	21.7(0.4)	16.4(0.3)	9.4(0.2)
108.2±12.5	172.8(1.6)	30.4(0.3)	19.5(0.2)	10.8(0.10)
216.2±14.6	280.8(1.3)	45.8(0.2)	22.9(0.1)	19.8(0.09)
430.0±20.6	324.0(0.8)	86.4(0.2)	25.8(0.06)	37.3(0.09)

961.2±70.5	1153.2(1.2)	576.6(0.6)	28.8(0.03)	76.9(0.08)

Table 1. IV Effect of Raid concentrations in feed on hepatic enzyme activity, reduced glutathione and glucose levels.

Raid concentrations Tissue In feed (μg/g)		Alk pase activity μgml^{-min-L}	GSH level mg/ml	Glucose level mg/g liver
430±20.2	Control	0.08±0.04	0.18±0.02	0.96±0.04
	Plasma	0.06±0.09	0.15±0.6	0.90±0.04
	Control	0.08±0.04	0.18±0.02	0.94±0.01
	Liver	0.06±0.02*	0.15±0.01	1.05±0.12
961.2±70.5	Control	0.09±0.05	0.19±0.05	0.96±0.52
	Plasma	0.06±0.01	0.11±0.05	1.09±0.52
	Control	0.08±0.08	0.19±0.02	0.96±0.06
	Liver	0.05±0.08*	0.09±0.03*	1.66±0.04

Data values are mean±SD

**Statistically significant $p< 0.05$*

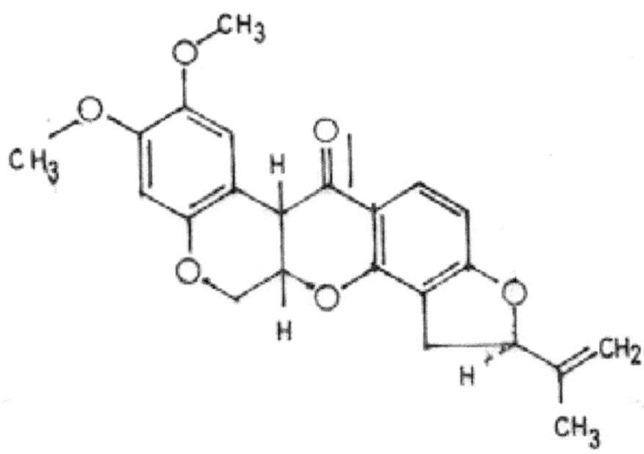

Fig. 3. Structure of rotenone.

Some other studies confirm that glutathione deficiency is associated with impaired survival in HIV disease (Herzenbery et al., 1997). Glutathione may be consumed by conjugation reaction, which mainly involve metabolism of zenobiotic agent. However, the principle mechanisms of hepatocyte glutathione turnover are known to be by cellular efflux (Sies et al., 1978). Glutathione reducatase is a known defense against oxidative stress, which in turn needs glutathione as co-factors. Catalase is an antioxidant enzyme which destroys H_2O_2 that can form a highly reactive radical in the presence of iron as catalyst (Gutter ridge, 1995).

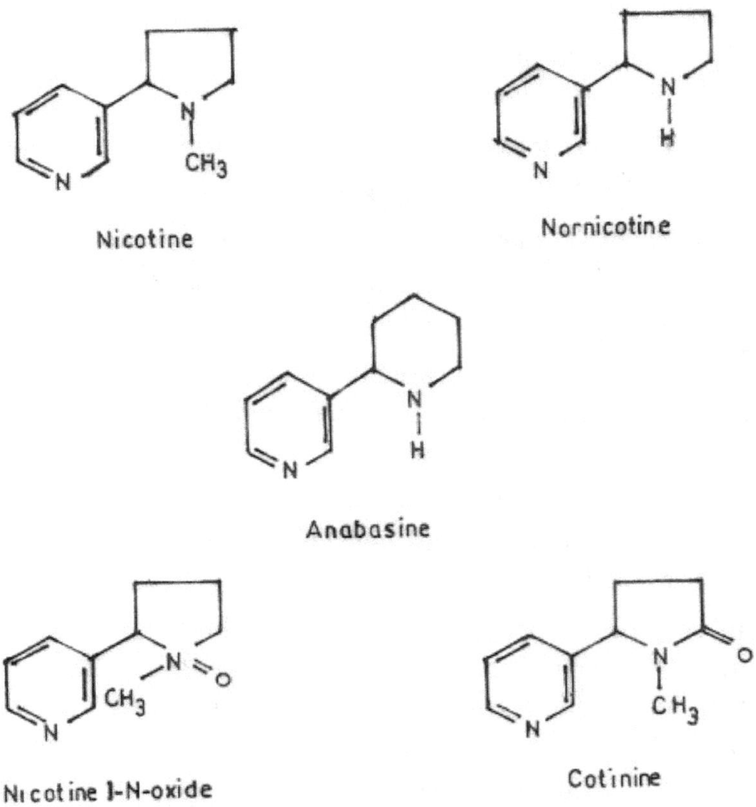

Fig. 4. Nicotine, narnicotine and anabasine with two important metabolites of nicotine.

Achudume et al., 2008 showed that bioaccumulation factor of insecticides raid was observed in lipid. Lipid peroxidation is a chemical mechanism capable of disrupting the structure and function of the biological membranes that occurs as a result of free radical attack on lipids. Some study confirms that insecticide raid increased lipid peroxidation, oxidative stress and hepatotoxicity due to reduced antioxidant system.

In addition, SOD is family of metalloid enzyme which is considered to be stress protein which decreases in response to oxidative stress (McCord, 1990). It is evident that decrease of SOD in the tissue is a confirmation of its protection from damage caused by insecticide Raid.

2. CLASSES OF INSECTICIDES

- The classification of insecticides is done in several different ways (Hayes, 1982),(Heam 1973, Lehman, 1954, Martin and 1977).
- Systemic insecticides are incorporated by treated plants. Insect ingest the insecticide while feeding on the plants.

- Contact insecticides are toxic to insects by direct contact. Efficacy is often related to the quality of pesticide application in aerosols which often improve performance.
- Natural insecticides, such as pincotine, pyrethrum and neem extracts are from plants as defences against insects.
- Inorganic insecticides are manufactured with metals e.g. Heavy metals
- Organic insecticides are synthetic chemicals which comprise the largest numbers of pesticides available.

Insecticides are pesticides used to control insects many of these insecticides are very toxic to insects and many others are relatively harmless to other organism except fish.

Insecticides decompose readily so the residues do not accumulate on crops or in the soil. Insecticides include ovicides and larvicides used against the eggs and larvae of insects respectively.

The use of insecticides is believed to be one of the major factors behind the increase in agricultural productivity (McLaughlin,1973, van Emden and Pealall, 1996). Nearly all insecticides have the potential to significantly alter ecosystems; many are toxic to humans; and others are concentrated in the food chain (WHO 1962, 1972). Selected inorganic metals are discussed in the next section followed by individual insecticides organ metals.

2.1 Barium

Barium is an alkaline earth metal in the same group as magnesium, calcium, strontium and radium. It valence is two. All are water-and acid soluble compounds. They are poisonous. Barium carbonate is a rat poison. It is used in ceramics, paints, enamels, rubber and certain plastics.

Absorption, Distribution, Metabolism Excretion (ADME): Barium carbonate is highly insoluble in water. It is partially solubilized by acid in the stomach. The danger of the insecticide is through ingestion. Various barium compounds can cause pneumoconiosis. It is absorbed from gastrointestinal tracts of rat rapidly and completely. It is stored in bone and in other tissues (Hayes 1982, Castagnou et al 1957, Dencker et al., 1976, 83). Excretion takes place rapidly in urine and feces in 24hr (Bauer et al. 1956).

Mode of action: Barium stimulates striated cardiac and smooth muscle, regardless of the innervation.

2.2 Chromium

Chromium is a metal somewhat like iron and separated in the periodic table by manganese. Only hexavalent chromium compounds (chromates) are important as pesticides. They are also the most toxic. Chromate is absorbed by the lung (Baetjer et al., 1959), gastrointestinal tract and skin. It is widely distributed in the liver, kidney, bone and spleen (Mackenzie et al 1958). Acute poisoning may produce death rapidly through shock or renal tubular damage and uremia (Steffee and Baetjer 1965).

2.3 Mercury

Mercury is toxic no matter what its chemical combination. It is widely distributed in the environment, and traces of it occur in food, water and tissues even in the absence of occupational exposure. Inhaled mercury vapour diffuses across the alveolar regions of the lung into the blood stream. Mercury vapour is a monatomic gas which is highly diffusible and lipid soluble (Berlin et al 1969a , Hush,1985). Once in the bloodstream mercury vapour enters the blood cells where it is oxidized to divalent inorganic mercury under the influence of catalase (Halbach and Clarkson 1978). Mercury is widely distributed with the highest concentrations in the kidney.

2.4 Thallium

Thallium stands between mercury and lead in the periodic table, and compounds of these metals show marked similarities. All of them may produce immediate local irritation followed by delayed effects in various organs, notable the nervous system. Thallium sulphate has been more widely used as pesticide than any other compound of thallium. It has produced many cases of poisoning and serves as good example of the toxicity of thallium generally (Lund, 1956b).

Thallium is easily absorbed by the skin as well as by the respiratory and the gastrointestinal tracts. Thallium accumulates in hair follicles and much less in those in the resting phase. Excretion is slow and is entirely by urine in humans but in rats via faeces (Barclay et al., 1953, Lund, 1956a)

2.5 Lead arsenate

Lead arsenate includes acid lead arsenate, dibasic lead arsenate, dilead orthoarsenate, diplumbic hydrogen arsenate, lead hydrogen arsenate and standard lead arsenate. Lead arsenate is used as an insecticide. it is used to control moths, leaf rollers and other chewing insects and in soil for the treatment of Japanese and Asian beetles in lawn. Absorption is generally via gastrointestinal. Dermal absorption is extremely small. Lead and arsenate are distributed separately in the body. lead is stored in highest concentration in the bone with much lower concentrations in soft tissues. Arsenic is stored in the liver and in some instances in the kidney at higher concentrations than those for lead (Fairhall and Miller, 1941. Lead is transferred to the fetus of animals humans (Heriuchi et al., 1959).

2.6 Antimony potassium tartrate

This compound serves as a poison in baits to control insects, especially thrips, and as an emetic in bait to control rodents. Ingestion of the compound usually leads to repeated vomiting. Excretion is mainly urinary (Fairhall and Hyslop, 1947).

2.7 Sodium selenate

Sodium selenate is an insecticide used in horticulture for control of mites, aphids and mealybugs. Various compounds of selenium are freely absorbed from the respiratory and gastrointestinal tracts. Dermal absorption is less important. Selenium is stored more in the liver, kidney, spleen, pancreas, heart and lung than in other organs (Underwood, 1977). Selenium is excreted chiefly in the urine but about 3-10% is metabolized and excreted by the lungs and through faecal excretion.

2.8 Sodium fluoride

Sodium fluoride is toxic to all forms of life. It has been used as an insecticide, rodenticide and herbicide and as fungicide for preservation of timber. Its toxicity to plants generally has restricted its use as an insecticide to bait formulations (who, 1970). Sodium fluoride concentrates more in the plasma and liver and is excreted in urine.

3. MISCELLANEOUS ELEMENTS

3.1 Boric acid

Boric acid and borax have been used as an insecticide, both mainly for the control of cockroaches. Boric acid also is known as boracic acid and as orthoboric acid. Absorption from the gastrointestinal tract is rapid and virtually complete. Its peak concentration is in brain and less in other tissues. Boric acid is excreted unchanged in the urine (wong et al., 1964).

3.2 Insecticides derived from living organisms and other sources:

Different groups of insecticide; derived from living organisms are entirely unrelated chemically and pharmacologically. They range from relatively simple alkaloids such as nicotine, with a molecular weight of only 162.2, through proteinaceous poisons to virulent living organism. They range in toxicity from harmless and fragile pheromones, which are used as a chemical warfare agent.

The distinction between synthetic compounds and those derived from living organisms is somewhat artificial. In practice, related compounds are assigned to one category or the other, depending on whether the particular compound of the group that was first known and used was of synthetic or of natural origin. For example, pyrethrum and later the naturally occurring pyrethriums were well known for years before the first synthetic pyrethroid was made; as a result, pyrethroids are thought of as various of natural compounds, even though they have not been found in nature and are unlikely to occur. By contrast, synthetic sodium fluoroacetate acquired a reputation as a rodenticids and was explored as a synthetic insecticide before it was realized that the potassium salt is the active principal of a poisonous plant. Thus pyrethroids are discussed extensively.

Perhaps the only unifying feature of the diverse array of poisons derived from living organism is the popular view that "natural" substances are harmless. On this matter of safety, an expert committee of the world health organisation pointed out that "all the most poisonous materials so far know are, in fact, of natural origin" (WHO,1967).

3.3 Pyrethrum and related compounds

The insecticidal properties of pyrethrum flowers (genus chrysanthemum) have been recognized since the middle of 1st century, when commercial sale of "insect powder" from Dalmatian pyrethrum flower heads began (McLaughlin, 1973). In addition to their insect- killing activity, their lack of persistence in the environment and rapid "knock down" activity whereby flying insects become uncoordinated and unable to fly makes it very useful. Pyrethrum used to be a major insecticide for both domestic and agricultural use despite its poor light stability. Its usefulness was extended by introduction of piperonyl butoxide and other compounds as synergists, which greatly reduced the unit cost of crop treatment. Development of synthetic pyrethroids with increased stability and insecticidal activity (Elliot 1977) reduced the use of pyrethrum. However, natural pyrethrins are now used mainly as domestic insecticides, while the synthetic pyrethroids represented 20-25% of the world foliar insecticide market in 1983 (Herve's 85) and the proportion is increasing steadily. Thousands of new synthetic pyrethroids have been synthesized, some showing complete divergence from the original pyrethrins (casida et al., 1973). Table 2.1 and Table 2.2.

4. MODE OF ACTION

Pyrethrum and the synthetic pyrethroid are sodium channel toxins which, because of their remarkable potency and selectivity, have found application in general toxicology (Lazdunski et al 1985). Their actions on the nerve membrane sodium channel are well understood. Pyrethroids have a very high affinity for membrane sodium channel, they have little effect on inactive sodium channels or close channels and produce subtle changes in their functions. After modification by prethroids, sodium channels continue in many of their normal functions, retaining their selectivity for sodium ions and link with membrane potential (Narahashi, 1986). The pyrethroids are thus known as open channel blockers. Detailed studies can be found in Narahishi 1986, Jacques et al., 1980, and Gray 1985.

4.1 Metabolism

The relative resistance of mammals to the pyrethroids is almost wholly attributable to their ability to hydrolyze the pyrethroids rapidly to their inactive acid and alcohol components, since direct injection into the mammalian CNS leads to susceptibility similar to that observed in insects (Lawrence and Casida, 1982). Some additional resistance of homoeothermic organisms can be attributed to the negative temperature coefficient of action of the pyrethroids (Van den Bercken et al., 1973) which are thus less toxic at mammalian body temperature but the major effect is metabolic.

The metabolic pathways for the breakdown of the pyrethroids vary little between mammalian species but vary somewhat with structure. This literature has been ably summarized by Leahy (1985), and further references to the metabolism of specific pyrethroids are given in the sections on individual compounds. Generally pyrethrum and allethrin are broken down mainly by oxidation, whereas for the other pyrethroids ester hydrolysis predominates. These reactions can take place in both liver and plasma and are followed by hydroxylation and conjugation to glucuronides or sulphates, which are then excreted in the urine (Gray 1985).

4.2 Individual insecticides

Other known insecticides pyrethroids under organophosphates are listed below only selective ones are discussed.

Table 4. I Other known insecticides.

Allethrin	Permethrin
Bifenthrin	
Cyhalothrin, Lambda-cyhalothrin	
Cypermethrin	Phenothrin
Cyfluthrin	Prallethrin
Deltamethrin	Resmethrin
Ftofenprox	Tetramethrin
Fenvalerate	Transfluthrin

4.3 Cypermethrin

Cypermethrin (R, S)-∞- cyano-3-pheno-xybenzyl-2, 2-dimethyl. There are eight isomeric forms. It was introduced commercially in 1977 as an emulsifiable concentrate to be used against a wide range of insect pest (Elliot, 1977).

4.4 Deltamethrin

Deltamethrin S-∞- cyano-3-phenoxybenzyl-(IR)-cis-3-(2, 2-dibromovinyl)2,2-dimethcyclo- propane carboxylate. It is a single isomer. It is used against a wide range of insect pests. It produces a typical type II motor symptom in mammals (Barnes and verschoyle, 1974. Metabolism of deltamethrin involves rapid ester cleavages and hydroxylation (Shono eta al; 1977).

4.5 Fenproponate

Fenproponate(∞-cyano-3-phenoxybenzyl-2,2,3,3-tetra-methylcyclopropanecarbo xylate). There are eight isomerism forms. Fenpropathrin is another common name, was first developed by sumitomo and commercialized in 1980 as an emulsifiable concentrate to be used against a wide range of insect pests. Fenproponate produces intermediate or mixed motor symptoms in mammals (Wright et al., 1988).

4.6 Fenvalerate

Fenvalerate (R,S)- ∞- cyano-3-phenoxy-benzyl (IR,IS)-2-(4-chlorophenyl)-3-methyl-1- butyrate. There are four isomeric forms. It should be noted that fenvalerate is not based on a cyclopropane ring structure. It was introduced commercially to be used against a wide range of insect pests fenvalrate produces typical type II motor symptoms in mammals (Verschoyle and Aldrige, 1980).

4.7 Phenothrin

Phenothrin (3-phenoxybenzy-(IR,IS)-cis,trans-3(2-methylprop-1-enyl)-2,2 dime-thyicyclo- propane carboxylate). There are four isomeric forms. It is used as a domestic insecticide in a partially resolved mixture rich in the IR isomer (Sumithrin) and for grain protection. Phenothrin produces typical type 1 moto symptoms in mammals (Lawrence and Casida, 1982)

4.8 Rotenone and related materials

Rotenone-bearing plants have longed being used as a fish poison by many ancient different indigenous people, nut their use as an insecticide is probably more than a century old. Plants known to produce rotenone and other rotenoids belong to at least 68 species of the family Leguminosae, the same as that for peas and beans. The genera most exploited so far are Derris, native to southeast Asia, and lonchocarpus to south America (shepard 1951).

Rotenone and other active principles often occur chiefly in the roots of rotenone bearing plants but may be in the leaves (as in *Tephrosia vogeli*), seeds (as in *Milletia pachycarpa*), or bark (as in *Mundulea serica*).

Regardless of the genus or the particular part of the plant involved, the active constituents of rotenone-bearing plants may be extracted with ether or acetone as resin.

Rotenone is (2R,6a 5,12a 5)-1,2,6,6a,12,12a-hexahydro-2-isopropenyl-8,9-dimethoxy- chromeno (3,4-b) furo(2,3-h) chromen-6-one. Its structure is depicted in fig. 3. Although rotenone generally is considered to be the active ingredient in all resins isolated, the other constituents show considerable insecticidal activity (Metcalf, 1955).

Rotenone is readily oxidized and racemized in the presence of light and the process is accelerated in alkaline solution (Cheng et al., 1972). It is active as a nonsytemic pesticide against a wide variety of insects, arachnids and molluscs. Its rapid photodecomposition means that it is active only for about 1 week on plants or 2-6 days in water and this limits its commercial use though still finds use as a domestic garden insecticide.

Rotenone is a highly potent mitochondrial poison, blocking NADH oxidation, this property dominates its actions in animals (Heikkila et al., 1985).

Rotenone is metabolized rather effectively by the liver in isolated rat liver mitochondria, the aerobic oxidation of pyruvate is almost completely inhibited by rotenone (Haley 1978).

4.9 Nicotine and related compounds

Three closely related compounds (nicotine, nornicotine and anabasine fig4) were commonly used as insecticides, although only the most potent, nicotine, is now used to any extent. Nicotine is usually obtained from the dried leaves of nicotiana tabacum, but it also occurs in *N. rustica* and *Duboisia*, another genus of the solanaceae, and in three other taxonomically diverse general, namely Asclepia (Asclepidaceae), Equisetaceae (Equisetaceae), and Lycopodium (Lycopodiaceae); Nicotine (S-3-(1-methyl pyrrolidin-2- yl) pyridine) is used as nicotine sulphate as a stomach poison for leaf eating insects (Haigh and Haigh 1980). Nicotine is rapidly absorbed from all mucosal surfaces, including those of the mouth, gastro-intestinal tract, and lung. Since nicotine readily forms salts in acid solution, its penetration through biological membranes is strongly pH dependent (Schievelbein, 1982).

The metabolism of nicotine is highly complex and reviewed by Gorrod and Jenner (1975) and schievelbein (1982). Metabolism mainly by cytochrome P.450 linked microsomal oxidative pathways in the liver. Cotinine (Fig4) is major metabolite, which then undergoes further oxidation. Nicotine stimulates the action of acetylcholine at nicotinic receptors in the central nervous system, autonomics ganglia and some pheripheral nerves. It central actions result in tremor and convulsions, stimulation and then depression of ventilation and induction of vomiting by a direct action on the medulla. Ventilation is stimulated by peripheral actions on the aortic and carotid chemoreceptors, and adrenal catecholamine. Secretion is increased at low doses. Heart rate and blood pressure are largely dominated by sympathetic effects and show increases compounded by adrenal catecholamines. The gastrointestinal tract is dominated by parasympathetic effects and shows hypersecretion followed by block as well as increased tone and peristalsis. Death is usually a result of block of neuromuscular transmission in the respiratory muscles or a consequence of seizures. In addition to its action on cholinergic transmission, nicotine can act at noncholinergic sites and also activate receptors on sensory nerve endings and vagal C fibers (Martin, 1986).

The carcinogenic potential of tobacco is well established, but there is debate about the role of nicotine, which, although probably not carcinogenic itself can be converted to carcinogens such as N'-nitrosonornicotine and 4-(methyl-nitrosamino)-1-(3-pyriyl)-1-butanone. The metabolites cotinine and nicotine 1-N-oxide are not carcinogenic although they do produce hyperplasia of the bladder epithelium (Hoffmann et al., 1985).

5. LIVING ORGANISMS AS PESTICIDES

The use of biological control agents has many potential advantages over chemical control, not least the possibility of high selectivity for the predators and other beneficial species. Several microorganisms or microbial products have been identified as potential insecticides (Miller et al, 1983). Most successful attempts have been directed against insects, as biological control of vertebrates has met with little success due to cross-infection problems. The world Health Organisation has investigated viruses, bacteria fungi and nematodes as potential insect control agents since all play a part in limiting the growth of natural insect populations.

5.1 Viral insecticides

Viral insecticides are still in the experimental stage but many are under investigation, as reviewed by Miller et al., 1983. Bacterial insecticides represent the largest and widest used group and reviewed by Burges (1982) and Lysenko(1985). All of those used are spore- formers, since the spores can be readily stored in dried form and applied by conventional means as wettable powders or dusts. Many form a crystalline toxin within the spore which enhances their pathogenicity to insects. The most widely used is *Bacillus thuringiensis*. A closely similar bacterium, *Bacillus papilliae* has been used against Japanese beetle. It has the advantage that once spores are introduced into the environment the bacterial population is sustained by reinfection of the insect hosts, but the disadvantage that spore production requires expensive in-vivo production using insect pupae and is now of declining importance. It is highly specific, does not infect vertebrates, and despite production of a crystal toxin is nontoxic to mammals by repeated oral administration (Burges, 1982).

5.2 Fungal insecticides

Fungal insecticides are commercially produced for a variety of specific applications. Their importance in controlling natural insect populations has been recognized since 1834, Aschersonia has been used to control Floridian white fly on citrus since the early 1900s. Fungi have the advantages of forming a stable population in the insect environment and are capable of infection through the insect cuticle, not by ingestion as bacteria. A disadvantage is their susceptibility to widely used fungicides. Examples include *Beauveria basiana* is marketed as Boverin and used against Colorado beetle and corn borer in Russia and China. *Metarhizium anisopliae* was used against a range of insects as metaquino. *Hirsutella thompsoni*, is used to control citrus rust in the united states as myear and *Vecticillium lecani* is used as vertalec or mycotal for aphid control in united kingdom. Some fungi such as *Beauveria bassiana* produce toxins which may be involved in their pathogenicity. *Culicinomyces clavosporus* and *lagenidium giganteum* are mosquito pathogens (Miller et al. 1983).

5.3 Nematate insecticides

Nematate insecticides have been isolated from mosquito larvae at low natural population densities. They are reared in vivo, which is expensive, and there some resistant mosquito population Nametodes are tolerant of many insecticides and insect growth regulators and can be used in combined malaria control programs and are rapidly broken down by human gastric juice (Gajana et. al; 1978).

6. CONCISE SUMMARY

The excitement of advocates for biological insect control may be unexpected, considering the minimal role these agents now play in pest management. Nonetheless, several challenges exist in maintaining a sufficiently big population of the control agent on crops, or, in the case of mosquitoes, at the water surface, as well as in agriculture due to the pronounced host specificity of certain agents. Fundamental issues include the potential risk posed by replicating agents that may proliferate in the environment and the possibility of gene transfer expressing toxins from invertebrate to vertebrate bacteria or viruses. Current experiences with biological control agents are notably promising, indicating their significant role in future integrated pest control systems (Laird, 1985).

Both animals and people may suffer negative consequences primarily from the absorption of active components; nevertheless, the impact of propellant chemicals should not be overlooked. Inflammatory activation may be a significant factor contributing to harmful consequences in the tissue (Mense et al., 2006). The contribution of propellant to the toxicity of the pesticide Raid remains ambiguous. An extensive evaluation of the risks linked to the environmental use of the pesticide

Raid was conducted across numerous tissues, as it impacts the fundamental biochemical molecules of cells. (Achudme et al., 2008).

7. REFERENCES

Achudume, A.C., Nwoha, P.C., Ibe, J.N. (2008) Toxicity and Bioaccumulation of insecticide "Raid" in Wistar Rats Inter Environ Toxicity 24(4); 357-361.

Barcley, R.K;peacock, W.C. and Karnofsky,G.A. (1953). Distribution and excretion of radioactive thallium in the chick embryo, rat and man.J. Phaemacol.Exp.Ther 107, 178-187.

Barnes, J.M. and Verschoyle, R.D. 1974 Toxicity of new Pyrethroid insecticide. Nature (London) 248-711.

Bauer, G.C.; Carlsson, A. and Lindquist, B. (1956). A comparative 45Ca in rats. Biochem. J. 63, 535-542.

Berlin, M.H; Nordbery, G.F. and Serenius, FR. (1969a). on the site and mechanism of mercuric nitrate Hg. 203.Arch.Environ. Health 18, 42-50.

Brodie, M.E. and Aldridge, W.N 1982 Elevated cerebellar cyclic GMP levels during the deltamethrin induced motor syndrome. Neurobehav. Toxicol. Teratol 4, 109- 113.

Burges, H.D. (1982). Control of insects by bacteria. Parasitology 84, (symp), 79-117.

Casida, J.E. (1973). Biochemistry of the pyrethrins. In "pyrethrum: The Natural insecticide" (j.E. casida, ed). Academic press, New York and London.

Castagnous, R; Paolett; C. And Larcebeau, S. 1957. Absorption and distribution of barium administered intravenously or orally to rats C.R. Hebd, seanes ser Acad Sci D 244, 2994-2996. (in French).

Cheng, P.Y., Buster, D., Hommock, B.D., Roe, R.M, and Alford, A.R. 1987. *Bacillus thuringiensis* van. Israelenisio-enctotosian Evidence of Neurotoxic action. Pestic Biochem Physiol. 27, 42-49.

Cremer, J.E and Seville, M.P. 1982-Comparative effects of two Pyrethroids, deltamethrin and cismethrin on plasma catecholamines and on blood glucose and lactate. Toxicol. Appl Pharmacol. 66, 124-133

Cremer, J.E., Cunningham, V.J. and Seville, M.P. 1983. Relationship between extraction and metabolism of glucose, blood flow, and tissue blood volume in regions of rat brain. J. Cereb. Blood Flow Metab. 3, 291-3002.

Dencker, L; Danielsson, B; khayal, A, and Lindren, A. (1983). Deposition of metals in the embryo and fetus. In "Reproductive and Developmental Toxicity of metals" (T.W. Clarkson, G.F Nordberg, and P.R. sage eds), pp 607-631. Plenum, New York.

Dencker, L; Nillson, A; Ronnback, C. And walinder, G. (1976). Nptake and retention of 133Ba and 140Ba-140La in mouse tissue. Acta Radiol: Ther; Phys; Biol (N.S) 15, 273- 28.

Elliot, M. 1977. "Synthetic Pyrethroids", ACS Symp. Ser. No. 42 Am. Chem. Soc., Washington, D.C.

Elliot, M., 1971. The relationship between the structure and the activity. Chem. Ind. (London) 24 776-791.

Fairhall, L.T. and Hyslop, F. (1947). The toxicology of antimony. Public Health Rep; suppl. 195.

Fairhall, L.T. and miller, J.W. (1941). A study of the relative toxicity of the molecular components of lead arsenate public Rep 56, 1610-1625.

Forshaw, P.J and Ray, D. E. 1990. A novel action of deltamethrin on membrane resistance in Mammalian skeletal muscle and non-myelinated nerve fibres. Neuropharmcol. 29, 75-81.

Gajana, A; kazimi, S.J; Bheemarao U.S., Suguna, S.G; and chandrahas, R.K 1978 studies on a nematode parasite (Romano-mermis sp; mermithidae) of mosquito larvae isolated in Pondicherry Indian J. Med Res. 68, 242-247.

Gorrod, J.W, and Jenner, P. 1975. The metabolism of tobacco alkaloids. Essays Toxical 6,35-78 schievelbein, H 1982.

Gray, A.J. 1985 Pyrethroid structure-toxicity relationships in mammals. Neurotoxicology 3, 25-35

Gray, A. J., Connors, T.A., Hoellinger, H. and Nguyen-Hoang-Nam. 1980. The relationship between the Pharmacolcinetics of intravenous cismethrin and bioresmethrin and their mammalian toxicity. Pestic. Biochem. Physiol. 13, 281-293.

Haigh, J.C. and Haigh, J.M. 1980, immobilizing drug emergencies in humans Vet Hum Toxical 22, 1-5.

Halbach, S. And Clarkson, T.W.(1978). Enzymic oxidation of mercury vapour by erythrocytes. Biochim Biophys Acta 523, 522-531.

Haley, T.J. 1978. A review of the literal of rotenone 1, 2,12,12a tetrahydro-8-9-dimethoxy-(2- (-1 methyl ethenyl)-1-benzo-pyrano (3,5-B)fluoro (2,3-H) (1)-benzo-pyran-6(6h), one J. Enviror pathol. Toxicol 1, 315-337.

Hayes, W.J., Jr (1982). "Pesticides Studies in Man". Williams & Wilkins, Baltimore, Maryland.

Hearn, C.E.D. (1973). A review of Agricultural Pesticide Incidents in Man in England and Wales, 1952-71, Br. J. Ind. Mect 30, 253 -258

Heikkila, E, Nicklas, W.J,vyas,I, and Duvoisin, R.C. 1985. Dopeminergic toxicity of rotenone and the 1 methyl-4-phenylpridium ion after stereotoxic administration to rats:implication for the mechanism of 1-methyl-4-phenyl-1,3,3,6-tetrahydropyridine toxicity. Neurosci. LeH. 62, 389-394.

Herve, J.J. (1985). Agricultural, public health health and animal health usage. In "The pyrethroid insecticides" (J.P. Leahey, ed) Taylor & Francis, London and Philadelphia.

Hoffmann, D; Laboie, E.J. and Hecht, S. 5,1985 Nicotine A precursor for carcinogens. Cancer letter, 26,67-75.

Horiuchi, K, Horiguchi S and suekane, M (1959). Studies on the industrial lead poisoning .I. Absorption, transportation, deposition and excretion of lead 6. The lead contents in organ tissues of the normal Japanese Osaka city med J. 5, 41-70.

Hursh, J.B.(1985). Partition coefficients of mercury (203Hg) vapour between air and biological fluids.J.Appl Toxicol 5,327-332.

Jacques, Y., romey, G., Cavey, M.T., Kartalovski, B. And Lazdunski, M. 1980. Interaction of Pyrethroids with the Na+ channel in mammalian neuronal cells in culture. Biochim. Biophys. Acta 600, 882-897.

Lawrence, L.J. and Caside, J.E. 1982 Pyrethroid toxicology: Mouse intracerebral structure- toxicity relationships. Pestic Biochem Physiol 18, 9-14.

Lazdunski, M., Barhanin, J., Borsotto, M., Frelin, C., Hugues, M.n Lombet, A., Pauron, D., Renaud, J., Schmid, A. and Vigne, P. 1985 Markers of membrane ionic channels In "Vascular Neuroeffector Mechanisms (J.A Bevan et al. Eds) Elsevier, Amsterdam.

Lehman, A.J, 1954. A toxiocological Evaluation of household insecticides Q.Bull-Assoc. Food Drug Off. 18, 3-13

Lund, A.(1956a). Distribution of thallium in the organism and its dimination. Acta pharmacol. Toxicol 12,251-259.

Lund, A. (1956b). The effect of various substances on the excretion and the toxicity of thallium in the rat. Acta Pharmacol. Toxicol 12, 251-259

Martin, B.R. (1986). Nicotine receptors in the central nervous system. In "The receptors (P.M. comm., ed), vol.3 pp. 393-415. Pergamon Oxford.

Martin, H. And Worthing, C.R; eds (1977). The Pesticide Manual," 5th ed. Br. Crop Rot. Counc; Molvern, Worcestershire, England.

McLaughlin, G.A. 1973. History of Pyrethrum. In "Pyrethrum: The Natural Insecticide" (J.E. Casida, ed), pp 3-15. Academic Press, New York and London.

Metcalf, R.L. (1955). "organic Insecticides" Wiley (intersciences), New York.

Miller, L.K., Lingg, A.J. and Bulla, L.A. 1983. Bacterial, viral and fungal insecticides science 219, 715-721.

Molaughlin, G.A. (1973). Histgory of Pyrethrum. In "Pyrethrum: The National Insecticide" (J.E Casida, ed), pp 3-15. Academic Press, New York and London.

Narahashi, R, 1986. Mechanisms of action of Pyrethroids on sodium and calcium channel gating. In "Neuropharmacology of Pesticide Action" (M.G. Ford, G.G. Lunt, R.C Reay, and P.N.B. Usherwood, eds), pp 36-40. Ellis Horword, Chichester, U.K.

Nebeker, A.V., Dunn, K.D, Griffis, W.H., Schuytema, G.S. 1994. Effects of dieldrin in food and growth and bioaccumulation in Mallard ducklings Arch Environ Contam. Toxicol 26; 29-32.

Parker, C.M., Albert, J.R., Van Gelder, G.A; Patterson, D.R. and Taylor, J.L. 1985. Neuropharmacologic and neuropathologic effect of Fenvalerate in mice and rats. Fundam.

Rickard, J. and Brodie, M.E. 1985 Correlation of blood and brain levels of the neurotoxic Pyrethroid deltamethrin with the onset of symptoms in rats. Pestic, Biochem. Physiol 23, 143-156.

Schieveibein, H. 1982, Nicotine, resorption and fate pharmacol. Ther 18, 233-248.

Shepard, H.H. 1951. "The Chemistry and action of Insecticides". 1st ed. McGraw-Hil, New York.

Shono, T., Ohsawa, K. And Casida, J.E. 1979. Metabolism of trans- and cis-cypermetrin and decamethrin by microsomal enzymes J. Agric Food Chem. 27, 316-325.

Staatz-Benson, C.G., and Hosko, M.J. 1986. Intataction of Pyrethroids with mammalian spinal neurons. Pestic. Biochem physiol 75, 19-30

Steffeee, C.H and Baetjer, A.M. (1965). Histopathologic effects of chromate chemical Arch. Environ.Health 11,66-75.

Underwood, E.J. (1977)."Trace elements in Human and Animal Nutrition". 4th ed. Academic press, New York.

Verschoyle, R.D. and Aldridge, W.N. 1980 Structure-activity relationship of some Pyrethroids in rats. Pestic. Biochem. Physiol. 2, 308-311.

Van den Bercken, J., Akkermann, L.M.A and van der Zalm, J.J, 1973. DDT. Like action of Allethrin in the sensory nervous system of Xenopus laevis. Eur. J. Pharmacol. 21, 95-106.

Van Emden HF, Pealall DB (1996) Beyond Silent Spring, Chapman & Hall, London, pp 322.

White, I.N.H., Verschoyle, R.D., Moradian, M.H., and Barnes, J.M. 1976. The relationship between brain levels of cismethrin and bioresmethrin in female rats and neurotoxic effects. Pestic. Biochem. Physiol. 6, 491-500.

WHO 1962 Accidental Food Poisoning with Agrosan. Communication to World Health, Organisation from S.A Raz Ali. WHO Inf. Circ. Toxic. Pestic. Man No 9, P. 23

WHO 1972, "IARC Monographs on the Evaluation of Carcinogenic Risk of Chemicals to Man," Vol. 1, Int. Agency Res. Cancer, Lyon, France.

Wong, L.C; Heimbach, M.D; Truscolt, D.R. and Duncan, B.D. (1964). Boric acid poisoning. Report of 11 cases. Can.med. Assoc J. 90, 1023.

World Health Organisation (WHO) (1970). "Fluuorides and Human Health". World Health organ. Geneva

World Health Organization (WHO) (19670) "safe use of pesticides in public Health", Who Tech. Rep ser. No 356 world Health Organ, Geneva.

Wright, C.D.P., Forshaw, P.J. and Ray, D.E. 1988 Classification of the actions of two Pyrethroid insecticides in the rat, using the trigeminal reflex and skeletal muscle as test systems. Pestic. Biochem, Physiol. 30, 79-86.

CHAPTER 2

Chlorfluazuron as a Reproductive Inhibitor

1. INTRODUCTION: AN OVERVIEW

Benzoyl phenyl ureas (BPUs) are insect growth regulators because they function as moult disruptors and inhibit chitin synthesis during growth and development in insects (Wright and Retnakaran, 1987; Binnington and Retnakaran, 1991). In a variety of situations, IGRs, such as dimilin, are effective against a significant spectrum of insect larvae and adults. The compound interferes with the synthesis of chitin, thereby disrupting the moulting process. Research has been conducted on the various aspects of dimilin, including its effects as a chitin synthesis inhibitor, toxicant, ovicide, disrupting adult emergence, and residual effects, with a variety of insect species. For example, Jakob (1973), James (1974), Qureshi et al. (1983), Naqvi and Rub (1985), Ganiev (1986), Khan and Naqvi (1988), Gupta et al. (1991), Tahir et al. (1992), and Nizam (1993). These pesticides have been reported to have a variety of modes of action. For instance, phagodeterrents and repellents (Abro et al., 1997), chitin synthesis inhibition (Hajjar and Casida, 1979), growth inhibition and abnormal development (Hashizume, 1988), ovicidal action (Hatakoshi, 1992), insecticidal effects on the reproductive system (Chang and Borkovec, 1980), and neurotoxic effects on insect behavior are a few examples. (Haynes, 1988).

1.1 Chlorfluazuron

Chlorfluazuron (Atabron®) is a benzyl phenyl urea (BPU) chitin-synthesis inhibitor (CSI) and insect growth regulator (IGR) is formed by Ishihara Sangyo Kaisha, Japan. Some important details concerning the insecticide chlorfluazuron are given below (provided by Ishihara Sangyo Kaisha, Japan) (Perveen, 2005):

Common name	Chlorfluazuron (proposed to ISO)
Other names	Atabron® or Helix® or Aimâ
Source	Ishihara Sangyo Kaisha Ltd., Tokyo, Japan
Code number	IKI-7899, CG-112913, pp-145
Formulation type	5% w/w (Emulsifiable concentration: EC)
Chemical name	[1-{3,5-dichlor-4-(3-chlor-5-trifluoromethyl-2-pyridyloxy) phenyl}-3-(2,6-ifluorobenzoyl) urea] (IUPAC nomenclature)

1.1.1 Salient physical and chemical properties (Perveen, 2005)

Appearance	Crystalline solid at 20 °C
Odor	Odorless
Melting point	222.0 – 223.9 °C (decomposes after melting)
Vapor pressure	<10 – 8 p_a, <10-10 Torr at 20 °C
Volatility	Relatively non-volatile
Specific gravity	1.4977 at 20 °C
Stability	No detectable decomposition over at least 3 months at 50 °C

1.2 Spodoptera litura

The *S. litura* is found in most of the Caroline and in the South Pacific Island including American Samoa. It also occurs in the northern two thirds of Australia. The moth is also widespread throughout India and recognized as quarantine pest in EU legislation. It is present in Mediterranean Europe and Africa. It is the most commonly intercepted in the UK, on imported ornamentals and their products. *Spodoptera litura* is also a destructive pest of subtropical and tropical agriculture, and has the potential to be a serious pest of glasshouse crops in northern Europe. It was found as feeding on impatience on Victoria Peak on Hong Kong Island and readily switched to (western) lettuce (Etman and Hooper, 1979). In 1974, Etman and Hooper initiated an investigation into the radiobiology of *S. litura*, and reported that it was a significant pest of cotton in the Ord River region, Australia (Etman and Hooper, 1979). Its larvae are a major cosmopolitan pest of a wide range of crops (Skibbe et al., 1995). Matsuura and Naito (1997) reported that *S. litura* causes serious widespread damage to many agricultural crops in the far southern of the Central Japan every year. They hypothesized that adult *S. litura* immigrate into Japan from overseas every year by long- distance migration. The larvae are destroyed many economically important crops such as *Gossypium hirsutum* L., *Brassica oleracea* L., *Spinacea oleracea* L., *Trifolium alexandrinum* L., *Medicago sativa* L., *Arachis hypogaea* L., *Phaseolus aureus* Roxb., *Phaseolus vulgaria* L. and *Nicotiana tabacum* L. during different seasons throughout the year in Pakistan (Younis, 1973). Their larvae eat nearly all types of herbaceous plants. Some examples of plants are: tobacco, *Nicotiana tabacum* L.; tomatoes, *Lycopersicum esculentum* Mill.; cauliflower, *Brassica botrytis* L.; beetroot, *Beta vulgaris conditiva* L.; silver beet (swiss chard) *Beta vulgaris cicla* (L.); peanuts, *Arachis hypogaea* L.; beans, *Phaseolus vulgaris* L.; banana, *Musa* paradisiaca L.; strawberry, *Fragaria* vesca L.; apple, *Malus pumila* Mill.; lettuce, *Lactuca sativa* L.; zinnia, *Zinnia elegans* Jacq.; dahlia, *Dahlia pinnata* Cav.; ape, *Alocasia macrorrhiza* (L.); geranium, *Pelargoniumx zonale*; St. John's lily, *Crinum asiaticum* L.; mangrove lily, *Crinum pedunculatum* (Fragrant); leek, *Allium porrum* (Leek); horsetail she oak,

Casuarina equisetifolia L.; *Fuchsia* and many other garden plants (Baloch and Abbasi, 1977). Several common names have been used for *S. litura*, for example defoliator cutworm, oriental leafworm, cluster caterpillar and common cutworm. The larvae are quite polyphagous for example eat all types of herbaceous plants and have been reaching the status of international pest. In 1968, a panel convened by the International Atomic Agency listed species of *S. litura* as a pest on which basic and applied research was needed in order to evaluate the potential of the sterile insect release method for control (Anonymous, 1969).

Eggs of *S. litura* are laid in batches, on plants and other surfaces such as pots, benches or glasshouse structures. Eggs are normally laid in the irregular furry masses covered with orange-brown hairs giving them a "felt-like" appearance on the underside of a leaf of a food plant similar to the egg of the brown locust, *Locustana pardalina* (Walk.) (Matthee, 1951). On hatching, larvae (caterpillars) are 2–3 mm long with white bodies and black heads and are very difficult to detect visually. If they emerge from eggs laid on glasshouse structures or hanging pots, they can reach the plants below by "parachuting" down on silken threads. The overall colour of the later-stages of the larvae can vary from light to dark brown, and the body is strongly speckled with tiny white spots. Initially, when larvae grow become a translucent green with a dark thorax. The young larvae are smooth-skinned with a pattern of red, yellow, and green lines, and with a dark patch on the mesothorax. Larvae initially eat only the flesh of their food leaves, leaving the veins intact. Later, as they grow, they eat whole leaves, and even flowers and fruit (Khuhro et al., 1986).

Many populations are extremely resistant to pesticides and, if they become well established, can be exceptionally difficult to control. In these cases, it is important that a comprehensive treatment programme is implemented, incorporating a range of reliable control methods, including physical destruction of insects. The *S. litura,* as it is the most common to be encountered in a UK nursery, but the larvae and adults of all noctuids are similar in appearance and are difficult to tell apart without laboratory examination (Khuhro et al., 1986). Larvae become brown with three thin yellow lines down the back, one in the middle and one on each side. A row of black dots runs along each side, and a conspicuous row of dark triangles decorates each side of the back. The last-instar larva is very dark, with four prominent yellow triangles on the mesothorax. When disturbed, the larvae curl into a tight spiral with the head protected in the centre. Larvae further develop characteristic markings on their backs. These include: a square of four yellow spots, each on a black patch, located just behind the head; a further pair of black patches just behind these, and another pair of black patches towards the end of the larva; typically, there are three orange-brown lines, punctuated with dashes of black and yellow along the back of the body. Depending on the background colour, these markings may be more evident on some larvae than others; the larvae ultimately grow up to 4.5 cm long; larvae are nocturnal, and during the day can be found at the base of plants or under pots. The feeding activity of young larvae causes "windows" in the leaves, while older larvae can completely defoliate plants if present in large numbers. Stems, buds, flowers and fruits may also be damaged. The larvae burrow into the soil below the plant for several centimeters and pupate there without a cocoon. As they do so, they produce a quantity of fluid, and drown in this if they pupate in captivity in an empty glass jar. They pupate successfully if 0.5 cm of sand is provided in the container. In January in Melbourne, the pupal stage lasts three weeks, but larvae that pupate at the end of summer emerge the following spring. The red-brown pupae are up to 2 cm long. The thoracical, ventral knobs found on covering of pupa (Khuhro et al., 1986). Adult moths with brown colour are up to 2 cm long with a wingspan of approximately 4 cm. The fore-wings are brown, with a large number of pale cream streaks and dashes and, when the adults are newly emerged, there may be a violet tint to the fore-wing. The hind-wings are a translucent white, edged with brown. The hind-wings are silvery white. It has a wingspan of about 4.0 cm. The males but not the females have a blue-grey band from the apex to the inner margin of each forewing. The pheromones of this species (specific sex-attractant scents used by females to attract males) have been elucidated. As the adults are nocturnal, light or pheromone traps should be used for monitoring purposes. Seek assurance from suppliers that plants are free from this pest as part of any commercial contract: carefully inspect new plants and produce on arrival, including any packaging material, to check for eggs and larvae and for signs of damage (Khuhro et al., 1986).

Early notification of the presence of this pest, will allow rapid implementation of a comprehensive treatment programme, and will help eradicate it quickly from nursery. Established outbreaks are very damaging and difficult to eradicate. Various methods of control of *S. litura* have been investigated. Biologically, it has been controlled by the nematode, *Steinernema carpocapsae* (Weiser) and parasitoid fly, *Exorista japonica* (Townsend). A baculovirus has also been used. Resistant species of plants are also grown to save the crops from this pest. Resistant tomatoes are most commonly cultivated (Khuhro et al., 1986).

2. EFFECTS OF CHLORFLUAZURN ON REPRODUCTION OF SPODOPTERA LITURA

Reproductive inhibition induced by BPUs has been reported the most widely when applied to adults or eggs of insect pests rather than to application to larvae or pupae (Fytizas, 1976). When these compounds were applied to females, males or both sexes of insect pests, BPUs induced a variety of effects on reproduction;

they caused a decrease in fecundity, fertility and/or hatchability. It has been reported that treatment of adult insect pests with diflubenzuron disrupts the secretion of adult cuticle (Hunter and Vincent 1974; Ker, 1977), and the production of peritrophic membrane in the grasshopper, *Locusta migratoria* (L.) (Clark et al., 1977) and the meal worm, *Tenebrio molitor* L. (Soltani, 1984; Soltani et al., 1987). In addition, topical treatments of male and female adult boll weevils, *Anthonomous grandis* Boheman; stable flies, *Stomoxys calcitrans* (L.) and *M. domestica* with TH-6040 [{N-(4- chlorophenyl)-N-26-difluorobenzoyl}urea] caused significant reduction of egg fertility and hatchability. It also causes inhibition in the fecundity of female adults of several species of insect pests (Holst, 1974; Taft and Hopkins, 1975; Crystal, 1978; Hajjar and Casida, 1979; Otten and Todd, 1979). Diflubenzuron applied to adult females caused a decrease in fecundity in the Mexican bean beetle, *Epilachna varivestis* Mulsant and Colorado potato beetle, *Leptinotarsa decemlineata* Say (Holst, 1974), and adversely affected egg viability in *St. calcitrans* and *M. domestica* (Wright and Spates, 1976). When 2 day-old female adults of the Japanese beetle, *Oryzae japonica* Willemse, were starved for 6 hours, and then allowed to consume 500 µg a.i. of diflubenzuron over another 6 hours, the fecundity of the treated females, in term of number of eggs laid per pod, was significantly decreased from controls. In controls, most pods gave an egg hatch of 82.5% but a hatch of only 8.5% hatched in the treated females (Lim and Lee, 1982). Similarly, treating eggs with diflubenzuron caused reduction in hatching in the mosquitoes, *Culex pipiens* Say, *C. quinquefascialus* Say (Miura et al., 1976), the almond moth, *Ephestia cautella* (Walk.) (Nickle, 1979), the two-spotted lady beetle, *Adalia bipunctata* (L.) and the seven-spotted lady beetle, *Coccinella septempunctata* L. (Olszak, 1994). Reports of inhibition of reproduction when larvae or pupae (instead of adults or eggs) were treated with BPUs are rare and little literature is available (Madore et al., 1983). Brushwein and Granett (1977), working with the spruce budworm, *Choristoneura fumiferana* (Celemens), demonstrated that certain moult-inhibiting IGRs such as EL-494 (Eli Lilly and Co., New York, USA) fed to sixth-instar larvae, caused reproductive failure in adults surviving after the larval treatment. Therefore, in this research newly ecdysed fifth- instar larvae and newly ecdysed pupae of *S. litura* were used as test materials. Chlorfluazuron, a comparatively new IGR and BPU that was discovered by Ishihara Sangyo Kaisha Ltd., Tokyo, Japan, that has been developed and sold commercially as Atabron[â], Helix[â] and Aim[â] in many countries, including Japan in cooperation with Novelties Co. Ltd, ICI-AGRO and Ciba Geigy. It is a relatively highly active chitin synthesis inhibitor and it is, therefore, an effective treatment for the control of major lepidopteran insect pests in crops such as cotton, fruits, tea, vegetables and where insect resistance to conventional insecticides is becoming a serious problem. Chlorfluazuron exhibits no activity against important beneficial insects (Haga et al., 1992). The highly selective insecticidal activity of chlorfluazuron is particularly suited to integrated pest management programmes. Although chlorfluazuron has contact toxicity at higher rates, the major route of toxicity to insects is ingestion, and it has no root, systemic or foliar translaminar activity. Like other BPUs, chlorfluazuron is believed to disrupt chitin formation and, thus, kills the insects when they moult. This mode of action necessarily means that it is effective only against immature insects and that it is relatively slow actions. When higher dosages of chlorfluazuron were applied to newly ecdysed fifth-instar larvae, it had a devastating effect on the *S. litura* population by killing them during larval, pupal and adult stages (Hashizume, 1988). In insect pest management, the purpose of research is to maintain the pest population below the economic injury level. The mode of action of chlorfluazuron, as a CSI is known to some extent. However, but the knowledge of its effects on reproduction are rare. Insect structure and physiology may vary considerably during growth and development, with certain stages being more susceptible to insecticides than others. For example, the cuticle varies in its composition during larval development and this has been related to changes in IGR susceptibility. The activity of various insecticides detoxifying enzymes, such as MFO, glutathione S-transferase and epoxide hydrase also fluctuate during the life cycle of an insect (Yu, 1983). For this purpose newly ecdysed fifth-instar larvae and newly ecdysed pupae were selected for the treatments for the present research. The main objective of the present research is to determine the effects of sublethal doses (LD_{10}: 1.00 ng larva^{-1}; 0.12 ng female pupa-1; 1.23 ng male pupa^{-1} or LD_{30}: 3.75 ng larva^{-1}) of chlorfluazuron on the reproduction (e.g., fecundity, fertility and hatchability) when ha been apply to newly ecdysed fifth-instar larvae and newly ecdysed pupae of *S. litura* (Perveen, 2000a).

2.1 Experimental procedures

2.1.1 Insect rearing

Experiments were conducted with *Spodoptera litura* (F.) (Lepidoptera: Noctuidae) taken from a stock that was established from eggs obtained from Aburahi Laboratory of Shionogi Pharmaceutical (Koga-Shiga-Pref., Japan). The larvae of *S. litura* were reared in the laboratory under controlled conditions on the artificial diet Insecta LF® (Nihon Nohsan- kohgyo, Kanagawa, Japan). The rearing temperature was maintained at 25±1 °C, with a L16:D8 hour photoperiod and 50-60% r.h. To facilitate observations, the dark period was set from 06:00 to 14:00 hours. Adults were fed on a 10% sucrose solution soaked in cotton. The eggs, which were laid on Rido® cooking paper (Lion, Tokyo, Japan), were collected every 3rd day and kept in 90 ml plastic cups (4 cm in diameter: 4×4 cm high) for hatching under the same environmental conditions (Perveen, 2000a).

2.1.2 Chlorfluazuron and its application

Sublethal doses, LD_{10} (1.00 ng larva^{-1}; 0.12 ng female pupa^{-1}; 1.23 ng male pupa^{-1}) or LD_{30} (3.75 ng larva^{-1}) were applied to newly ecdysed fifth-instar larvae and newly ecdysed pupae. These LD_{10} and LD_{30} values were calculated based on interpret alone of the results of the toxicity data of larval tests at adult emergence. The treated and untreated insects, at all developmental stages including fifth- and sixth-instar larvae, pupae and adults, were weighed separately, on different developmental days, using an analytical balance (Sartorius Analytical AC-2105, Tokyo, Japan) to a precision of 0.001 mg, to determine the effect of chlorfluazuron on the body weight. The duration of each developemental stage was also strictly recorded (Perveen, 2000a).

2.1.3 Mating

After both larval and pupal treatments, females and males that emerged between 2 and 8 hour (most adults emerged in the dark photoperiod) on the same day were collected at 0200–1000 hour. These adults were considered as 0 day old and paired just before the dark photoperiod (12 hour old) of the next day. Each female and male pair was kept separately in a plastic cup (430 cm3; height: 8.0 cm; diameter: 9.5 cm) for the whole life-span. The cup was padded with Rido cooking paper on its wall and with a disc of 70 mm filter paper on the bottom. The pairs were fed throughout their life by cotton wool soaked in 10% sugar solution in small plastic cups. All *S. litura* were examined daily (Perveen, 2000a).

To determine the effects of sublethal doses of chlorfluazuron on reproductivity, seven different mating combinations of female and male crosses were established. These were: (1) Untreated female mated with untreated male (${}_{U}$♀′${}_{U}$♂); (2) LD_{10}-treated female mated with untreated male (${}_{LD10}$♀′${}_{U}$♂); (3) Untreated female mated with LD_{10}-treated male (${}_{U}$♀′${}_{LD10}$♂); (4) LD_{10}-treated female mated with LD_{10}-treated male (${}_{LD10}$♀′${}_{LD10}$♂); (5) LD_{30}-treated female mated with untreated male (${}_{LD30}$♀′${}_{U}$♂); (6) Untreated female mated with LD_{30}-treated male (${}_{U}$♀′${}_{LD30}$♂); (7) LD_{30}-treated female mated with LD_{30}-treated male (${}_{LD30}$♀′${}_{LD30}$♂). For the fecundity, fertility and hatchability experiments, 15–30 pairs were used for each cross. Eggs were laid on the cooking paper after 24 hours and were collected during 0800–1000 hour, cut out and kept in cups (90 cm3) for hatching. Eggs laid were hatched within 84 hours. Observation of oviposition continued until the death of female. Four days after each collection of eggs, the fecundity, fertility and hatchability of the laid eggs were assessed. After the natural death of females, the spermatophores were separated from the bursa copulatrix with a fine forceps in 0.9% NaCl (Saline or Ringer's solution: Barbosa, 1974) under the binocular microscope (10′magnification) (Olympus Co. Ltd., Tokyo, Japan) (Perveen, 2000a).

2.1.4 Data analysis

Data for the effects of sublethal doses of chlorfluazuron on reproductivity and viability were analyzed using analysis of variance, one way ANOVA (Concepts, 1989; Minitab, 1997; Walpol and Myers1998) at P<0.0001 and Scheffe's F-test (multiple range) (Scheffe, 1953) at 5%. Hatchability percentage values were normalized by arcsin transformation before statistical analysis (Anderson and McLean, 1974).

2.2 Results

When the LD_{10} (1.00 ng larva^{-1}; 0.12 ng female pupa-1; 1.23 ng male pupa-1) and the LD_{30} (3.75 ng larva^{-1}) of chlorfluazuron were applied to newly ecdysed fifth-instar larvae or newly ecdysed pupae, it was observed that the fecundity of the resulting adults and the fertility and hatchability of their eggs, was significantly reduced {P<0.0001 (for larval treatment); P<0.05 (for pupal treatment)}, compared with untreated adults, but no significant differences were observed between larval and pupal treatments (P<0.02) (Tables 1 and 2). When chlorfluazuron was applied to newly ecdysed fifth-instar larvae at sublethal doses, the number of eggs oviposited by a treated females mated with an untreated male ($_{T}$♀′$_{U}$♂) was suppressed to the same degree as an untreated female mated with a treated male ($_{U}$♀′$_{T}$♂) or a treated female mated with a treated male ($_{T}$♀′$_{T}$♂) (Table 1). The mean female fecundity was 2250±198 eggs when both male and female were untreated (control), i.e., ($_{U}$♀′$_{U}$♂), (Table 2). When the female was treated either by the LD_{10} or LD_{30} and mated with an untreated male, the fecundity was 1462±353 ($_{LD10}$♀′$_{U}$♂) and 1266±237 ($_{LD30}$♀′ $_{U}$♂), respectively. When the male was treated with the LD_{10} and mated with an untreated female, the fecundity was 1407±334 ($_{U}$♀′$_{LD10}$♂). However, in the same cross when the male was treated with LD_{30} instead of the LD_{10} ($_{U}$♀′$_{LD30}$♂), the fecundity was 1270±215. When both sexes were treated with the LD_{10} ($_{LD10}$♀′$_{LD10}$♂), it was 1330±295 and when both sexes were treated with the LD_{30} concentration ($_{LD30}$♀′$_{LD30}$♂), it was 1331 ±295. In all the crosses, the fecundity was significantly reduced when compared with the control cross ($_{U}$♀′$_{U}$♂) (Table 1) (Perveen, 2000a).

Table 1. Effects of sublethal doses of chlorfluazuron on fecundity, fertility and hatchability after topical application of newly ecdysed-fifth instar larvae of *Spodoptera litura* (Source: Perveen, 2000a).

Mating pairs[a] (female′male)	na	Fecundity[b,c] (mean±SD)	Fertility[b,c] (mean±SD)	Hatchability%[c,d] (mean±SD)
$_{U}$♀′$_{U}$♂	30	2250±198a	1984±208a	88.4±6.6a
LD10♀′U♂	30	1462±353b	1010±315b	68.9±11.8b
U♀′LD10♂	28	1407±334b	688±317c	48.3±17.1c

Mating pairs[a] (female′male)	na	Fecundity[b,c] (mean±SD)	Fertility[b,c] (mean±SD)	Hatchability%[c,d] (mean±SD)
LD10♀′LD10♂	30	1330±295b	643±265c	48.6±14.2c
LD30♀′U♂	29	1266±237b	828±206b	65.8±11.4b
U♀′LD30♂	30	1270±215b	36±155d	28.8±11.2d
LD30♀′LD30♂	29	1331±295b	33±121d	27.1±9.1d

a LD10: 1.00 ng larva^{-1}; LD30: 3.75ng larva^{-1}; U: untreated (control); ♀: female adults; ♂: male adults; n: number of pairs used

b Number of eggs oviposited during the whole life of female adult were counted (fecundity) and from the eggs number of hatched that larvae were counted (fertility).

c Data were analyzed using one way ANOVA (Concepts, 1989) at P<0.0001). Means within columns followed by different letters indicate significant differences by Scheffe`s F-test (Scheffe, 1953) at 5%. dHatchability % values were normalized by arcsin transformation before statistical analysis (Anderson and McLean 1974).

When the LD_{10} of chlorfluazuron was applied to newly ecdysed pupae and the resulting adults were paired, the control fecundity was 2170±175. It was not significantly reduced (P<0.02) compared with the larval treatment. The fecundity in the treated pupal cross ($_{LD10}$♀′$_{U}$♂) was 640±83. It was not significantly reduced (P<0.02) compared with the same cross with treated larvae. In the pupal treatment, the fecundity was suppressed to a similar degree in the crosses $_{LD10}$♀′$_{U}$♂, $_{U}$♀′$_{LD10}$♂, $_{LD10}$♀′ $_{LD10}$♂. This reduction was significant (P<0.0001) when compared with the control cross (Table 1) (Perveen, 2000a). There was no significant reduction (P<0.02) between the larval and pupal treatments with respect to fecundity (Tables 1 and 2) (Perveen, 2005).

The mean fertility of females was 1984±208 larvae when both the male and female were untreated, i.e. the control ($_{U}$♀′$_{U}$♂) (Table 1). When the female was treated with either the LD_{10} or LD_{30} and mated with an untreated male, the fertility was ($_{LD10}$♀′$_{U}$♂: 1010±315; $_{LD30}$♀′ $_{U}$♂: 828±206, respectively), i.e., significantly reduced compared with the control cross ($_{U}$♀′ $_{U}$♂). When the male was treated with the LD_{10} and mated with an untreated female, the fertility ($_{U}$♀′ $_{LD10}$♂: 688±317) was significantly reduced compared with the crosses, i.e., $_{LD10}$♀′$_{U}$♂ and $_{LD30}$♀′$_{U}$♂. However, in the same cross when the male was treated with the LD_{30} instead of LD_{10}, the fertility was ($_{U}$♀′ $_{LD30}$♂: 368±155) significantly lower than the $_{U}$♀′$_{LD10}$♂ cross. When both sexes were treated with the LD_{10}, the fertility was ($_{LD10}$♀′$_{LD10}$♂: 643±265) not significantly different from the $_{U}$♀ ′$_{LD10}$♂ cross. Similarly, when both sexes were treated with the LD_{30}, the fertility was ($_{LD30}$♀′$_{LD30}$♂: 333±121) not significantly different from the $_{U}$♀′$_{LD30}$♂ cross (Table 1) (Perveen, 2000a).

Table 2. Effects of a sublethal doses of chlorfluazuron on fecundity, fertility and hatchability after topical application of newly ecdysed pupae of *Spodoptera litura* (Source: Perveen, 2005).

Mating pairsa (female´male)	na	Fecundityb,c (mean±SD)	Fertilityb,c (mean±SD)	Hatchability%c,d (mean±SD)
$_U$♀´$_U$♂	15	2170±175a	2123±177a	97.8a
LD10♀´U♂	15	1640±83b	1090±79b	66.5b
U♀´LD10♂	15	1580±75b	827±49c	52.3c
LD10♀´LD10♂	15	1524±76b	751±51c	49.3c

a LD10: 0.12 ng female pupa-1; 1.23 ng male pupa-1; n: number of pairs used

b Number of eggs oviposited (fecundity) the during whole life of female adults were counted and from the number of eggs that hatched larvae were counted (fertility).

c Data were analyzed using one-way ANOVA (Concepts, 1989) at P < 0.001. Means within a column followed by different letters indicate significant differences according to Scheffe's F-test (Scheffe, 1953) at 5%.

d Hatchability% values were normalized by arcsin transformation before statistical analysis (Anderson and McLean 1974).

When the LD_{10} of chlorfluazuron was applied to newly ecdysed pupae and resulting adults were paired, the fertility of the control cross, $_U$♀´$_U$♂ was 2123±177, which was not significantly reduced (P<0.02) than that in the larval treatment. In the same way, with the $_{LD10}$♀´$_U$♂ cross, the fertility was 1090±79, which was not significantly reduced (P<0.02) than the same cross in the larval treatment. The fertility was reduced 61 –65% in the $_U$♀´$_{LD10}$♂ cross, which was significantly reduced (P<0.02) than the $_{LD10}$♀´$_U$♂ cross with a fertility reduction of 46–49%. The fertility of the $_U$♀´$_{LD10}$♂ cross was not significantly reduced (P<0.02) than the $_{LD10}$♀´$_{LD10}$♂ cross, which was reduced 65–68% for both larval and pupal treatments. There were no significant reductions (P<0.02) between larval and pupal treatments with respect to fertility (Tables 1 and 2). The mean hatchability during a female life-span was 88.4(±6.6)% when both male and female were untreated, i.e. the control ($_U$♀´$_U$♂) cross (Table 1). When the female was treated, either by the LD_{10} or LD_{30} and mated with an untreated male, the hatchabilities were 68.9 (±11.8)% ($_{LD10}$♀´$_U$♂) and 65.8(±11.4)% ($_{LD30}$♀´$_U$♂), respectively significantly reduced compared with the control cross. When the male was treated with the LD_{10} and mated with an untreated female, the hatchability, 48.3(±17.1)% ($_U$♀´$_{LD10}$♂) was significantly reduced than the $_{LD10}$♀´$_U$♂ and $_{LD30}$♀´$_U$♂ crosses. However, in the same cross when the male was treated with LD_{30}, instead of LD_{10}, hatchability, 28.8(±11.2)% ($_U$♀´$_{LD30}$♂), significantly reduced than the $_U$♀´$_{LD10}$♂ crosses. When both sexes were treated with LD_{10}, the hatchability was 48.6(±14.2)% ($_{LD10}$♀´$_{LD10}$♂), not significantly different from the $_U$♀´$_{LD10}$♂ cross. Similarly, when both sexes were treated with LD_{30}, the hatchability was 27.1(±9.1)% ($_{LD30}$♀´$_{LD30}$♂), which was not significantly different from the $_U$♀´$_{LD30}$♂ cross (Table

3.1). The hatchability for the $_{U}$♀′$_{U}$♂ cross was 88.4% and 97.8%, respectively, for the larval and pupal treatments. The larval-treated cross, $_{LD10}$♀′$_{U}$♂, was 68.9%, which is not significantly reduced (P<0.02) than the same cross of the pupal treatment in which the hatchability was 66.5%. The hatchability was reduced 48.7% and 52.7% in the $_{U}$♀′$_{LD10}$♂ cross, respectively, for the larval and pupal treatments. In the $_{LD10}$♀′$_{LD10}$♂ cross, it was reduced to 48.6% and 49.3%, respectively, for these treatments. There was no significant reduction (P<0.02) between larval and pupal treatments with respect to hatchability (Tables 1 and 2) (Perveen, 2000a).

2.3 Discussion

When chlorfluazuron was applied to newly ecdysed fifth instars at sublethal doses, LD_{10} (1.00 ng larva^{-1}) or LD_{30} (3.75 ng larva^{-1}), it was observed that the fecundity of resulting adults as well as the hatching rate of their eggs was suppressed. The hatching rate of eggs oviposited by an untreated female mated with a treated male was suppressed to the same degree as that ofeggs oviposited by a treated female mated with a treated male. However, Madore et al. (1983) studied the effects when different concentrations of sublethal doses of the UC-62644 (chlorfluazuron-25) fed to sixth instar larvae of spruce budworm. Homologous crosses between adults of the 0.01, 0.025 and 0.034 ppm treatments showed 0, 69 and 97% reduction, respectively, in the numbers of eggs laid per 30 pairs of moths when compared with control. Emam et al. (1988) reported the fecundity of *S. littoralis* adults decreased significantly from 977.64 eggs in control to 421.75 eggs, a decrease of about 56%, for adults feeding 10% honey solution containing 0.5 p.p.m. chlorflufluazuron. The corresponding fertility inhibition amounted to 32%. In the present case the fertility was significantly different when only the female was treated or only the male was treated. It is obvious from the results that the fertility and hatchability were affected more when the male was treated in comparison with the female, as also reported by Abro et al. (1997), who found that males were more sensitive to insecticides than females, when five concentrations of cyhalothrin and fluvalinate were tested against fourth instar larvae of *S. litura*.

2.4 Concise Summary

To clarify the sublethal effects of chlorfiuazuron on reproductivity of common cutworm, *Spodoptera litura*, experiments were conducted under laboratory conditions. Reduction in the body weight was observed in the larvae and pupae when treated with a sublethal dose (LD_{30}: 3.75 ng larva^{-1}) and in the adults when treated with sublethal doses (LD_{10}: 1.00 ng larva^{-1}; LD_{30}: 3.75 ng larva^{-1}) as newly ecdysed fifth instar larvae of *S. litura*, although the number of matings per female and life span of adult females and males remained unaffected by the same treatments. When sublethal doses were applied only to females or only to males, or both sexes, the average fecundity reduction was up to 35–44%. When only females were treated with sublethal doses, fertility was reduced by 49–58%; when only males were treated fertility was reduced by 65–81% and when both sexes were treated, fertility was reduced by 68±83%. Hatchability was reduced by 22–26% when only females were treated, by 44–66% when only males were treated and by 45–72% when both sexes were treated with LD_{10} or LD_{30} doses as newly ecdysed fifth instars. The results from these observations suggest that the fecundity was reduced to a similar degree when only females or only males or both sexes were treated with LD_{10} or LD_{30} doses as newly ecdysed fifth instars. However the fertility and hatchability were affected more when only males were treated with LD_{10} and much more when treated with LD_{30}. Currently, work is in progress to find out the main reasons for the sublethal effects of chlorfluazuron on reproductivity and viability.

3. EFFECTS OF CHLORFLUAZURN ON FEMALE REPRODUCTIVE SYSTEM OF SPODOPTERA LITURA

In many insects oviposition requires the development of the ovary, egg maturation, mating and, in some insects, feeding of the females. Ovarian development, which includes oöcyte growth and vitellogenesis, is under the hormonal control, of either juvenile hormone or ecdysteroid (Engelmann, 1979). In many insects, juvenile hormone (JH) regulates the biosynthesis and uptake of vitellogenin by the oöcytes. Among Lepidoptera, e.g., the tobacco hawkmoth, *Manduca sexta* L. (Sroka and Gilbert, 1971; Nijhout and Riddiford, 1974) and the large white butterfly *Pieris brassica* L. (Karlinsky, 1963 and 1967; Benz, 1969), juvenile hormone is required for full development of the ovaries in adults, whereas in the silkworm *Bombyx mori* L. (Chatani and Ohnishi, 1976), giant silk moth, *Hyalophora cecropia* (L.) (Williams, 1952; Pan, 1977), ailanthus silkmoth, *Samia cynthia* (Drury) (Takahashi and Mizohata, 1975) and ricemoth, *Corcyra cephalonica* (Stainton) (Deb and Chakarvorty, 1981) ovarian development occurs as part of adult development initiated by ecdysteroid. Juvenile hormone or juvenile hormone analogue (JHA) application at a critical period, however, induces abnormal development of the ovary as well as other tissues, although juvenile hormone analogues can replace natural juvenile hormone in regulating oöcyte maturation (Nomura, 1994). In the normal state, the ovary develops during one day before and after eclosion in the presence of juvenile hormone (as described above) and a haemolymph factor stimulates the ovary to start oviposition. When S-71639 was applied to pupae, it inhibited adult emergence when a relatively a high dose was applied. If adults did emerge, they could not oviposit through inhibition of the haemolymph factor, hatchability was also reduced (Hatakoshi and Hirano, 1990). The effects of diflubenzuron on fecundity resulted from treatment of adult females by contact or ingestion (Leuschner, 1974;

Fytizas, 1976). When *O. japonica*, adults females were fed diflubenzuron, it retarded the maturation of oöcytes (Lim and Lee, 1982). In *T. molitor*, diflubenzuron reduced mealworm longevity (Soltani et al., 1987), the number of oöcytes per ovary, the duration of the oviposition period and the fecundity (Soltani, 1984). Diflubenzuron, topically applied (0.5 µg insect-1) to codling moth, *Cydia pomonella* L. on pupal ecdysis, inhibited the growth and development of oöcytes. It delayed adult emergence and caused a decrease in both the thickness of the follicular epithelium and the size of the basal oöcytes during pupal development. On the other hand, the size of basal oöcytes, the protein content per ovary and the number of oöcytes per ovary recorded in newly emerged adults were significantly reduced by the diflubenzuron treatment. These results, together with observations in several other species, indicated that the reduction in fecundity and egg viability was probably due to interference by diflubenzuron with vitellogenesis (Soltani and Mazouni, 1992). Under laboratory conditions, effects of topical application of sublethal doses of chlorfluazuron (LD_{10}: 1.00 ng larva^{-1} or LD_{30}: 3.75 ng larva^{-1}) on newly ecdysed fifth-instar on fecundity, fertility and hatchability have been investigated. Thus, it is investigated the causes of the decrease in these parameters. To obtain more information, sublethal doses of chlorfluazuron topically have been applied to newly ecdysed fifth-instar larvae of *S. litura* and the effects on female reproductive system during ovarian development and oögenesis have been observed.

3.1 Experimental procedure

3.1.1 Ovary measurement

Experimental *S. litura* were rared in the same way as mentioned in Section 2.1.1. Sublethal doses, LD_{10} (1.00 ng larva^{-1}; 0.12 ng female pupa^{-1}; 1.23 ng male pupa^{-1}) or LD_{30} (3.75 ng larva^{-1}) were applied to newly ecdysed fifth-instar larvae same mentioned in Section 2.1.2. To determine the effects of chlorzuazuron on the ovaries, control and treated batches of insects, collected from the fifth day after pupation to the seventh day after adult emergence, were used, depending on the experiment requirements. Ovaries were dissected from these insects in Ringer's solution under a binocular microscope (10´magnifcation: Nikon, Nippon Kogaku, Tokyo, Japan), and the lengths of pedicle, vitellarium, and germarium of each ovariole were measured. The number of mature oöcytes in each of the ovarioles was also counted. The Ringer's solution was removed, and the freshly dissected ovaries were placed in a small covered container that had been preweighed. The dissected ovaries were weighed on an analytical balance (AC-205, Sartorius Analytical, Tokyo, Japan) and kept in the same container in the oven for 24 h at 62±1 °C for evaporation of water. The dried ovaries were reweighed (Perveen and Miyata, 2000).

3.1.2 Histology

The procedure to stain the nuclei for cell density from the germarium during female adults required age was adapted from the method described by He (1994). First, the germarium of the female were removed and kept on a microscope slide and carefully crushed with micro forceps until it was extended and roughly evenly distributed over the slide. Second, several drops of 3:1 methanol-acetic acid were introduced to the slide to fix the preparation for 15 min and then the excess fixing solution was absorbed with a filter paper after the fixation. Third, several drops of a 2–5% Giemsa solution dissolved by Sorensen-Gomori buffer solution (monobasic and dibasic sodium phosphate, 0.07 M, pH 6.8) were introduced to the slide for 10–30 min to stain the preparation. Then after the staining, the slide was washed gently and carefully and dried in the air. Finally, the air-dried preparation was checked with a phase-contrast microscope at 20´magnification (Perveen and Miyata, 2000).

The length and width of the basal oöcytes were measured from the 5th d after pupation to 0 day after adult emergence in control and treated insects. Oöcyte measurements were made on three to four basal oöcytes per pair of ovaries that were taken from 9-10 insects. Measurements were made with a graduated slide under the phase contrast microscope at 400´magnification (BH_2, Olympus, Tokyo, Japan). The size of basal oöcytes was calculated by the formula used by Loeb et al. (1984) for the size of a prolate spheroid, 4/3π(ab2), where *a* is the radius of the long and *b* is the short dimension of the same oöcyte (Perveen and Miyata, 2000).

The thickness of the follicular epithelium of basal oöcytes was observed by making a parafilm microtomy conducted according to the method described by Yoshida (1994). Basal oöcytes were fixed in Carnoy's solution for 3 hour, washed in 70% ethanol for 2 hour, and dehydrated in an alcohol series 70, 80, 90, and 95% and twice in 100%, followed by a benzene and ethanol solution (1:1), each for 30 min. Incubation was done three times at 60°C in benzene and paraffin (1:1), and then in parafin only, each for 30 min. Five-mm microtome sections were cut into a rolling ribbon. It was stained in xylene I (10 min), xylene II (5 min), followed by an ethanol series of 100, 90, 80, and 70%, each for 5 min, Mayer's hematoxylin for 15 min, and washed under running water. Microtome sections were mounted in 1% eosin (10 min), distilled water (2 sec), followed by an alcohol series 50, 70, 80, and 90%, and twice in 100% (each for 1-5 min), xylene and 100% ethanol (1:1; 5 min), xylene I (5 min), xylene II (5 min). Finally, the sections were embedded on microscopic graduated slide in a drop of Canada balsam. The microscopic graduated slide was covered with a glass cover slip. The thickness of follicular epithelium was measured under a phase contrast microscope (BH_2, Olympus, Tokyo, Japan) at 400´magnifcation (Perveen and Miyata, 2000).

3.1.3 Data analysis

Data were analyzed using analysis of variance, one way ANOVA (Concepts, 1989) at $P<0.01$ and Scheffe's F-test (Scheffe, 1953) at 5%.

3.2 Results

The morphology of the adult female reproductive system of *S. litura* is shown in Figure 1.

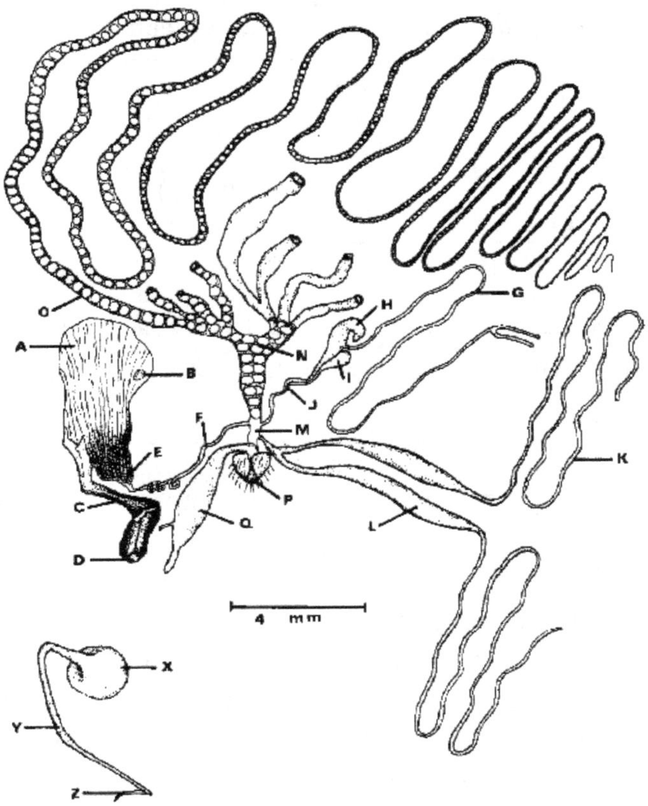

Fig. 1. The morphology of the female reproductive system of *S. litura*: A: corpus bursae; B: signum, C: ductus bursae; D: ostium bursae; and E: diverticulum of bursa copulatrix; F: ductis seminalis; G: spermathecal gland; H: utriculus I: lagena of spermatheca; J: ductus receptaculi; K: accessory gland (paired); L: accessory gland reservoir (paired); M: vestibulum; N: calyx of the unpaired oviductus communis; O: one of four ovarioles of ovary (paired); P: papillae anales; Q: rectum; X: corpus, Y: collum, and Z: frenum of spermatophores; (Source: Etman and Hooper, 1979).

3.2.1 Effects on ovarian development

Sublethal doses of chlorfluazuron (LD_{10}: 1.00 ng larva^{-1}; LD_{30}: 3.75 ng larva^{-1}), applied to newly ecdysed fifth-instar larvae significantly ($P<0.0001$) reduced the body weight, fresh ovarian weight and dry ovarian weight in newly emerged adults when compared with the controls (Table 3). Significant reductions were not observed in fresh body weight (P=0.0567), fresh ovarian weight (P=0.7788) and dry ovarian weight (P=0.5757), when the LD_{10} and LD_{30} treatments were compared. Similarly, ratios of fresh ovarian/fresh body weight (31.0%), dry ovarian/fresh ovarian weight (28.0%) and dry ovarian/fresh body weight (9.0%), were not significantly different (Table 3) (Perveen and Miyata, 2000).

Table 3. Effects of sublethal doses of chlorfluazuron on the ovarian and body weight of newly emerged adults after topical application to newly ecdysed fifth-instar larvae of *Spodoptera litura* (Source: Perveen and Miyata, 2000).

	FBW[a,b]	FOW[a,b]	DOW[a,b]	% R=FOW	% R=DOW	% R=DOW
	Ta	na	(M±SD)	(M±SD) mg (M±SD) mg /FBW[a,b]	/FBW[a,b]	/FOW[a,b]
mg			(M±SD)		(M±SD)	(M±SD)
C	30	255±11.6a 81.1±9.2a	23.6±1.4a	31.5±2.9a	28.9±2.4a	9.2±0.7a
LD_{10}	30	229±3.2b 72.0±3.4b	20.0±1.4b	31.2±1.5a	28.1±1.2a	9.0±0.8a
LD_{30}	30	224±5.9b 70.9±4.2b	19.9±1.3b	31.3±2.4a	28.0±0.5a	9.1±0.8a

a C: control; T: trearment; LD10: 1.00 ng larva^{-1}; LD30: 3.75 ng larva^{-1}; n: number of insects used; FBW: fresh body weight; FOW: fresh ovarian weight; DOW: dry ovarian weight; % R: percent ratio

b Data were analyzed using one-way ANOVA (Concepts, 1989) at P<0.001. Means within a column followed by different letters indicate significant differences according to Scheffe's F-test (Scheffe, 1953) at 5%.

Ovaries are small on the 8th day after pupation. From the 8th day after pupation to the day before adult emergence, ovarian weight slowly increased; after that it increased sharply until the day of adult emergence, and then, increased gradually until the 2nd day after adult emergence, when it reached maximum (120±19.4 mg). Then, in the controls it decreased gradually until the 7th day after adult emergence. The pattern of changes in fresh ovarian weight in the LD_{10-} or LD_{30-}treated females was similar as observed in the controls during various developmental days of pupae and adults. The fresh ovarian weight was significantly reduced on the 8th day after pupation (P<0.0001); on the 9th day after pupation (P<0.0001); on the 1st day after adult emergence (P<0.0003); on the 2nd day after adult emergence (P<0.0001); on the 3rd day after adult emergence (P<0.0001); on the 4th day after adult emergence (P<0.0001); on the 5th day after adult emergence (P<0.0001); on the 6th day after adult emergence (P<0.0001); on the 7th day after adult emergence (P<0.0001) in the LD_{10}- or LD_{30}-treated females compared with the controls, but no significant reduction was observed (P=0.0979–0.970) between the LD_{10}- or LD_{30}-treated females during ovarian development (Figure 2) (Perveen and Miyata, 2000).

In newly emerged the LD_{10-} or LD_{30-}treated adults, the total length of the ovariole was significantly reduced (P<0.0001) compared with the control, but there were no significant reductions (P=0.0508) between the LD_{10-} or LD_{30-}treatments. In the LD_{10-} or LD_{30-}treated insects, the germarium (immature oögonia) was significantly longer (P<0.0001) than that of the pedicle (fully mature ova) and the vitellarium (under developing oöcytes) compared with the controls in which the vitellarium was significantly longer (P<0.0001) than the germarium and pedicle (Figure 3; Table 4) (Perveen and Miyata, 2000; Perveen, 2011).

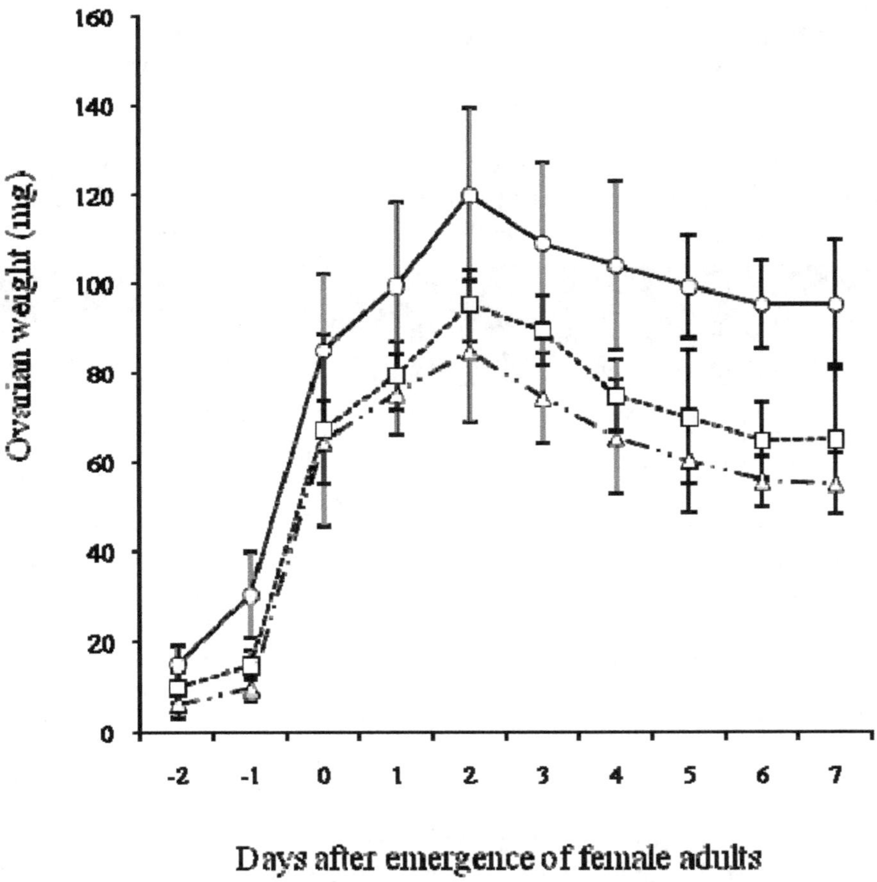

Fig. 2. Effect of sublethal doses of chlorfluazuron on ovarian weight during different developmental days (post pupal and 1 to 7 day after adult emergence). For control (○; n = 10), LD_{10} (1.00 ng larva^{-1}) treated (□; n = 9), and LD_{30} (3.75 ng larva^{-1}) treated (Δ; n = 9) after topical application to newly ecdysed fifth instars of *Spodoptera litura*. Data were analyzed using one-way ANOVA (Concepts, 1989) at *P*<0.0001 and Scheffe's *F*-test (Concepts, 1989) at 5%. Vertical bars indicate SD; (Source: Perveen and Miyata, 2000).

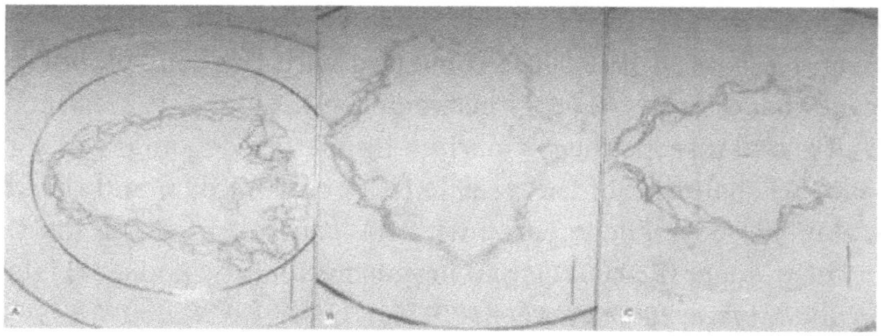

Fig. 3. A comparison of ovarian morphology of newly emerged adult *Spodoptera litura*, in A: untreated (control); B: treated with the LD_{10} dose and C: treated with the LD_{30} dose of chlorfluazuron. Bars in photographs indicate 100 µm (Source: Perveen, 2011).

Table 4. Effects of sublethal doses of chlorfluazuron on ovarian development in newly emerged adults after topical application to newly ecdysed fifth instars of *Spodoptera litura* (Source: Perveen and Miyata, 2000).

Treatmentsa	n_1	n_2	TLObc (M±SD)mm	LPbc (M±SD) mm	LVbc (M±SD) mm	LGbc (M±SD) mm	R (M±SD) mm	P:V:G
Control	10	80	104.8±5.1a[3]	33.5±3.4a	45.2±4.4a	26.1±4.2a	32:43:25	
LD_{10}	9	72	91.5±5.5b	22.1±5.0b	30.5±2.1b	39.0±2.7b	24:33:43	
LD_{30}	9	72	88.7±9.6b	20.4±4.4	28.6±6.7b	40.0±8.9b	23:32:45	

a LD10, 1.00 ng larva^{-1}; LD30, 3.75 ng larva^{-1}; n1: number of insects used; n2: number of ovariole measured; TLO: total length of ovariols; LP: length of pedicle; LV: length of vitellarium; LG: length of germarium; RP:V:G: Ratio of pedicle:vitellarium:germanium

b In Spodoptera litura, the paired ovaries are composed of 8 ovarioles. Four ovarioles are found on each side of the body cavity, forming several loops. Each ovarioles differentiated into 3 parts: (1) pedicle (fully matured eggs), (2) vitellarium (oöcytes and trophocytes), (3) germanium (oögonia) (Etman and Hooper 1979).

c Data were analyzed using one-way ANOVA (Concepts, 1989) at P<0.0001. Means within a column followed by different letters indicate significant differences according to Scheffe's F-test (Scheffe, 1953) at 5%.

When ratios of the length of the pedicle, vitellarium and germarium were compared, they were 32:43:25 for the controls, 24:33:43 for the LD_{10} and 23:32:45 for the LD_{30}. There was a significant reduction when the %ratios of the LD_{10} and LD_{30} were compared with the controls, but there was significant reduction between the LD_{10} and LD_{30} treatments. When ovarian maturation was observed untreated females had mature ova with an occasional one or two being absorbed (solid ova) in the ovarioles. In the LD_{10-}treated females, the spacing in the ovarioles and the absorption of ova different from the control. In LD_{30-}treated females, besides the spacing and absorption , sometimes only immature ova (germarium) were found in the ovarioles (data is not presented) (Perveen and Miyata, 2000). Mature ova were not observed in the pupae during the 2nd day before adult emergence, but a few mature ova were found the day before adult emergence. The number of mature ova sharply increased until the 1st day after adult emergence, and then gradually increased until the 2nd day after adult emergence. The maximum number of mature ova 725±2.0 was found on the second day after adult emergence. On the same day, the number of mature eggs was significantly reduced (P<0.0001) in the LD_{10-} or LD_{30-}treated females as compared with the controls, but no significant reduction was observed (P=0.0984) between the LD_{10} and LD_{30} treatments. From the 2nd day to the 7th day after adult emergence, absorption of mature ova started gradually in the controls, and the LD_{10-} or LD_{30-}treated females. The pattern of maturation of ova in ovaries was similar in the controls, LD_{10} and LD_{30} treatments (Figure 4) (Perveen and Miyata, 2000).

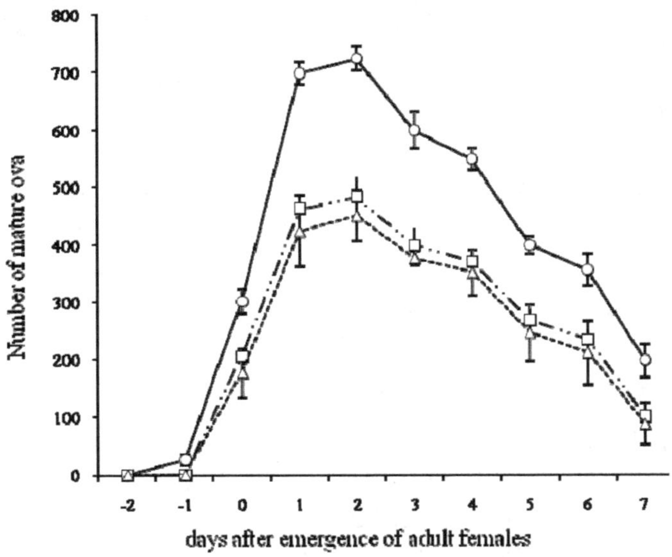

Fig. 4. Effect of sublethal doses of chlorfluazuron on number of mature eggs in the ovaries during different developmental days (post pupal and 1 to 7 day after adult emergence). For control (○; n = 13), LD_{10} (1.00 ng larva^{-1}) treated (□; n = 11), and LD_{30} (3.75 ng larva^{-1}) treated (Δ; n = 13) after topical application to newly molted fifth instars of *Spodoptera litura*. Data were analyzed using one-way ANOVA (Concepts, 1989) at *P*<0.0001 and Scheffe's *F*-test (Scheffe, 1953) at 5%. Vertical bars indicate SD; (Source: Perveen and Miyata, 2000).

The cell density, expressed as number of nuclei per mm2, was determined at various days during sexual maturation in the germaria of the controls, LD_{10-} or LD_{30-}treated females (Table 4.3). In the controls, on the 2nd day of adult emergence, the density was 1636±9.17 nuclei mm2-1. The cell density increased until the 3rd day, when it was 1829±8.87 nuclei mm2-1 and decreased thereafter. On the 4th day of adult emergence, the cell density was 1323±56.20 nuclei mm2-1. In the LD_{10-} or LD_{30-}treated insects, the patterns of the cell density change in the germarium were the same as in the controls but the values were significantly (P<0.01) lower in the LD_{10-} treated females and more were significantly decreased in the LD_{30-}treated females compared with the controls, during the days the 2nd, 3rd and 4th of adult female development. There was also a significant reduction (P<0.05) in the cell density between the LD_{10-} and LD_{30-} treated females during adult development (Table 5) (Perveen, 2011).

Table 5. Effects of sublethal doses of chlorfluazuron on the cell density in the germarium after topical application to newly ecdysed fifth-instar larvae of *Spodoptera litura* **(Source: Perveen, 2011).**

Treatments[a]	n_1	n_2	Cell density number of nuclei $(mm^2)^{-1}$ in the germarium during female adults age (M±SD)[b,c]		
			2 day-old	3 day-old	4 day-old
Control	5	10	1636±9.17a	1829±8.87a	1323±56.20a
LD_{10}	5	10	1570±42.50b	1753±49.91b	1235±9.50b
LD_{30}	5	10	1489±8.60c	1644±7.68c	1089±61.42c

a LD_{10}: 1.00 ng larva^{-1}; LD30: 3.75 ng larva^{-1}; n1: number of insects used; n2: number of ovariole measured

b The age of female adults was taken from the day of adult emergence.

c Data were analyzed using one-way ANOVA (Concepts, 1989) at P<0.01. Means within a column followed by different letters indicate significant differences according to Scheffe's F-test (Scheffe, 1953) at 5%.

3.2.2 Effects on oöcytes development

In the controls, the basal oöcytes were tiny on the 5th day after pupation, but increased sharply until the 8th day, after which they increased slowly until adult emergence. The maximum size of the basal oöcytes on the day of adult emergence was significantly reduced (P<0.0002) in the LD_{10-} or LD_{30-}treated females, but there was no significant difference (P=0.9976) between LD_{10-} and LD_{30-}treated females (Figure 5) (Perveen and Miyata, 2000).

The thickness of the follicular epithelium of the basal oöcytes gradually increased and reached a maximum on the 8th day after pupation, after which it sharply declined. On the 8th day after pupation, it was significantly reduced (P<0.005) in the LD_{10-} or LD_{30-}treated females when compared with control females, but there was no significant difference (P=0.8686) between the LD_{10} or LD_{30} treatments. The reduction was approximately 34% in the LD_{10-} and 39% in the LD_{30-}treated females. The patterns of the development of the follicular epithelium of basal oöcytes were not similar in the LD_{10-} or LD_{30-}treated females compared with the controls. However, pattern was similar between the LD_{10-} and LD_{30-} treated females. In the LD_{10-} or LD_{30-}treated females, the follicular epithelium reached maximum on the 9th day after pupation, and then declined. The development of the follicular epithelium was delayed by one day in the LD_{10-} or LD_{30-}treated females compared with the controls. On the 9th day, it was thicker in the LD_{30-}treated females and thickest in LD_{10-}treated females than in control females, but significant differences were not observed (P=0.7611) among these three groups, i.e. controls, LD_{10-} and LD_{30-}treated females (Figure 6) (Perveen and Miyata, 2000).

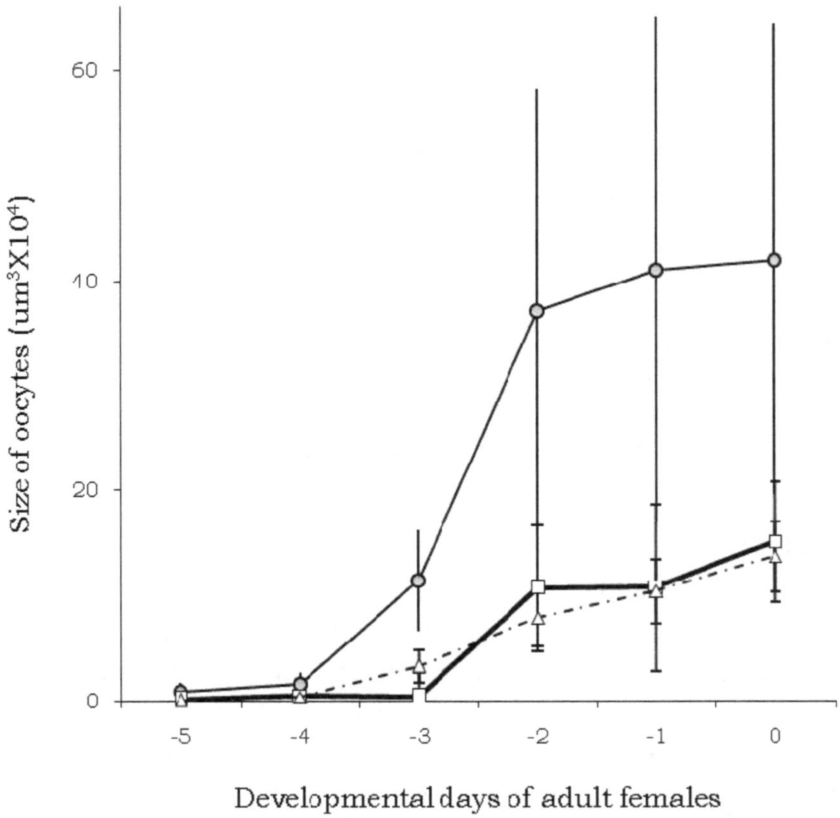

Fig. 5. Effect of sublethal doses of chlorfluazuron on size of basal oöcytes during different developmental days (5 to 9 day after pupation pupae and newly emerged adults). For control (○), LD_{10} (1.00 ng larva^{-1}) treated (□), and LD_{30} (3.75 ng larva^{-1}) treated (Δ) after topical application to newly ecdysed 5th instars of *Spodoptera litura*. Data were analyzed using one-way ANOVA (Concepts, 1989) at *P*<0.0002 and Scheffe's *F*-test (Scheffe, 1953) at 5%. Verticle bars indicate SD (n = 9-10); (source: Perveen and Miyata, 2000).

Data were analyzed using one-way ANOVA (Concepts, 1989) at *P*<0.005 and Scheffe's *F*-test (Scheffe, 1953) at 5%. Vertical bars indicate SD (n = 9-10); (Source: Perveen and Miyata, 2000).

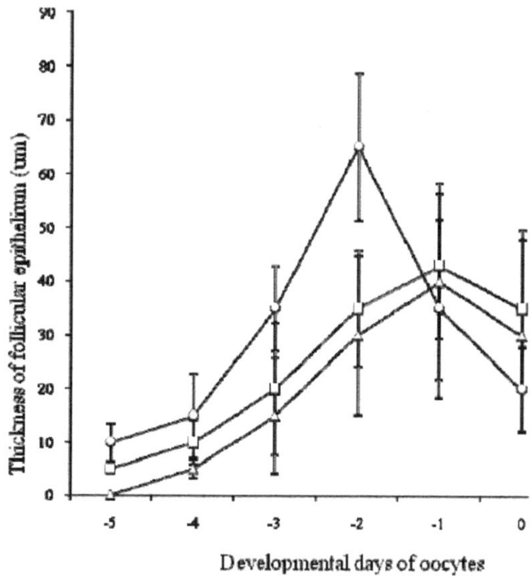

Fig. 6. Effect of sublethal doses of chlorfluazuron on thickness of follicular epithelium of basal oöcytes during different developmental days (5 to 9 days after pupation and newly emerged adults). For control (○), LD_{10} (1.00 ng $larva^{-1}$) treated (□), and LD_{30} (3.75 ng $larva^{-1}$) treated (Δ) after topical application to newly ecdysed 5th instar larvae of *Spodoptera litura*.

3.3 Discussion

Topical application of sublethal doses of chlorfluazuron (LD_{10}: 1.00 ng per larva or LD_{30}: 3.75 ng per larva) on newly ecdysed fifth instars had an effect on the reproduction of *S. litura* by reducing fecundity, fertility, and hatchability (Perveen, 2000a). Thus, this study was conducted to establish the causes of the reduction in these parameters. It was found that topical appication of sublethal doses of chlorfuazuron had an affect on ovarian development and oögenesis by decreasing the weight of ovaries during postpupal and adult developmental days of LD_{10} or LD_{30} treated females. The basic factors responsible for the reduction in ovarian weight were reduction in the length of different parts of the ovarioles, decrease in the number of mature ova, reduction in the size of basal oöcytes and thickness of their follicular epithelium, and reduction in protein content of ovarian constituents as compared with the controls. In this study, topical application of sublethal doses of chlorfuazuron signifcantly reduced the ovarian weight to that of the controls in postpupal and adult developmental days (Figure 2; Table 4.3). The ratios of fresh ovarian/fresh body weight, dry ovarian/fresh ovarian weight, and dry ovarian/ fresh body weight were the same among the newly emerged control, LD_{10} or LD_{30} treated females, indicating a reduction in the body, fresh, and dry ovarian weights with the same degree of reduction in treated females and controls. However, Soltani and Pickens and DeMilo (1977) reported that 0.5 mg DFB, when topically applied at pupal ecdysis to *Cydia pomonella* (L.), did not cause the ovarian weight be significantly reduced ($P<0.05$) between control and treated newly emerged adult females. Nor did Hatakoshi (1992) observe any signifcant reduction ($P<0.05$) in the ovarian weight between control and treated last day of pupae to third day after adult emergence of *S. litura,* when 0.3 ng pyriproxyfen was topically applied at pupal ecdysis. The differences in results observed in these experiments may be related to the pesticides, the kind of insect used and their developmental stages. *Spodoptera litura* has paired ovaries that branch into four polytrophic meroistic ovarioles located on the ventral side of the body cavity, making several loops of ovarioles, with all basal oöcytes developing simultaneously each ovariole is differentiated into tree portions according to the developmental stages of the oöcytes: (1) the yellowish green pedicle, where fully matured ova are stored; (2) the reddish orange vitellarium, which contains the developing oöcyte and trophocyte follicles which undergo accumulation of yolk proteins, and choriogenesis; and (3) the whitis germarium, which contains oögoia, from which germ cells proliferate and follicles are formed. Similar observations were reported by Riakhel and Dhadialla (1992) and Etman and Hooper (1979), which were confirmed here.

As in other Lepidopterous species, the ovaries of *S. litura* start to differentiate, and develop at the pupal stage. Indeed, in controls, the thickness of follicular epithelium of basal oöcyte reached its maximum size on the eighth day after pupation. This coincided with the start of follicular epithelium resorption. Histological examination on *S. litura* showed that topical application of sublethal doses of chlorfuazuron to newly ecdysed ffth-instar larvae affected growth and development of oöcytes during pupal and adult stages by affecting size and thickness of follicular epithelium (Figures 5 and 6). However, in *C. pomonella,* (in controls) the basal oöcytes reached their maximum size 7 day after pupation. In this insect, this coincided with the start of follicular epithelium resorption. Hence, a 0.5-mg dose of DFB applied topically to newly ecdysed pupae affected the growth and development of oöcytes by causing a decrease in both the thickness of the follicular epithelium and size of basal oöcytes during the pupal development (Soltani and Mazouni, 1992). Lim and Lee (1982) reported that 2-d-old adult females of *O. japonica,* starved for 6 h and consumed 500 mg (AI) of DFB with two maize discs. The females were found to have retarded ovarian development, caused by a delay of oöcytes development, and an increased percentage of oöcytes resorption. This caused a decrease in fecundity and egg viability of the females. However, significant reduction was not observed either in the number of ovarioles or in the length of basal oöcytes in treated insects. Differences in these results might be a result of the use of different BPUs. Also, the doses used by Lim and Lee (1982) were very high compared with those used in present study or in the Soltani and Mazouni (1992) experiments.

In newly emerged treated adults, the germarium was much longer than the pedicle and vitelarium as compared with the controls in which the vitelarium was longer than the germarium and pedicle (Table 4). This shows that maturation of oöcytes was delayed in treated adult females as compared with the controls. The maximum thickness of the follicular epithelium of basal ooctes was observed on the 9 day after pupation in treated females, whereas it was on the 8 d after pupation in the controls (Figure 6). Subsequent to the 8th or 9th day, resorption of follicular epithelium started in control and treated females, respectively. When ovarian maturation was scored, as depicted in Figure 4.4, a maximum number of matured oöcytes were found in the second day after adult emergence in the controls. From this day, resorption of mature oöcytes started. The chlorfuazuron-treated females showed the same pattern of mature oöcyte resorption up to the seventh day after adult emergence as in the controls. However, Hatakoshi (1992) reported that when 0-day pupae of female *S. litura* were topically treated with pyriproxyen (0.3 ng per pupa), few or no mature oöcytes were found in newly emerged females, but controls had mature oöcytes with one occasionally being resorbed. The maturation of insect eggs dependent, among other factors, on the materials taken up from the surrounding hemolymph (Telfer et al., 1981), and by materials synthesized by the ovary in situ (Indrasith et al.,1988). These materials include proteins, lipids, and carbohydrates, all of which are required for the embryogenesis (Kunkel and Nordin 1985, *Kanost* et al., 1990). Difubenzuron also caued a decrease in ovarian protein content in *C. pomonella* (Soltani and Mazouni, 1992). Decrease in the ovarian protein content suggests an interference of BPUs with vitellogenesis. It has been reported that DFB could affect ecdysteroid secretion from other organs, such as the epidermis, in *T. molitor* (Soltani, 1984), ovaries in *C. pomonella* (Soltani et al., 1989a; 1989b), and the concentration of hemolymph constituents in *T. molitor* (Soltani, 1990). Future studies should clarify the biochemical mechanism. Moreover, this work dose not clarify why signiÞcant differences were not observed ($P<0.0001$) between effects of LD_{10} (1.00 ng larva^{-1}) and LD_{30} (3.75 ng larva^{-1}) treated females, although LD_{30} dose was much higher than LD_{10} dose. Further studies are needed to obtain more knowledge about the effects of chlorfuazuron on oögenesis. Currently, the biochemical mechanism involved has been explored.

3.4 Concise Summary

In comparison to the controls, the ovarian weight and number of mature eggs in pupae and adults were significantly reduced by sublethal doses of chlorfiuazuron (LD10: 1.00 ng larva^{-1} or LD30: 3.75 ng larva^{-1}) when applied topically to freshly ecdysed fifth instars of S. litura. The ratios of fresh ovarian/fresh body weight, dry ovarian/fresh ovarian weight, and dry ovarian/fresh body weight were consistent among controls, LD10, and LD30 treated newly emerged adults. In comparison to the controls, the germarium was significantly longer in treated adults than the pedicle and vitelarium, while the vitelarium was significantly longer in controls than the germarium and pedicle. This suggests that the maturation of ovarioles in treated cutworms is delayed. The size of basal oöcytes and the thickness of follicular epithelium are also substantially impacted by these doses, which disrupt the growth and development of oöcytes. In comparison to the controls, the maximum size of basal oöcytes recorded on the day of adult emergence was markedly reduced in females treated with LD10 or LD30. The controls attained a maximum thickness of the follicular epithelium of basal oöcytes on the eighth day, while the treated females reached a maximum thickness on the ninth day after pupation. The reduction in fecundity that is believed to be caused by sublethal exposure to chlorfiuazuron is believed to be the result of the effects of chlorfiuazuron on ovarian development and oögenesis.

4. EFFECTS OF SUBLETHAL DOSES OF CHLORFLUAZURN ON MALE REPRODUCTIVE SYSTEM OF SPODOPTERA LITURA

The deep yellow-coloured testes of *S. litura* are distinctly paired in larvae and they appear as a single round organ in adults. The testes of *S. litura* resemble those of other lepidopterans, being enclosed in a common membrane called the scrotum. The testes lie dorsally and appear to be held in place by trachea and strands of basement membrane-like material (Amaldoss, 1989). Although reports of several layers surround the testes of lepidopteran have been made, a single capsule and follicular layers are present in *S. litura* (Amaldoss, 1989). Chase and Gilliland (1972) described the *tunica externa* and *interna* as nothing more than basement membranes over the capsule and follicular layers. The intra-follicular layer is divided into eight incomplete compartments in the tobacco leafminer, *Phthorimaea operculella* (Zeller). This layer is similar to that in the larger canna leafroller, *Calpodes ethlius* (Stoll), and tobacco moth, *Ephestia elutella* (Hübner). It also bears the pigments responsible for the bright yellow-coloured testes. It is clear that spermatogenesis persists in the adult testis.

A characteristic feature of the testicular follicles is the presence of large cells or a nucleated mass of protoplasm in the apex of the germarium. This is known as an apical cell or versonian cell. This is the region where there are successive stages of development of the germ cells occur. The upper part contains the primary spermatogonia and is known as the germarium. This is followed by a region called the zone of growth. The region or zone of growth is where spermatogonia multiply and usually become encysted. The maturation zone, where maturation takes place follows. Finally, is the zone of transformation where the spermatocytes develop into spermatids (spermiogenasis) completing spermatogenesis (Amaldoss, 1989). Two distinct types of spermatozoa are produced in the Lepidoptera: eupyrene (nucleated) spermatozoa which can fertilize the egg; and apyrene (anucleated) which are smaller and completely lacking in nuclear material, and do not appear to play any role in activation of the eggs (Doncaster, 1911; Goldschmidt, 1916). The eupyrene sperm can easily be counted in the male tract because they remain in bundles until they are transferred during mating, but the apyrene sperm are dispersed shortly after they leave the testis. Like eupyrene sperm, the apyrene sperm are produced in large numbers, usually contributing over half the sperm complement, and are transferred to the females with the eupyrene sperm during mating. It was thought that the apyrene sperm did not appear to play any role in activation of the eggs (Friedlander and Gitay, 1972). Their function has remained unclear ever since their discovery by Meves (1902), although several hypotheses concerning the function of the apyrene sperm have been proposed (Silberglied et al., 1984). Holt and North (1970 b) proposed that, in the cabbage looper, *Trichoplusia ni* (Hüebner), apyrene sperm might aid the transport of eupyrene sperm from the male reproductive tract to the female reproductive tract. Katsuno (1977 a) reported that the apyrene sperm in the *B. mori*, might facilitate the migration of eupyrene sperm through the cellular barrier, separating the testis from the efferent ducts. Gage and Cook (1994) reported that nutritional stress seriously affected the number and size of eupyrene and apyrene sperm production in the Indian meal moth, *Plodia interpunctella* (Hübner). However, sperm development in Lepidoptera takes place in the larvae (Munson, 1906; Machida, 1929; Garbini and Imberski, 1977). Usually, spermatogenesis starts in the late larval instars and proceeds on a schedule well correlated with the insect's metamorphosis. Studies *in vitro* and *in vivo* indicated that high titre of juvenile hormone inhibits spermatogenesis, and that sperm mitosis and meiosis require sufficient ecdysteroid titre. During the post-embryonic development of eupyrene and apyrene sperm bundles, when the insect is going to pupation, the juvenile hormone titre declines (Leviatan and Friedlander, 1979). Other factors have also been reported to promote spermatogenesis *in vitro* and *in vivo* (Dumser, 1980 a). In adult males *S. litura*, both apyrene and eupyrene sperm appear in bundles in the testis, but in the vas deferens only the eupyrene sperm were still in bundles, as reported for *B. mori* (Katsuno 1977 b), *T. ni* (Holt and North, 1970 a) and the army worm, *Pseudaletia separata* (Walk.) (He, 1994). The testes of early larvae contain a large number of spermatogonial cells near the outer border of the follicles. There is a preponderance of spermatocytes containing primary spermatocytes during the penultimate and early last-instar larvae. Secondary spermatocytes are present in the early last-instar larvae, persisting through to the middle of the last-instar larvae. They then begin to differentiate into spermatids. In the process of elongation and maturation of spermatids, the spermatocytes assume an elliptical shape. The sperm bundles formed as a result of maturation of the spermatids are seen abundantly in adults. Spermiogenesis is, however, not synchronous, and spermatozoa in various stages of differentiation can be detected in the testes of freshly emerged adults (Sridevi et al., 1989a). In the pupa, the apyrene sperm bundles emerge from the testicular follicle into the vas efferens earlier than the eupyrene sperm bundles and the bundles separate when they pass through the basement membrane of the testis (Katsuno, 1977b). The apyrene spermatozoa migrate from the vas efferens into the seminal vesicle through the vas deferens during the pupa (Katsuno, 1977c). In the post-pupal period, however, the eupyrene sperm bundles and apyrene spermatozoa migrate simultaneously through the same way (Katsuno, 1977d). The effects of chlorfluazuron have been examined on male reproductive system during testicular development and spermatogenesis when sublethal doses have been topically applied to newly ecdysed fifth-instar larvae of *S. litura*.

4.1 Experimental procedure

4.1.1 Histology of testis

Testes from newly molted sixth instar larvae to 5-day-old virgin adult males (treated and control), were dissected in 0.9% of NaCl under a binocular microscope. The length and width of each testis were measured by the same procedure as used for the oöcytes measurement. Testis volume was calculated for larval testes using the formula 4/3π (length′width2), assuming that the testis is a prolate spheroid (Loeb et al., 1984). For the fused pupal and adult testes, the formula 4/3πr3 was used, with r as the radius of the globular gonad. The treated and control weight and sheaths thickness of testes were measured by the same procedure used for the ovaries as described above (Perveen, 2000b). The thickness of treated and control testes sheaths or vas deferenti of untreated and treated relevant stages of insect was observed by making a parafilm microtomy conducted according to the method used by Yoshida (1994) and the procedure to stain the nuclei of sperm was adapted from the method by He (1994) (Perveen, 2000b).

4.1.2 Spermatogenesis

A staining method was used for determining number of the cysts, eupyrene and apyrene sperm for treated and control (He et al., 1995). First, the testis was transferred to a microscopic grid slide (each square= 1mm2) and crushed until it was evenly distributed on the slide. Secondly, several drops of methanol-acetic acid solution (3 : 1; v/v) were added to the slide to fix the reparation for 15 min, and the excess fixing solution was absorbed with filter paper. Third, several drops of 2-5% Giemsa solution dissolved in Sorensen-Gomori buffer solution (monobasic and dibasic sodium phosphate, 0.07 M, pH 6.8) were added to the slide to stain the preparation for 10-30 min. The slide was washed with water and air dried after staining. Finally, the air-dried preparation was observed for counting of bundles and cysts under a phase contrast microscope at 20´mignification. Cysts were classified into the following six developmental stages as described by Chaudhury and Raun (1966): (1) spermatogonia; (2) primary spermatocytes; (3) secondary spermatocytes; (4) spermatids; (5) elongated cysts with maturing sperm and (6) bundles with fully matured sperm. The length and width of sperm bundles were measured with a calibrated ocular micrometer a phase contrast microscope at 400´magnification (Perveen, 2000b).

4.1.3 Data analysis

Data were analyzed using analysis of variance, one way ANOVA (Concepts, 1989) at $P<0.0001$ and Scheffe's *F*-test (Scheffe, 1953) at 5%.

4.2 Results

The structural morphogenesis was seen in sixth-instar larvae of *S. litura* during development. When fifth-instar larvae ecdysed into newly sixth-instar larvae (N: 0 day), larvae remained unchanged for upto 2 hour. Then, they changed to a slender surface during 1st day (S). After that, they changed to being puffy during 2nd day (early-last-instar stage: P). Then, they changed to digging stage during 3rd day (mid-last-instar stage: D). Then, they changed to early burrow during 4th day (pre-late-last-instar stage: B_1). After that, it changed to late burrow during 5th day (post-late-last-instar stage: B_2). The morphogenesis for phase variations used for convenient of observations during experiments is given in Table 6. The morphology of the adult male reproductive system of *S. litura* is shown in Figure 7 (Perveen, 2000b).

Table 6. Structural morphogenesis during five developmental days of sixth-instar larvae of *Spodoptera litura* (Source: Perveen, 2005).

Developmental category (days)	Symbols for phase variations[a]	Name of the phases[a]	Duration (hour day^{-1})
0	N	newly ecdysed	0–2
1	S	slender surface	1st
2	P	puffy (early last instar stage)	2nd
3	D	digging (mid last instar stage)	3rd
4	B1	early burrowed (pre late last instar stage)	4th
5	B2	late burrowed (post late last instar stage)	5th

a Symbols for phase variations observed during developmental days of sixth-instar larvae were used for the convenience of observations.

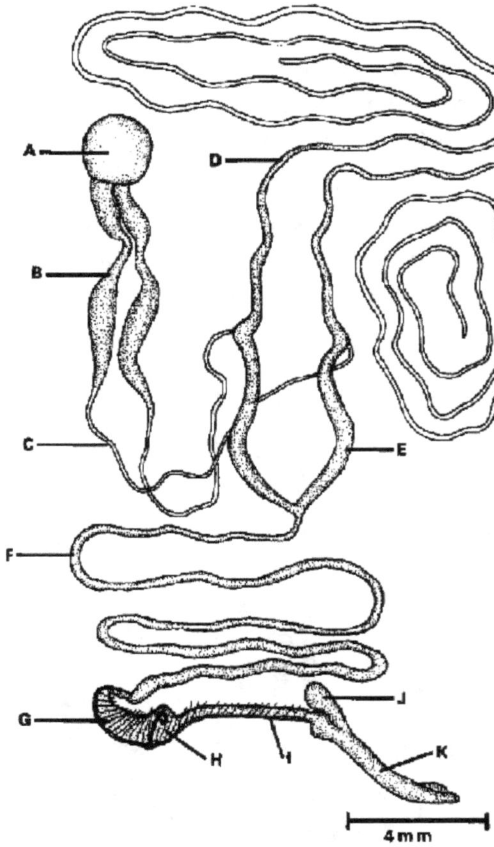

Fig. 7. The morphology of the male reproductive system of *Spodoptera litura*: A: testis; B: seminal vesicle (paired); C: vas deferens (paired); D: accessory glands (paired); E: ductus ejaculatorious duplex; F: primary segment of ductus ejaculatorious simplex; G: muascular area; H: area of frenum formation; and I: area of collum formation of the cuticular secondary segment of the ductus ejaculatorious simplex; J: caecum of aedeagus; K: aedeagus (Source: Etman and Hooper, 1979).

4.2.1 Effects on testicular development

The testes of *S. litura* show the three dimentional measureable structure. Each mature testis consists of four follicles or lobes, each separated by an inner layer of sheath cells. An outer sheath cell layer further surrounds all follicles. In sixth-instar larvae, the volume and weight of the testes gradually then rather sharply increased until the 4th day of sixth-instar larvae. The volume weight decreased when two larval testes fused on the 5th day (last day) after moulting of sixth-instar larvae. They again sharply increased in size, reached a maximum (6.21±1.31 mm3 and 11.94±0.42 mg, n=30, respectively) on the 0 day of pupation and gradually declined until the 5th day after adult emergence. Sublethal doses of chlorfluazuron rapidly disrupted the development of testes by decreasing the volume and weight of testes compared with the controls. The weight and size of testes were significantly reduced (P<0.001) in the LD_{10}-treated and more significantly reduced (P<0.0001) in LD_{30}-treated males compared with the controls from newly ecdysed sixth-instar larvae to the 5th day after adult emergence. Testes reached their maximum size in treated males (LD_{10}: 4.16±1.54 mm3 and 8.0±0.83 mg; LD_{30}: 2.79±1.00 mm3 and 4.88±1.05 mg; n= 10, respectively) on the same day as the controls. The patterns of development of the testes with respect to the volume and weight were the similar in the controls and the LD_{10}- or LD_{30}-treated males (Figures 8a and b) (Perveen, 2000b).

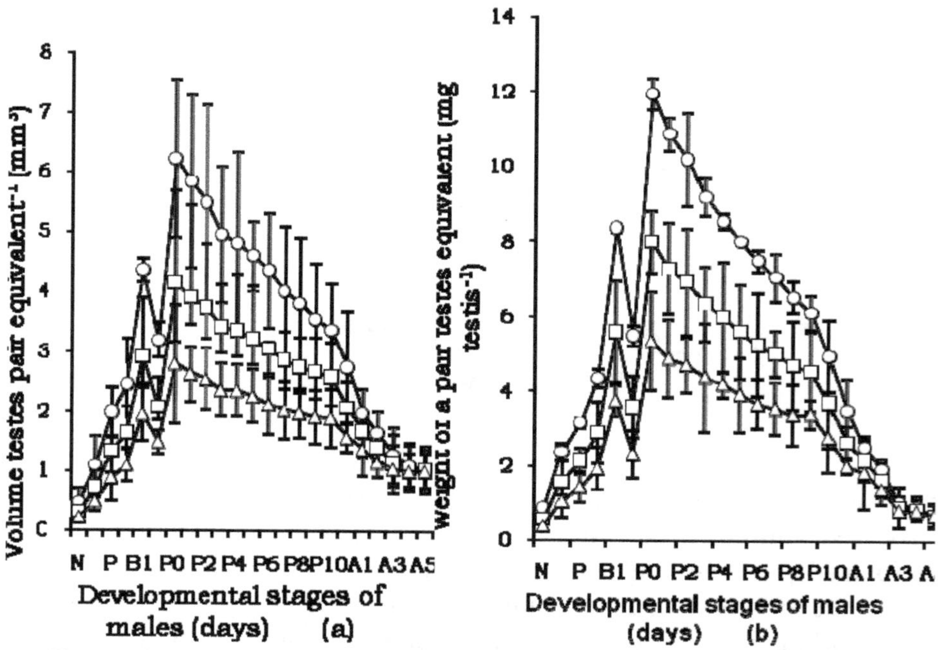

Fig. 8. Effects of sublethal doses of chlorfluazuron (LD_{10}: 1.00 ng larva^{-1}; LD_{30}: 3.75 ng larva^{-1}) on the testis volume (a) and weight (b) of *Spodoptera litura* during newly ecdysed sixth-instar larvae to 5th day after adult emergence; controls: O; LD_{10}: □; LD_{30}: Δ; data analyzed using one-way ANOVA (Concepts, 1989) at *P*<0.0001 and Scheffe`s *F*-test (Scheffe, 1953) at 5%; vertical bars: SD; N, S, P and D: larval (Table 5.1), P: pupal and A: adult developmental days; n = 10 for each point; paired larval testes and fused single pupal or adult testis were considered as testes pair equivalent; (Source: Perveen, 2000b).

The thickness of the testis sheath gradually and then sharply increased until the 4th day of moulting of sixth-instar larvae. It decreased when the two larval testes fused on the 5th day (last-day) of the sixth-instar larvae; it again increased and reached to a maximum [(6±0.51)´10-2 mm; n=10] on the 0 day of pupation and gradually declined in the newly emerged adults. The thickness remained constant until the 2nd day after adult emergence. Sublethal doses rapidly disrupted the development of testis by significantly decreasing (P<0.0001) the thickness of the testes sheath as compared with that of the controls. This reduction occurred from newly ecdysed sixth-instar larvae to the 0 day of pupation in the LD_{10}-treated and the 1st day after pupation in the LD_{30}-treated males. The thickness of the testes sheath in chlorfluazuron-treated males reached a maximum [(5.9±0.67)′10-2 mm] in the LD_{10}-treated males on the 1st day and [(5.9±1.1)′10-2 mm] in the LD_{30}-treated males on the 2nd day after pupation whereas, in the controls, it was on the 0 day of pupation. This result shows that attainment of the maximum thickness of the testes sheath was delayed by one day in LD_{10}- and by two days in LD_{30}-treated males compared with the controls. However, no significant reduction was observed in the maximum thickness of the testes sheath among the control and LD_{10}- or LD_{30}-treated males. The developmental pattern of the testes sheath in the LD_{10}- or LD_{30}-treated males was similar to that of the controls (Figure 9) (Perveen, 2000b).

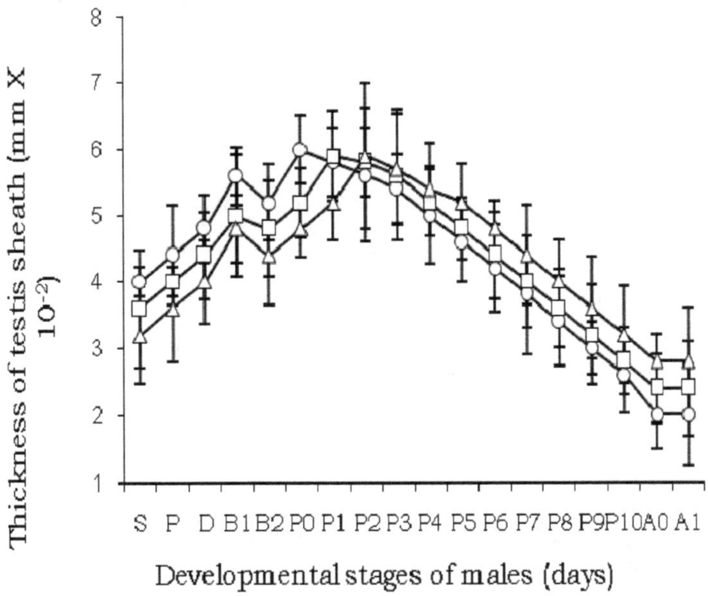

Fig. 9. Effects of sublethal doses of chlorfluazuron (LD_{10}: 1.00 ng larva^{-1}; LD_{30}: 3.75 ng larva^{-1}) on the thickness of testis sheath of *Spodoptera litura* during newly ecdysed sixth-instar larvae to 5th day after adult emergence; controls: O; LD_{10}: □; LD_{30}: Δ; data analyzed using one-way ANOVA (Concepts, 1989) at *P*<0.0001 and Scheffe`s *F*-test (Scheffe, 1953) at 5%; vertical bars: SD; S, P, D and B_1: larval (Table 5.1), P: pupal and A: adult developmental days; n = 10 for each point; paired larval testes and fused single pupal or adult testis were considered as a testes pair equivalent; (Source: Perveen, 2000b).

4.2.2 Effects on spermatogenesis

When the different developmental stages of cysts were observed in testes during spermatogenesis on the 1, 3 and 5 day-old sixth-instar larvae, the number of spermatogonia, primary and secondary spermatocytes was significantly reduced (P<0.001) in the LD_{10}- and even more significantly reduced (P<0.0001) in the LD_{30}-treated males as compared with those of controls. Spermatids, elongated cysts with mature sperm and eupyrene sperm bundles were not found in the controls and LD_{10}- or LD_{30}-treated larval testes (Table 7) (Perveen, 2000b).

Table 7. Effect of sublethal doses of chlorfluazuron on spermatogenesis in the testes of sixth- instar larvae during 1st, 3rd and 5th days of development after topical application to newly ecdysed fifth-instar larvae of *Spodoptera litura* **(Source: Perveen, 2000b).**

TS[a]	T[a]	n[a]	*Sg[b] (M±SD)	*PS[b] (M±SD)	*SS[b] (M±SD)	St (M±SD)	ECMS (M±SD)	ESB (M±SD)
1st	C	13	2447±18a	4893±35a	1397±18a	0.0±0.0	0.0±0.0	0.0±0.0
	LD_{10}	11	1960±7b	3920±7b	1120±6b	0.0±0.0	0.0±0.0	0.0±0.0
	LD_{30}	10	1320±6c	2641±5b	754±4c	0.0±0.0	0.0±0.0	0.0±0.0
3rd	C	13	1311±8a	4280±9a	3146±8a	0.0±0.0	0.0±0.0	0.0±0.0
	LD_{10}	11	1050±5b	3430±7b	2520±7b	0.0±0.0	0.0±0.0	0.0±0.0
	LD_{30}	10	707±5c	2311±7c	1689±7c	0.0±0.0	0.0±0.0	0.0±0.0
5th	C	13	437±8a	3574±5a	4717±7a	0.0±0.0	0.0±0.0	0.0±0.0
	LD_{10}	11	342±5b	2870±7b	3780±6b	0.0±0.0	0.0±0.0	0.0±0.0
	LD_{30}	10	224±6c	1933±8c	2546±9c	0.0±0.0	0.0±0.0	0.0±0.0

a TS: treatment stages; T: treatments; n: number of males used; C: control; LD10: 1.00 ng larva^{-1}; LD30: 3.75 ng larva^{-1}; Sg: spermatogonia; PS: primary spermatocytes; SS: secondary spermatocytes; St: spermatids; ECMS: elongated cysts with mature sperm; ESB: eupyrene sperm bundles; larvae ecdysed usually during 0200 to 0800 hour and collected between 0800 to 1000 hour

b Data were analyzed using one-way ANOVA (Concepts, 1989) at P<0.0001. Means within columns followed by different letters are significantly different by Scheffe`s F-test (Scheffe, 1953) at 5%.

When different developmental stages of the cysts were observed in testis during spermatogenesis in newly ecdysed pupae, on the 5th and 10th day after pupation, the number of spermatogonia, primary and secondary spermatocytes, were decreased. However, the spermatids and elongated cysts with mature sperm gradually increased in controls. Eupyrene sperm bundles were not present on the 0-5 day-old pupae. However, they were found on the 10th day after pupation (mean: 1002±3.0 numbers). The pattern of spermatogenesis was the same in the controls, LD_{10}- and LD_{30}-treated male pupae. However, the developmental stages of the sperm were significantly reduced (P<0.001) in the LD_{10}-treated and even more significantly reduced (P<0.0001) in the LD_{30}-treated males compared with the controls (Table 8) (Perveen, 2000b).

Different developmental stages of cysts were observed in testes during spermatogenesis in newly emerged, 1 and 2 day-old adults. The spermatogonia were not present in newly emerged and 1 day-old adults. Primary spermatocytes were not found in 1 and 2 day-old adults, but they (mean: 174±4.0 number) were found in newly emerged adults. The secondary spermatocytes, spermatids and elongated cysts with mature sperm were present, but gradually decreased in number in the controls. Eupyrene sperm bundles gradually increased in number in the controls. The pattern of spermatogenesis was the same in the controls, LD_{10}- and LD_{30}-treated male pupae. However, the stages of sperm development were significantly reduced ($P<0.001$) in the LD_{10}- and even more significantly reduced ($P<0.0001$) in LD_{30}-treated males as compared with the controls (Table 9) (Perveen, 2000b).

Table 8. Effect of sublethal doses of chlorfluazuron on spermatogenesis in the testes of pupae during 1st, 5th and 10th days of development after topical application to newly ecdysed fifth-instar larvae of *Spodoptera litura* (Source: Perveen, 2000b).

TS[a]	T[a]	n[a]	*Sg[b] (M±SD)	*PS[b] (M±SD)	*SS[b] (M±SD)	St (M±SD)	ECMS (M±SD)	ESB (M±SD)
1st	C	13	262±6ac	1747±5a	2710±7a	3670±6a	349±8a	0.0±0.0
	LD_{10}	11	205±6b	1400±4b	2170±6b	2940±7b	280±5b	0.0±0.0
	LD_{30}	10	142±5c	943±6b	1462±5c	1980±4c	189± 5c	0.0±0.0
3rd	C	13	175±7a	1047±8a	2010±4a	4805±5a	697±11a	0.0±0.0
	LD_{10}	11	140±4b	840±5b	1610±3b	3805±5b	560±5b	0.0±0.0
	LD_{30}	10	94±5c	566±6c	1085±3c	2593±7c	377±4c	0.0±0.0
5th	C	13	86±4a	436±5a	698±3a	2969±6a	3581±3a	0.0±0.0
	LD_{10}	11	70±3b	350±4b	560±6b	2660±4b	3339±3b	0.0±0.0
	LD_{30}	10	47±4c	236±3c	377±5c	1792±3c	2263±4c	0.0±0.0

a TS: treatment stages; T: treatments; n: number of males used; C: control; LD10: 1.00 ng larva^{-1}; LD30: 3.75 ng larva^{-1}; Sg: spermatogonia; PS: primary spermatocytes; SS: secondary spermatocytes; St: spermatids; ECMS: elongated cysts with mature sperm; ESB: eupyrene sperm bundles; pupation occurred usually during 0200 to 0800 hour and collected between 0800 to 1000 hour.

b Data were analyzed using one-way ANOVA (Concepts, 1989) at P<0.0001. Means within columns followed by different letters are significantly different by Scheffe's F-test (Scheffe, 1953) at 5%.

Table 9. Effect of sublethal doses of chlorfluazuron on spermatogenesis in the testes of adults during 0, 1st and 2nd day of development after topical application to newly ecdysed fifth-instar larvae of *Spodoptera litura* (Source: Perveen, 2000b).

TS[a]	T[a]	n[a]	*Sg[b] (M±SD)	*PS[b] (M±SD)	*SS[b] (M±SD)	St (M±SD)	ECMS (M±SD)	ESB (M±SD)
1st	C	13	0±0a	174±4a	524±3a	1048±4a	2097±5a	4893±3a
	LD_{10}	11	35±3b	140±7b	385±3b	840±3b	1680±6b	3920±4b
	LD_{30}	10	24±2c	94±3b	259±2c	566±6c	1132±3c	2641±2c
3rd	C	13	0±0a	0±0a	174±6a	611±3a	1047±6a	6902±5a
	LD_{10}	11	14±3b	28±3b	98±1b	490±5b	840±3b	5530±4b
	LD_{30}	10	9±2c	19±2c	66±2c	330±4c	566±2c	3728±5c
5th	C	13	0.0±0.0	0.0±0.0	43±3a	219±4a	437±6a	8037±6a
	LD_{10}	11	0.0±0.0	0.0±0.0	35±3b	175±3b	350±5b	6440±5b
	LD_{30}	10	0.0±0.0	0.0±0.0	24±2c	118±4c	236±4c	4338±4c

a TS: treatment stages; T: treatments; n: number of males used; C: control; LD10: 1.00 ng larva^{-1}; LD30: 3.75 ng larva^{-1}; M: mean; Sg.: spermatogonia; PS: primary spermatocytes; SS: secondary spermatocytes; St.: spermatids; ECMS: elongated cysts with mature sperm; ESB: eupyrene sperm bundles; adults emerged usually between 2300 to 0200 hour and and between collected 0800 to 1000 hour.

b Data were analyzed using 1-way ANOVA (Concepts, 1989) at P<0.0001. Means within columns followed by different letters are significantly different by Scheffe`s F-test (Scheffe, 1953) at 5%.

Table 10. Effect of sublethal doses of chlorfluazuron and comparison of the number of eupyrene and apyrene sperm bundles in the testis of newly emerged unmated adults after topical application to newly ecdysed fifth-instar larvae of *Spodoptera litura* (Source: Perveen, 2000b).

T[a]	n[a]	NESB[b] (M±SD)	NASB[b] (M±SD)	Ratios (%)=ESB:ASB
Control	30	4893±546a	4697±520a	51.1:48.9
LD_{10}	30	3920±426b	3763±466b	51.0:48.9
LD_{30}	30	2641±161c	2386±271c	52.5:47.5

a LD_{10}: 1.00 ng larva^{-1}; LD_{30}: 3.75 ng larva^{-1}; T: treatments; n: number of males used; NESB: number of eupyrene sperm bundle; NASB: number of apyrene sperm bundle; ESB: eupyrene sperm bundle; ASB: apyrene sperm bundle

b Data were analyzed using one-way ANOVA (Concepts, 1989) at P<0.0001. Means within columns followed by different letters are significantly different by Scheffe`s F-test (Scheffe, 1953) at 5%.

In the testis of newly emerged LD_{10}- treated adults, the number of eupyrene and apyrene sperm bundles was significantly reduced (P<0.001), and even more significantly reduced (P<0.0001) in LD_{30}- treated adults compared with the controls. When males were treated with the LD_{10} or LD_{30}, the ratio of eupyrene to apyrene sperm bundles was not significantly changed; apyrene sperm bundle comprised about half of the total sperm complement (Table 10).

Table 11. Effect of sublethal doses of chlorfluazuron on the size of various developmental stages of sperm observed in the testes of sixth-instar larvae and puape after topical application to newly ecdysed fifth-instar larvae of *Spodoptera litura* (Source: Perveen, 2000b).

Developmental stages of sperm	T[a]	n[a]	Size (dm in µm)	
			(M±SD)[b]	ranges (min–max)
Spermatogonia (S)	Control	10	7.1±0.7a	6–8
	LD_{10}	10	5.9±0.8b	5–7
	LD_{30}	10	4.1±0.9c	3–5
Primary spermatocytes (D)	Control	10	15.1±0.7a	14–16
	LD_{10}	10	13.0±1.3b	11–14
	LD_{30}	10	11.4±1.5c	10–48
Secondry spermatocyte (B_2)	Control	10	31.0±1.8a	28–33
	LD_{10}	10	29.2±1.6b	27–31
	LD_{30}	10	25.0±1.2c	24–27
Spermatids (P_{10})	Control	10	4.1±0.9a	3–5
	LD_{10}	10	2.9±0.87b	2–4
	LD_{30}	10	1.9±0.7c	1–3

a LD_{10}: 1.00 ng larva^{-1}; LD_{30}: 3.75 ng larva^{-1}; T: treatments; n: number of males used; for S, D and B2 refer to Table 5.1; P10: ten day old pupae

b Data were analyzed using one-way ANOVA (Concepts, 1989) at P<0.0001. Means within columns followed by different letters are significantly different by Scheffe's F-test (Scheffe, 1953) at 5%.

Table 12. Effect of sublethal doses of chlorfluazuron on size of various developmental stages of sperm observed in the testes of newly emerged adults after topical application to newly ecdysed fifth-instar larvae of *Spodoptera litura* (Source: Perveen, 2000b).

Developmental stages of sperm	Ta	na	sizes of various developmental stages			
			length (µm)[b]		width (µm)[b]	
			M±SD	Ranges (min–max)	M±SD	Ranges (min–max)
Elongated spermatocytes	Control	10	45.5±2.4a	(42–50)	40.0±2.0a	(35–42)
	LD_{10}	10		(40–44)	37.2±1.3b	(35–39)
	LD_{30}	10	M±SD	Ranges	35.2±1.6c	(34-38)

Developmental stages of sperm	Ta	na	sizes of various developmental stages			
			length (μm)[b]		width (μm)[b]	
			M±SD	Ranges (min–max)	M±SD	Ranges (min–max)
Eupyrene sperm bundles	Control	10	98.0±4.4a	(92–105)	33.2±1.9a	(30–36)
	LD_{10}	10	94.7±1.4b	(93–97)	31.3±1.4b	(29–33)
	LD_{30}	10	91.5±1.0c	(90–93)	29.1±1.2c	(28–31)
Apyrene sperm bundles	Control	10	24.9±1.2a	(23–27)	16.1±1.7a	(14–19)
	LD_{10}	10	94.7±1.4b	(21–25)	14.2±1.3b	(12–16)
	LD_{30}	10	20.3±0.9c	(19–22)	12.2±1.2c	(11–14)

a LD_{10}: 1.00 ng larva^{-1}; LD_{30}: 3.75 ng larva^{-1}; T: treatments; n: number of males used

b Data were analyzed using one-way ANOVA (Concepts, 1989) at P<0.0001. Means within columns followed by different letters are significantly different by Scheffe's F-test (Scheffe, 1953) at 5%.

The size of the spermatogonia, primary spermatocytes, secondary spermatocytes and spermatids was significantly reduced (P<0.001) in LD_{10}-treated and more significantly reduced (P<0.0001) in LD_{30}-treated insects compared with the controls (Table 11) in newly emerged adults (Perveen, 2000b).

In newly emerged adults, the width and length of elongated cysts with mature sperm, eupyrene and apyrene sperm bundles were significantly reduced (P<0.001) in LD_{10}- and more significantly reduced (P<0.0001) in LD_{30}- treated insects compared with the controls (Table 12) (Perveen, 2000b).

Table 13. Effect of sublethal doses of chlorfluazuron on the number of eupyrene sperm bundles in the vas deferens during different developmental days of adults after topical application to newly ecdysed fifth-instar larvae of *Spodoptera litura* **(Source: Perveen, 2000b)**

T[a]	n[a]	Eupyrene sperm bundles in vas deferens		
		Pre-adult[b] (M±SD)	Newly emerged adult[b] (M±SD)	One day old adult[b] (M±SD)
Control	30	102±29a	1002±116a	2513±407a
LD_{10}	30	0b	23±4.9b	1621±159b
LD_{30}	30	0b	0c	1080±75c

a LD_{10}: 1.00 ng larva^{-1}; LD_{30}: 3.75 ng larva^{-1}; T: treatments; n: number of males used

b Data were analyzed using one-way ANOVA (Concepts, 1989) at P<0.0001. Means within columns followed by different letters are significantly different by Scheffe's F-test (Scheffe, 1953) at 5%.

In the vas deferens of male pre-adult controls, the mean number of eupyrene sperm bundles was 102±29, but no sperm bundles observed in the LD_{10}- or LD_{30}-treated males of the same age (Table 13) (Perveen, 2000b).

In newly emerged control males, the mean number of eupyrene sperm bundles was 1002±116, and in LD_{10}-treated adult males, 23 ± 4.9. In LD_{30}-treated adults male there was no sperm bundles were observed. Moreover, in 1 day-old LD_{10}-treated adult males, the number of eupyrene sperm bundles was significantly (P<0.001) reduced and more significantly (P<0.0001) reduced in LD_{30}-treated males compared with the controls (Table 14) (Perveen, 2000b).

In the testis and vas deferens of newly emerged, LD_{10}-treated males, the total number of eupyrene sperm bundles was significantly reduced (P<0.001) and more significantly reduced (P<0.0001) in LD_{30}-treated males had no sperm bundles in the vas deferens compared with the controls (Table 14) (Perveen, 2000b).

Table 14. Effect of sublethal doses of chlorfluazuron on the total number of eupyrene sperm bundles in the testis and vas deferens of newly emerged adults after topical application to newly ecdysed fifth-instar larvae of *Spodoptera litura* (Source: Perveen, 2000b)

T[a]	n[a]	NESB in testis[a,b] (M±SD)	NASB in vasa deferens[a,b] (M±SD)	TNESB (testis+ vasa deferens)[a,b] (M±SD)
Control	30	4893±546a	1002±116a	5791±640a
LD_{10}	30	3920±426b	23±4.9b	3943±425b
LD_{30}	30	2641±161c	0±0c	2641±161c

a LD_{10}: 1.00 ng larva^{-1}; LD_{30}: 3.75 ng larva^{-1}; T: treatments; n: number of males used; NESB: number of eupyrene sperm bundle; NASB: number of apyrene sperm bundle; ESB: eupyrene sperm bundle; ASB: apyrene sperm bundle; TNESB: total number of eupyrene sperm bundle

b Data were analyzed using one-way ANOVA (Concepts, 1989) at P<0.0001. Means within columns followed by different letters are significantly different by Scheffe's F-test (Scheffe, 1953) at 5%.

4.3 Discussion

Topical application of sublethal doses of chlorfuazuron (LD_{10}: 1.00 ng larva^{-1}; LD_{30}: 3.75 ng larva^{-1}) has an effect on reproduction of *S. litura* by reducing the fecundity, fertilityand hatchability. Fecundity was reduced to a similar degree (35±44%) when females, males or both sexes were treated with LD_{10} or LD_{30}. Fertility was reduced by 42% or 52% when females were treated with LD_{10} or LD_{30}, respectively, and by 60% or 63%, respectively, when males or both sexes were treated with LD_{10}. Fertility was reduced by 78% or 80% when males or both sexes were treated with LD_{30}. The hatchability was reduced by 20% or 23% when females were treated with LD_{10} or LD_{30}, respectively, and by 37% or 39%, respectively, when males or both sexes were treated with LD_{30}. Hatchability was reduced by 55% or 56% when males or both sexes were treated with LD_{30} (Perveen, 2000a). Thus, this study was conducted to determine the causes of the reduction in these reproductive parameters. Topical application of sublethal doses of chlorfuazuron has an effect on testis development by decreasing the volume and weight of testes and its sheath thickness, and also on spermatogenesis by decreasing the number of cysts, eupyrene and apyrene bundles during different times in development. Reduction in the testes volume and weight might be caused by a reduction in the number of cysts, number and size of sperm bundles, thickness of testes sheath, and/or the reduction in protein content of testes constituents. When females are treated with LD_{10} of chlorfuazuron, no signifiant reduction in the inseminated eupyrene sperm number is observed compared with controls. However, LD_{10} and LD_{30} treatment of males significantly reduces (65.8% and 88.6%) the number of inseminated eupyrene sperm. Moreover, no significant (P<0.0001) differences in the reduction are observed on the inseminated eupyrene sperm number when males or both sexes are treated either with LD_{10} or LD_{30} (Perveen, 2008). Therefore, the main cause of the reduction in the fecundity, fertility and hatchability is a decrease in inseminated eupyrene sperm numbers. In larval *S. litura*, the bright yellow-coloured testes are distinctly paired, reniform and situated between the 5th and 6th abdominal segments. Each of the lateral testicular lobes is made up of four follicles. The testicular lobes are enclosed within two thick, double-layered peritoneal sheaths. Each of these sheaths comprises two layers of epithelial cells. The external sheath is made up of lightly stained cuboidal cells resting on a basement membrane. This forms a common envelope to all the follicles of the testis and is penetrated by tracheal branches. The inner sheath is made up of more darkly stained elliptical cells, within which muscle fibres are distinguisable. Ingrowths from this sheath in the form of double-walled septa penetrate between the follicles to separate them from one another. The testes sheath reaches its maximum thickness at day 0 of pupation in the controls, whereas in LD_{10-} treated males maximum thickness occurs later at 1 day. In LD_{30}-treated males maximum thickness occurs at 2 day after pupation. Difubenzuron affects ecdysteroid secretion from the epidermis in Tenebrio molitor (Soltani, 1984), ovaries of *Cydia pomonella* (Soltani et al., 1987; Soltani et al., 1989b) and also the concentrations of haemolymph constituents in *T. molitor* (Soltani, 1992). Ecdysteroids have been reported to stimulate spermatogenesis in many insect species (Dumser, 1980b; Gelman and Hayes, 1982). In Mamestra brassica (Shimizu et al., 1985), *Heliothis virescens* (Loeb, 1986) and *L. dispar* (Loeb et al., 1988) testes synthesize ecdysteroids. The treatment with chlorfuazuron effects ecdysteroid production by the testis sheet remains to be investigated. Gelman and Hayes (1982) and Gelman et al. (1988) observed in *Ostrinia nubilalis*, the size and weight of separate testes paired. Topical application of the sublethal dose, LD_{10} of chlorfuazuron significantly reduces (P<0.0001) the weight and size of testes, and this is even greater in LD_{30}-treated males. However, the topical application of similar sublethal doses of chlorfuazuron significantly reduced ($P<0.0001$) ovarian weight in post- pupal and developing pharate adult females as compared with that of the controls (Perveen and Miyata, 2000). However, significant differences are not observed in ovarian weight when adult females are treated with LD_{10} or LD_{30} doses. The maturation of insect testes depends, among other factors, upon the materials that are taken up from the surrounding haemolymph and by materials synthesized by the testes *in situ*. These materials include protein, lipid and carbohydrate, all of which are required for development of the genital tract (Kunkel and Nordin, 1985; Kanost et al., 1990). In newly emerged males from LD_{10-} or LD_{30}-treated larvae the eupyrene and apyrene sperm bundles are significantly ($P<0.0001$) smaller in size and number compared with those of controls. Spermatozoa descend regularly from the testis through the vas deferens into the seminal vesicles, which fill with eupyrene sperm bundles with cysts and individual apyrene sperm. Yoshida (1994) also reported that the number of eupyrene sperm bundles in the testis and vas deferens of newly emerged treated (LD_{10}) males is reduced by 36%, and the initiation of sperm movement from testis to seminal vesicle was delayed. The present results show that initiation of sperm movement from testis to vas deferens is delayed after chlorofuazuron treatment and this is caused by the delay of spermatogenesis. When males or both sexes are treated on the first day of pairing, few or none mate. However, seven to nine pairs mate in control or female treated crosses. During the next day, seven to 10 pairs mate in 10±13 pairs of seven combinations of crosses (Perveen, 2008). These results suggest that the delay of the first mating is caused by the delayed spermatogenesis. The effect of chlorfuazuron on testicular development and spermatogenesis is one of the factors responsible for the reduction in fecundity, fertility and hatchability caused by the sublethal doses of chlorfuazuron. More work is in progress to determine the biochemical mechanism of these effects in *S. litura*.

4.4 Concise Summary

When sublethal doses (LD10: 1.00 ng larva^{-1} or LD30: 3.75 ng larva^{-1}) of chlorfluazuron are administered topically to the cuticle of newly ecdysed fifth instars of Spodoptera litura, the physiological mechanism of action on testicular development and spermatogenesis is described. In comparison to the controls, these dosages disrupt the growth and development of testes by reducing the volume and weight of testes and the thickness of the testes sheath. Furthermore, these dosages disrupt spermatogenesis by decreasing the quantity and size of eupyrene and apyrene sperm bundles in the testis. In contrast to controls, there are either negligible or no eupyrene sperm bundles observed in the vas deferens of pre- and newly ecdysed adults. This finding indicates that the transfer of sperm bundles from the testes to the vas deferens is delayed in males who have been treated. The reduction in fecundity, fertility, and hatchability that sublethal concentrations of chlorfluazuron induce is believed to be attributable to the effects of chlorfluazuron on testicular development and spermatogenesis.

5. REFERENCES

Abro, G. H., Memon, A. J. and Syed, T. S. (1997). Sublethal effects of cyhalothrin and fluvalinate on biology of *Spodoptera litura* F. Pakistan J. Zool. 29: 181–184.

Anderson, V. L. and McLean, R. L. (1974). Design of experiments: a realistic approach. Dekker, New York. p. 70.

Amaldoss, G. (1989). Ultrastructure and physiology of the ductus ejaculatorius duplex and seminal vesicle of male *Spodoptera litura* (Fabricius) (Lepidoptera: Noctuidae). Proc. Indian Acad. Sci. (Anim. Sci.) 98 (6): 405–418.

Anonymous, (1969). Summary of panel meeting in "Sterile male technique for eradication or control of harmful insect". Intl. Atom. Ener. Com. Vienna. pp. 80

Baloch, S. H. and Abbasi, Q. D. (1977). Investigation of the ecology and biology of the cutworm (Noctuidae: Lepidoptera) in the Hyderabad region. First Annu. Res. Report, Pakistan Sci. Found. Islamabad. p. 80.

Barbosa, P. (1974). Dissecting fluids and saline solutions. *In*: "Manual of basic techniques in insect' histology". (Mass, A. D. ed.), Oxford Publisher, London. p. 243.

Bargers, A. and Raw, F. (1967). Soil biology. Academic Press. New York. pp. 53. Beeman, R. W. (1982). Advances in mode of action of insecticides. Ann. Rev. Entomol. 27: 253- 281.

Benz, G. (1969). Influence of mating, insemination and other factors on oÖgenesis and oviposition in the moth *Zeiraphera diniana*. J. Ins. Physiol. 15: 55–71.

Binnington, K. and Retnakaran, A. (1991). Epidermis: A biologically Active target for metabolic inhibitors. *In*: "Physiology of the insect epidermis" (Binnington, K. and Retnakaran, A. eds.) CSIRO Publications, Australia. p. 307–334.

Brushwein, J. R. and Granett, J. (1977). Laboratory tests of two moulting disruptants, El-494 and dimilin, against the spruce budworm (Lepidoptera: Tortricidae): mortality and reproductive effects. Me. Life Sci. Agr. Exp. St. Mis. Report. p. 192.

Chase, J. A. and Gilliland, F. R. J. (1972). Testicular development in Thet tobacco budworm. Ann. Entomol. Soc. Amer. 65: 901- 906.

Chatani, F. and Ohnishi, E. (1976). Effects of ecdysone on theovarian development of *Bombyx* silkworm. Develop. Gro. Diff. 18: 48–484.

Chaudhury, M. F. B. and Raun, A. S. (1966). Spermatogenesis And testicular development of the European corn borer, *Ostrinia nubilalis* (Lepidoptera: Pyraustidae). Ann. Entomol. Soc. Amer. 59: 1157–1159.

Clark, L., Temple, G. H. R. and Vincent, J. F. V. (1977). The effects of a chitin inhibitor, dimilin on the reproduction of the peritrophic membrane in the locust, *Locusta migratoria*. J. Insect Physiol. 24: 241–246.

Concepts, A. (1989). Super ANOVA. Abacus Concepts, Berkeley, *CA*. pp. 120.

Crystal, M. M. (1978). Diflubenzuron-induced decrease of egg hatch of screw worms (Diptera: Calliphoridae). J. Med. Entomol. 15: 52–56.

Deb, D. C. and Chakaravorty, S. (1981). Effects of a juvenoid on the growth and differentiation of the ovary of *Corcyra cephalonica* (Lepidoptera). J. Insect Physiol. 27: 103–111.

Doncaster, L. (1911). Some stages in the spermatogenesis of Abraxas *grossulariata* and its variety *lacticolor*. J. Gent. 1: 179–184.

Dumser, J. B. (1980 a). The regulation of spermatogenesis in insects. Ann. Rev. Entomol. 25: 341–369.

Dumser, J. B. (1980 b). *In vitro* effects of ecdysterone on the spermatogonial cell in *Locusta*. Intl. J. Invert. Rep. 2: 165–174.

Emam, A. K., El-Refal, S. A. and Degheele, D. (1988). Effect of Sublethal dosages of four chitin synthesis inhibitors on the reproduction potential and F-1 generation of the Egyptian cotton leafworm *Spodoptera litoralis* (boisd.) (Lepidoptera: Noctuidae). Med. Van de Facland. Rij. Gent. 53: 249–254.

Engelman, F. (1979). Insect vitellogenin: identification, bio-synthesis, and role in vitellogenesis. Adv. Insect Physiol. 27: 49–108.

Etman, A. A. M. and Hooper, G. H. S. (1979). Developmental and reproductive biology of *Spodoptera litura* (F.) (Lepidoptera: Noctuidae). J. Aus. Entomol. Soc. 18: 363–372.

Friedlander, M. and Gitay, H. (1972). The fate of the normal-a-nucleated spermatozoa in inseminated females of the silkworm, Bombyx mori. J. Morphol. 138: 121–129.

Fytizas, E. (1976). L'action du TH6040 sur la metamorphose de *Dacus oleae* Gemel. (Diptera, Trypetidae). *Z. Angew. Entomol. 31*: 440–444. (Abstract in English read).

Gage, M. J. G. and Cook, P. A. (1994). Sperm size or number? Effects of nutritional stress upon eupyrene and apyrene sperm production strategies in the moth, *Plodia interpunctella* (Lepidoptera: Pyralidae). Func. Ecol. 8: 594–599.

Ganiev, S. G. (1986). Residual amount of dimilin in the plantand soil of paistaccio stands in SW Kirgizia. *Lesn. Khoz. 10*: 63–70.

Gelman, D. B. and Hayes, D. K. (1982). Methods and markers for synchronizing the maturation of fifth-stage diapause-bound and non diapause-bound larvae, pupae and pharate adults of the European cornborer, *Ostrinia nubilalis* (Lepidoptera: Pyralidae). Ann. Entomol. Soc. Amer. 75: 485–493.

Gelman, D. B., Woods, C. W. and Borvokovec, A. B. (1988). Ecdysteroid profile for haemolymph and testes from larvae, pupae, and pharate adults of the European cornborer, *Ostrinia nubilalis.* Arch. Ins. Biochem. Physiol. 7: 267–279.

Goldschmidt, R. (1916). The function of the apyrene spermatozoa. Science 44: 544–546.

Gupta, H. C., Bareth, S. S. and Sharma, S. K. (1991). Bioefficacy of edible and non-edible oils against pulse, *Callosobrochus chinensis* L. on stored pulses and their effect on germination. Agri. Biol. Res. 7: 101–107.

Haga, T., Toki, T., Tsujii, Y. and Nishiyama, R. (1992). Development of an insect growth regulator, chlorfluazuron. J. Pestic. Sci.17: S103–S113.

Hajjar, N. P. and Casida, J. E. (1979). Structure-activity relationships of benzoyl phenyl urea as toxicants and chitin synthesis inhibitors in *Oncopeltus fasciatus.* Pestic. Biochem. Physiol. 11: 33–45.

Hashizume, B. (1988). Atabron® 5E, a new IGR insecticide (chlorfluazuron). Jpn. Pestic. Inform. 58: 32–34.

Hatakoshi, M. (1992). An inhibitory mechanism over oviposition in the tobacco cutworm, *Spodoptera litura* by juvenile hormone analogue pyriproxyfen. J. Insect. Physiol. 38: 793–801.

Hatakoshi, M. and Hirano, M. (1990). Effects of S-71639 on the reproduction of *Spodoptera litura.* Adv. Invert. Rep. 5: 429–434.

Haines, L. C. (1981). Changes in colour of a secretion in the reproductive tract of adult males of*Spodoptera littoralis* (Boisduval) (Lepidoptera: Notuidae) with age and mated status. Bull. Entomol. Res. 71: 501–598.

Haynes, K. F. (1988). Sublethal effect of neurotoxic insecticides on insect behaviour. *Ann. Rev. Entomol. 33*: 149–168.

He, Y. (1994). Reproductive strategies in the armyworm, *Pseudaletia separata* (Lepidoptera: Noctuidae): with special reference to rearing density. Ph. D. Thesis. Graduate School of Agricultural Sciences, Nagoya University, Japan. pp. 122.

He, Y., Tanaka, T. and Miyata, T. (1995). Eupyrene and apyrene sperm and their numerical fluctuations inside the female reproductive tract of the armyworm, *Pseudaletia separata.* J. Insect. Physiol. 41: 689–694.

Holst, H. Z. (1974). Die fertility ätsbeeinflussende Wirkung des neuen insektizids P D 60-40 bei *Epilachna varivestis* Muls (Col. occinellidae) und *Leptinotarsa decemlineata* Say. (Col. Chrysomelidae). Z. Pflkranzk. Pflschutz. 81: 1–7. (Abstract in English read)

Holst, H. Z. (1974). Die fertility ätsbeeinflussende Wirkung des neuen insektizids P D 60-40 bei *Epilachna varivestis* Muls (Col. occinellidae) und *Leptinotarsa decemlineata* Say. (Col. Chrysomelidae). Z. Pflkranzk. Pflschutz. 82: 1–7. (Abstract in English read)

Holt, G. G. and North, D. T. (1970a). Spermatogenesis in the cabbage looper, *Trichoplusia ni.* (Lepidoptera: Noctuidae). Ann. Entomol. Soc. Amer. 63: 501–507.

Holt, G. G. and North, D. T. (1970b). Effects of gamma Irradiation on the mechanisms of sperm transfer in *Trichoplusia ni.* J. Insect. Physiol. 16: 2211–2222.

Hunter, E. and Vincent, J. F. V. (1974). The effects of a novel insecticide on insect cuticle. Experientia 4: 484–490.

Indrasith, L. S., Sasaki, T., Yaginuma, T. and Yamashita, O. (1988). The occurrence of a premature form of egg-specific protein in vitellogenic follicles of *Bombyx mori.* J. Comp. Physiol. 158: 1–7.

Jakob, W. L. (1973). Development inhibition of mosquitoes and house fly by urea analogues. *J. Med. Entomol. 10*: 540–543.

James, E. W. (1974). Laboratory and field evaluation of TH-6040, against House fly and stable fly. J. Econ. Entomol. 67: 746–747.

Kanost, M. R., Kawooya, J. K., Law, J. H., Ryan, R. O., Heusden, V. M. C. and Ziegler, R. (1990). Insect haemolymph proteins. *In: "Advances Insect physiology"* (Evans, P. D. and Wigglesworth, V.B. eds.), Academic Press, London. v. 22, p. 299–396.

Karlinsky, A. (1963). Effects de l'ablation des corpora allata imaginaux sur le developement ovarien de *Pieris brassicae* L. (Lepidoptera). *C. R.* Acad. Sci. Paris. *256*: 4101–1103. (Abstract in English read)

Kasuga, H., Osani, M., Yonezawa, Y. and Aigaki, T. (1985). An energy supply device for spermatozoa in the spermatophore of the silkworm. Develop. Gro. Diff. 27: 514– 522.

Katsuno, S. (1977a). Studies on eupyrene and apyrene spermatozoa in the silkworm *Bombyx mori* L. (Lepidoptera: Bombycidae). I. The intra testicular behaviour of the spermatozoa at various stages from the 5th instar to the adult. Appl. Entomol. Zool. 12: 142–147.

Katsuno, S. (1977 b). Studies on eupyrene and apyrene spermatozoa in the silkworm, *Bombyx mori* L. (Lepidoptera: Bombycidae). II. The intra testicular behaviour of the spermatozoa after emergence. Appl. Entomol. Zool. 12: 148–153.

Katsuno, S. (1977c). Studies on eupyrene and apyrene spermatozoa in the silkworm, *Bombyx mori* L. (Lepidoptera: Bombycidae). III. The post testicular behaviour of spermatozoa at various stages from the pupa to adult. Appl. Entomol. Zool. 12: 142–247.

Katsuno, S. (1977d). Studies on eupyrene and apyrene spermatozoa in the silkworm *Bombyx mori* (Lepidoptera: Bombycidae). IV. The behaviour of the spermatozoa in the internal reproductive organs of female moths. Appl. Entomol. Zool. 12: 352–359.

Ker, F. (1977). Investigation of locust cuticle using the insecticide diflubenzuron. *J. Insect Physiol. 23*: 39–48.

Khan, M. F. and Naqvi, S. N. H. (1988). Toxicological studies of diflubenzuron and nicotine cyanurate on house flies larvae. Pakisrtan J. Entomol. 3: 95–106.

Khuhro, R. D., Abbasi, Q. D., Abro, G. H., Baloch, S. H. and Soomor, A. H. (1986). Population dynamics of *Spodoptera litura* (F.) as influenced by ecological conditions and its host plant preference at Tandojam. Pakisrtan J. Entomol. 18: 351–357.

Kunkel, J. G. and Nordin, J. H. (1985). Yolk proteins. *In: "Comprehensive Insect Biochemitry, Physiology and Pharmacology"* (Kerkut, G. A. and Gilbert, L. I. eds.), Pergamon Press, Oxford. v. 1, p. 83–111.

Leuschner, K. (1974). Effects of the new insecticide PDD 60-40 on nymphs and adults of the cabbage bug, *Eurydema oleraceum* L. (Heteroptera, Pentatomidae). Z. Pflkrankh. Pflschutz. 81: 8–12.

Leviatan, R. and Friedlander, M. (1979). The eupyrene-apyrene dichotomous spermatogenesis of Lepidoptera. I. The relationship with post embryonic development and the role of the decline in juvenile hormone titre towards pupation. Develop. Biol. 68: 515–525.

Lim, S. J. and Lee, S. S. (1982). Toxicity of diflubenzuron to the grass hopper, *Oxy japonica*; effect on reproduction. Entomol. Exp. Appl. 31: 154–158.

Loeb, M. J. (1986). Ecdysteroids in testis sheaths of *Heliothis Virescens* larvae: An immunological study. Arch. Insect Biochem. Physiol. 3: 173–180.

Loeb, M., Brand, E. P. and Birnbaum, M. J. (1984). Ecdysteroid production by testes of budworm, *Heliothis virescens*, from last-larval instar to adult. *J. Insect Physiol. 30: 375–381.*

Loeb, M. J., Brandt, E. P., Woods, C. W. and Bell, R. A. (1988). Secretion of ecdysteroids by sheaths of testis of the gypsy moth, *Lymantria dispar*, and its regulation by testis ecdysiotropin. J. Exp. Zool. 248: 94–100.

Machida, J. (1929). Feine experimentelle untersuchung über die apyrene spermatozoen des seidenspinners *Bombyx mori* L. Zur. Beitru. Wissen. 9: 466–510. (Abstract in English read)

Madore, C. D., Drion, G. B. and John, B. D. (1983). Reduction of reproductive potential in spruce budworm (Lepidoptera: Tortricidae) by a chitin-inhibiting insect growth regulator. J. Econ. Entomol. 76: 708–710.

Matsuura, H. and Naito, A. (1997). Studies on the cold-hardiness and overwintering of *Spodoptera litura* F. (Lepidoptera: Noctuidae) VI. possible overwintering areas predicted from meteorological data in Japan. Appl. Entomol. Zool. 32: 167–177.

Matthee, J. J. (1951). The structure and physiology of the egg of *Locustana pardalina* (Walk.). Bull. Develop. Agri. For. Un. S. Afr. no 316: p. 83.

Mellanby, K. (1976). Pesticides, the environment and the balance of nature. *In: "Pesticides and Human Welfare"* (Gunn, D. I. and Steven, J. G. R. eds) Oxford University Press, p. 217–227.

Meves, F. (1902). Über oligpyrene und apyrene spermien und über ihre entstehung, nach beobachtugen an *Paludina* und *Pygaera. Arch. Mikro. Anatom. 61*: 1–84. (Abstract in English read)

Minitab, Inc. (1979). Statistics software. Release 12. Minitab Inc., Stat College, USA. p. 16801. Mitsuhashi, J. (1995). A continuous cell line from pupal ovaries of the common cutworm, *Spodoptera litura* (Lepidoptera: Noctuidae). Appl. Entomol. Zool. 30: 75–82.

Munson, J. P. (1906). Spermatogenesis of the butterfly, *Papilio rutulus*. Proc. Boston Soc. Natu. His. 33: 43–124.

Naqvi, S. N. H. and Rub, A. (1985). Effect of dimilin on the activity of melanization enzyme of *Musca domestica* L. (PCSIR strain). J. Sci. Karachi Univ. 13: 171–177.

Nickle, D. A. (1979). Insect growth regulators: New protectants against the almond moth in stored inshell peanuts. J. Econ. Entomol. 72: 816–819.

Nijahout, M. M. and Riddiford, L. M. (1974). The control of egg maturation by juvenile hormone in the tobacco hornworm, *Manduca sexta*. Biol. Bull. 146: 377–392.

Nizam, S. (1993). Effects of allochemicals against 3rd instar larvae of *Musca domestica* L. (Malir strain). Ph. D. Thesis, Department of Zoology, University of Karachi. pp. 250.

Nomura, M. (1994). Physiological effects of pyriproxyfen against the common cutworm, *Spodoptera litura. M. Sc.* Thesis, Laboratory of Applied Entomology, Nagoya University. pp. 125.

Olszak, R. W., Pawlik, B. and Zajac, R. Z. Z. (1994). The influence of some insect growth regulators on mortality and fecundity of the aphidophagous coccinellids, *Adalia bipunctata* L. and *Coccinella septempunctata* L. (Col., Coccinellidae). J. Appl. Entomol. 117: 58–63.

Ottens, R. J. and Todd, J. W. (1979). Effect of diflubenzuron on reproduction and development of *Graphognathus peregrinus* and *G. leucoloma*. J. Econ. Entomol. 72: 743–746.

Pan, M. L. (1977). Juvenile hormone and vitellogenin synthesis in the *Cecropia* silk moth. Biol. Bull. 153: 336–345.

Perveen, F. (2011) Effects of Sublethal Doses of Chlorfluazuron on Ovarioles in the Common Cutworm, *Spodoptera litura* (F.) (Lepidoptera: Noctuidae). Journal of Life Science, USA (Accepted).

Perveen, F. (2000a). Sublethal effects of chlorfluazuron on reproductivity and viability of *Spodoptera litura* (F.) (Lep., Noctuidae). J. Appl. Entomol. 124: 223–231.

Perveen, F. (2000b). Effects of sublethal dose of chlorfluazuron on testicular development and spermatogesis in the common cutworm, *Spodoptera litura*. J. Physiol. Entomol. 25: 315–323.

Perveen, F. (2005). Effects of chlorfluazuron on the growth of common cutworm, *Spodoptera litura* (Lepidoptera: Noctuidae) with special reference to the morphology and histology of the reproductive system. Ph.D. Thesis, Department of Zoology, University of Karachi.

Perveen, F. (2008). Effects of sublethal doses of chlorfluazuron on the number of inseminated sperm in *Spodoptera litura* (Lepidoptera: Nocutidae). J. Entomol. Sci. 11: 111–121.

Perveen, F. and Miyata, T. (2000). Effects of sublethal dose of chlorfluazuron on ovarian development and oögenesis in the common cutworm, *Spodoptera litura* (F.) Lepidoptera: Noctuidae). Ann. Entomol. Soc. Ame. 93: 1131–1137.

Qureshi, R. A., Qadri, S., Anwarullah, M. and Naqvi, S. N. H. (1983). Effect of neopesticides (JHA) on the morphology, emergence and fertility of *Musca domestica* L. (PCSIR strain) and its relation to phosphatases. Zool. Ang. Entomol. 95: 304–309.

Riakhel, A. S. and Dhadialla, T. S. (1992). Accumulation of yolk proteins in insect oÖcytes. Ann. Rev. Entomol. 37: 217–251.

Ross, D. C. and Brown, T. M. (1982). Inhibition of larval growth in *Spodoptera frugipera* by sublethal dietary concentrations of insecticides. J. Agr. Food Chem. 30: 193–196.

Scheffe, H. (1953). A method for judging all contrasts in the analysis of variance. Biometrika 40: 87–104.

Shimizu, T., Moribayashi, A. and Agui, N. (1985). *In vitro* analysis of spermatogenesis and testicular ecdysteroids in the cabbage armyworm, *Mamestra brassicae* L. (Lepidoptera: Noctuidae). Appl. Entomol. Zool. 20: 56–61.

Silberglied, R. E., Shepherd, J. G. and Dickinson, J. L. (1984). Eunuchs: the role of apyrene sperm in Lepidoptera? Amer. Natur. 123: 255–265.

Skibbe, J. T., Challaghan, P. T., Eccles, C. D. and Laing, W. A. (1995).Visulization of pH in the larval midgut of *Spodoptera litura* using 31P-NMR Microscopy. J. Insect Physiol. 42: 777–790.

Soltani, N. (1984). Effects of ingested diflubenzuron on the longevity and peritrophic membrane of adult mealworms (*Tenebrio molitor* L.). Pestic. Sci. 15: 221–225.

Soltani, N., Quennedey, A., Delbecque, J. P. and Delchambre, J. (1987). Diflubenzuron-induced alterations during *in vitro* development of *Tenebrio molitor* pupal integument. Arch. Insect Biochem. Physiol. 5: 201–209.

Soltani, N., Delachambre, J. and Delbecque, J. P. (1989a). Stage-specific effects of diflubenzuron on the ecdysteroid titres during the developmental of *Tenebrio molitor*: Evidence for a change in hormonal source. Gen. Comp. Endocrinol. 76: 350–356.

Soltani, N., Mauchamp, B. and Delbeque, J. P. (1989b). Effects of diflubenzuron on the ecdysteroid titres in two insect species. Akad Landwirtsch Wiss. Tag. Ber., DDR, Berlin. 274: 171–177.

Soltani, N. and Mazouni, N. S. (1992). OÖgenesis in mealworms: cell density of germarium, thickness of chorion and ecdysteroid production. Effects of regulators. Med. Fac. Landbouww. Univ. Gent. 62: 565–571.

Sridevi, R., Gupta, A. D. and Ramamurty, P. S. (1989a). Spermatogenesis in *Spodoptera litura* (Lepidoptera: Noctuidae). Entomon 14: 01–09.

Sroka, P. and Gilbert, L. I. (1971). Studies on the endocrine control of post emergence ovarian maturation in *Manduca sexta*. J. Insect Physiol. 71: 2409–2419.

Taft, H. M. and Hopkins, R. A. (1975). Boll weevils: field populations controlled by sterilizing emerging over winter females with a TH-6040 sprayable bait. J. Econ. Entomol. 68: 551–554.

Tahir, S., Anwer, T. and Naqvi, S. N. H. (1992). Toxicity and Residual effects of novel pesticides against rice weevil, *Sitophilus oryzae* (L.) (Coleoptera: Curculionidae). Pakistan J. Zool. 24: 111–114.

Takahashi, S. and Mizohata, H. (1975). Critical period necessary for the imaginal moulting and egg maturation of the eri-silkworm, *Samia cynthia* Ricini. Zool. Mag. 84: 283- 295.

Telfer, W. H., Rubenstein, E. and Pan, M. L. (1981). How the ovary makes yolk in *Hylophora. In: "Regulation of Insect Development and Behaviour"* (Wroclaw, M. K. ed.) Technical University Press, p. 637–654.

Walpole, R. E. and Myers, R. H. (1998). Probability, statistics for engineer scientist, 6th eds. Prentice-Hall, USA.

Williams, C. M. (1952). Physiology of insect diapause. IV. The brain and prothoracic glands as an endocrine system in the *Cecropia* silk. Biol. Bull. 103: 120–138.

Wright, J. E. and Spates, G. E. (1976). Reproductive inhibition activity of the insect growth regulator TH-6040 against the stable and house fly: Effect on hatchability. J. Econ. Entomol. 69: 365–368.

Wright, J. E. and Retnakaran, A. (1987). "Chitin and Bezoyl phenyl Ureas" (Junk, K. eds.) The Hague. v. 1, pp. 225.

Yoshida, K. (1994). Studies on the physiological effects of chlorfluazuron: with special reference to reproductive effects of chlorfluazuron on *Spodoptera litura*. M. Sc. Thesis, Graduate School of Agricultural Sciences, Nagoya University. p. 69.

Younis, M. (1973). Losses of cotton crop due to insect pests. *Proc. Crop. Produc. Sem. Organ. ESSO (Pak.) Ltd*. p. 110.

Yu, S. J. (1983). Age variation in insecticide susceptibility and detoxification capacity of fall armyworm (Lepidoptera: Noctuidae) larvae. J. Econ. Entomol. 76:

CHAPTER

3

Glucose Homeostasis and Organophosphorus Insecticides

1. INTRODUCTION: AN OVERVIEW

The agricultural practices revolution that have given rise to a significant increase in agricultural productivity has been a significant factor in the modern world's prosperity. Pesticides are potentially one of the most critical and effective strategies of the green revolution. Pesticides are the sole category of toxic substances that are designed to be intentionally released into the environment in order to achieve a greater benefit, a decision that significantly outweighs their toxicological concerns. One of the most frequently employed categories of insecticides is organophosphorus insecticides (OPI). Chemically, they are derivatives of phosphoric (H3PO4), phosphorous (H3PO3), or phosphinic acid (H3PO2) (Abou-Donia, 2003). Initially, the OPI were introduced as alternatives to the highly persistent organochlorine insecticides (Galloway & Handy, 2003). OPI are employed as pest control agents in a variety of situations due to their systemic, contact, and fumigant effects. OPI are extensively employed in agricultural practices to safeguard food and commercial commodities from a variety of insects. Furthermore, OPI are employed in domestic settings to mitigate the presence of hazardous insect species. They are not chemically or biochemically stable and are degraded in soil, detritus, and surface water. It is possible that the pervasive and indiscriminate use of these agents is a result of their instability, which has exposed both human and animal life to a variety of health hazards. The widespread use of these substances has resulted in a variety of ecotoxicological issues, and it is widely believed that exposure to OPI is a significant cause of morbidity and mortality in numerous countries.

OPI exposure is a significant toxicological threat that may impact the health of humans and animals due to its diverse toxicities, including neurotoxicity, endocrine toxicity, immunotoxicity, reproductive toxicity, and genotoxicity. Additionally, it has the potential to induce organ damage, disrupt glucose homeostasis, and alter cellular oxidative balance. A substantial body of scientific evidence supports this finding. In fact, the data on OPI residue levels in various sources reported from India does raise significant concerns regarding their toxic effects. Raw and bottled water samples were reported to contain a variety of OPI residues, some of which were significantly higher than the recommended levels (Mathur et al., 2003). OPI residue levels in breast milk samples from India have been reported by Sanghi et al. (2003). It has been hypothesized that infants may ingest malathion at levels 4.1 times higher than the average daily intake levels recommended by the World Health Organization, as indicated by the levels of OPI residues. In the same vein, human blood samples were reported to contain residues of phosphamidon, malathion, chlorpyrifos, and monocrotophos. (Mathur et al., 2005). Therefore, OPI pose a plausible environmental hazard that has the potential to impact a variety of human health conditions..

2. TOXICITY OF ORGANOPHOSPHORUS INSECTICIDES

The toxicity of active OPI is attributed to their ability to inhibit acetylcholinesterase (AChE, choline hydrolase, EC 3.1.1.7), an enzyme that catalyses the hydrolysis of the neurotransmitter acetylcholine (ACh), leading to cholinergic stress as a result of stimulation of muscarinic and nicotinic ACh receptors (Fukuto, 1990; Sogorb & Vilanova, 2003; Abou- Donia, 2003). The inhibition of AChE by an OPI takes place via a chemical reaction in which the serine hydroxyl moiety (of the active site) is phosphorylated. The phosphorylated enzyme is highly stable and, depending on the groups attached to the central 'P' atom of the OPI molecule, may be irreversibly inhibited.

There are several factors that determine the toxicity of OPI. Important of these are route and levels of exposure, structure of the substance and its interaction with the biotransformation/detoxification system of the body. The metabolic fate of OPI is basically the same in insects and animals. Following absorption, the distribution of OPI is variable. Blood half-lives are usually short, although plasma levels are in some cases maintained for several days. OPI undergo extensive biotransformation, which is complex and involves several metabolic systems in different organs, with simultaneous oxidative biotransformation at a number of points in the molecule, utilizing the cytochrome P-450 isoenzyme system. Metabolism occurs principally by oxidation, hydrolysis by esterases, and by transfer of portions of the molecule to glutathione. Oxidation of OPI may result in more or less toxic products. Most mammals have more efficient hydrolytic enzymes than insects and, therefore, are often more efficient in their detoxification processes. Numerous conjugation reactions follow the primary metabolic processes, and elimination of the phosphorus-containing residue may be via the urine or faeces. Some bound residues remain in exposed animals. Binding seems to be to proteins, principally, since there are limited data showing that incorporation of residues into DNA (Eto, 1974).

2.1 Neurotoxicity

Based on structure-function relationships, OPI are essentially neurotoxicants. Most important of their neurotoxicities is their 'cholinergic toxicity', which is a consequence of acetylcholinesterase (AChE) inhibition by OPI leading to accumulation of ACh and cholinergic stress. Signs of cholinergic toxicity include miosis, muscle fasciculation, excessive glandular secretions, nausea and vomiting (Namba, 1971). In addition, OPI are known to exert two other forms of neurotoxicities- Organophosphorus ester-induced delayed neurotoxicity (OPIDN) and Organophosphorus ester-induced chronic neurotoxicity (OPICN). OPIDN is a neurodegenerative disorder characterized by delayed onset of prolonged ataxia and upper motor neuron spasticity as a result of single or multiple exposures. OPICN refers to other forms of neurotoxicity that is distinct from both cholinergic toxicity and OPIDN. OPICN is characterized by neuronal degeneration and subsequent neurobehavioral and neuropsychological consequences (Abou-Donia, 2003).

2.2 Oxidative stress

Numerous studies provide evidence for the propensity of OPI to disrupt oxidative balance leading to oxidative stress (Soltaninejad & Abdollahi, 2009). Increased lipid peroxidation, protein carbonylation, depletion of cellular antioxidant pools and alterations in enzymatic antioxidant status appear to be chief mechanisms of OPI-induced oxidative stress that often results in pathophysiological changes and organ damage. Several studies have demonstrated usefulness of antioxidant intervention in alleviating oxidative stress and pathophysiological changes induced by OPI (Kamath et al., 2008, Soltaninejad & Abdollahi, 2009). These studies lend unequivocal support to view that oxidative stress mediates as one of the chief mechanisms of OPI toxicity.

3. ORGANOPHOSPHORUS INSECTICIDES AND GLUCOSE HOMEOSTASIS: MECHANISTIC INSIGHTS

In addition to neurotoxicity and oxidative stress, alterations in glucose homeostasis often culminating hyperglycemia is increasingly being reported as characteristic outcome of OPI toxicity. Meller et al., (1981) have described two cases of human subjects who were hospitalized with many complications including hyperglycemia. With no pseudocholinesterase detected, patients were given pralidoxime (AChE activator), which improved their condition and normalized hyperglycemia. Investigations revealed that they may have been exposed to malathion sprayed in their area. This case presents a classic case of hyperglycemic outcome following exposure to OPI as patients also exhibited miosis and muscle twitching. Numerous experiments have been conducted with experimental animals that reveal hyperglycemia as a characteristic outcome of OPI poisoning. A recent review by Rahimi & Abdollahi (2007) provides an exhaustive account of investigations revealing hyperglycemia in cases of OPI exposure.

There are certain characteristic features of alterations induced by OPI in glucose homeostasis. In cases of exposure to single dose of OPI, hyperglycemia appears to set in rapidly and peak changes are often followed by a trend of normalization. High dose of diazinon has been reported to cause hyperglycemia in mice that follows a trend of normalization (Seifert, 2001). Acute exposure of rats to malathion resulted in hyperglycemia with peak increase occurring at 2.2h after administration followed by decrease after 4h (Rodrigues et al., 1986). A similar case of reversible hyperglycemia has been reported by Lasram et al., (2008) following administration of a single dose of malathion to rats.

Biochemical changes associated with hyperglycemia serve as useful tools to understand etiology of OPI-induced hyperglycemia. Malathion has been reported to cause hyperglycemia in fasted rats. Interestingly, these hyperglycemic responses were not associated with hepatic glycogen depletion. The reversible phase of hyperglycemia was associated with increased glycogen deposition in liver, indicating that glucose may have come from gluconeogenesis (Gupta, 1974). Malathion induced hyperglycemia was associated with AChE inhibition in pancreas. More importantly, the trend of reversibility coincided with spontaneous reactivation of inhibited AChE (Lasram et al., 2008), indicating involvement of AChE-inhibition in hyperglycemia. Increase in blood glucose induced by sub chronic exposure of rats to malathion has been reported to be associated with increased glycogen phosphorylase and phosphoenolpyruvate carboxykinase activities, indicating involvement of both glycogenolytic and gluconeogenic processes. Increase in blood glucose levels induced by sub chronic exposure of rats to acephate has been reported to be associated with decrease in hepatic glycogen content (Deotare & Chakrabarti, 1981).

3.1 Pancreatic dysfunctions

Acute pancreatitis is also a well known complication of OP poisoning (Dressel et al., 1979; Frick et al., 1987; Hsiao et al., 1996), and epidemiological findings indicate that the incidence of pancreatitis is high in OPI intoxication based on various pathophysiological reports (Gokalp et al., 2005). The precise mechanisms underlying OPI-induced pancreatitis are still undefined, although it is believed to involve obstruction of pancreatic ducts and /or enhanced reactive oxygen species (Dressel et al., 1982; Sevillano et al., 2003, Sultatos, 1994). Involvement of oxidative stress following acute exposure to OPI has been reported recently (Banerjee et al., 2001) and it has been demonstrated unequivocally that lipid peroxidation is one of the molecular mechanisms involved in OPI-induced cytotoxicity (Akhgari et al., 2003; Ranjbar et al., 2002; Abdollahi et al., 2004b).

In view of the above, we attempted to understand the potential of repeated oral doses of dimethoate (DM) (at 20 and 40mg/kg b.w/d for 30days; doses corresponding to 1/20 and 1/10LD50 values) to cause alterations in glucose homeostasis and the associated biochemical alterations in pancreas of rats. We observed distinct signs of glucose intolerance among rats administered DM (**Fig. 1**) at time points at which un-treated rats showed normal glucose tolerance after an oral glucose load (3g/kg b.w.). We also observed that DM at both doses caused significant increase in blood glucose levels with concomitant inhibition of acetylcholinesterase activity and depletion of reduced glutathione contents in pancreas (**Table 1**) (Kamath & Rajini, 2007).

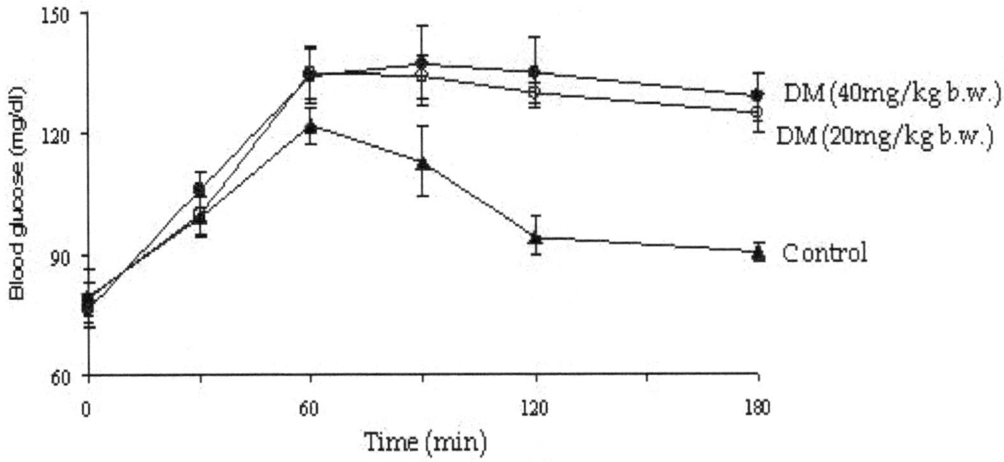

Fig. 1. Oral glucose tolerance at the end of 30 days in control (CTR) and Dimethoate (DM) treated rats.

Table 1. Blood glucose, acetylcholinesterase (AChE) and reduced glutathione (GSH) levels in pancreas of rats administered oral doses of Dimethoate (DM) for 30 days.

Treatment	Blood glucose (mg/dl)		AChE (nmoles substrate hydrolyzed /min/mg protein)	GSH (mg/g tissue)
Dimethoate (mg/kg b.w.)	Initial	Final		
0	85.33 ± 3.85	91.33 ± 2.41	4.96 ± 1.47	1.11 ± 0.02
20	87.34 ± 5.23	105.28 ± 3.57a	2.94 ± 1.75	0.99 ± 0.05a
40	85.00 ± 5.30	138.67 ± 5.70,b	0.43 ± 0.21a,b	0.91 ± 0.07a,b

Values are mean ± SEM (n=6);

a Comparison of control and other groups;

b Comparison of DM (20mg /kg b.w.) group with DM (40 mg/kg b.w.) group

Further, DM also caused significant pancreatic damage as reflected by increased amylase (2- 3 folds) and lipase (20 & 38%) activities in serum (**Fig 2**). These changes were sharply paralleled by significant damage in pancreatic milieu. There was a dose-related elevation in ROS levels in pancreas of treated rats. While the increase at the lower dose was 66%, a dramatic (150%) increase was evident at the higher dose. Concomitantly, a dose-related increase in TBARS (lipid peroxidation index) levels was observed in the pancreas of DM treated rats. There was 2.5 and 3.7 fold increase in TBARS level at lower and higher doses of DM respectively (**Fig. 3**). Activities of selected antioxidant enzymes were significantly elevated in the pancreas of treated rats compared to that of control rats. (**Table 2**) (Kamath & Rajini, 2007). These results are in accordance with the study of Hagar et al., (2002) who had earlier reported increased blood glucose levels and hyerinsulinemia with concomitant histochemical and ultramicrostructural changes in pancreas of rats following chronic exposure to dimethoate.

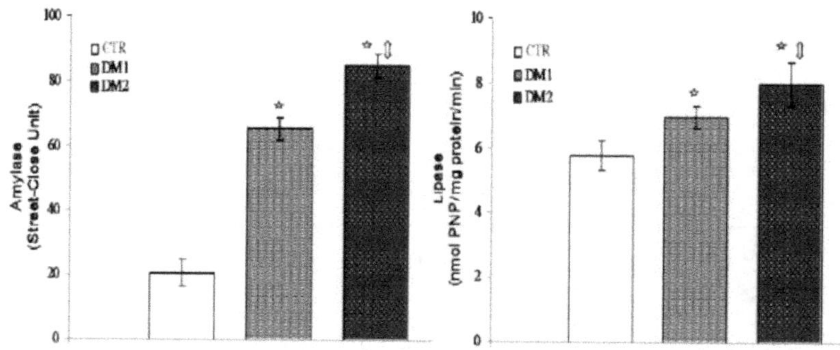

Fig. 2. Changes in pancreatic damage markers in rats induced by Dimethoate after 30 days (DM1: 20 mg/kg b.w/d; DM2: 40 mg/kg b.w/d). Values are mean ± SEM (n=6); * Comparison of control and other groups (*P* < 0.01), Comparison of DM1 with DM2 (*P* < 0.01)

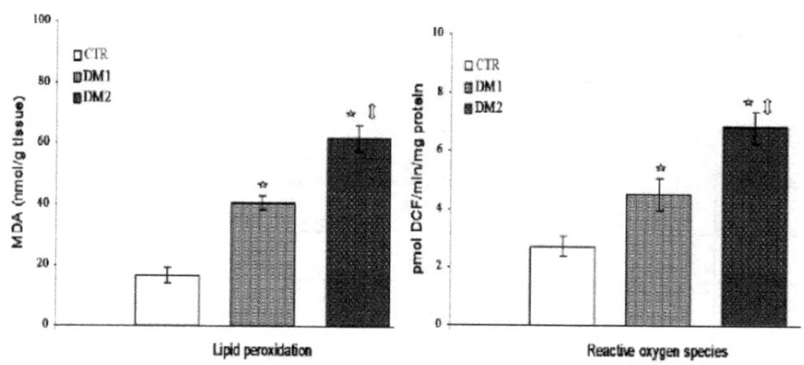

Fig. 3. Extent of lipid peroxidation and ROS levels in pancreas of control (CTR) and Dimethoate treated rats (DM1: 20 mg/kg b.w/d; DM2: 40 mg/kg b.w/d). Values are mean ± SEM (n=6); * Comparison of control and other groups (*P* < 0.01), ⇳ Comparison of DM1 with DM2 (*P* < 0.01).

Several studies have demonstrated pancreatitis after exposure to OPI (Dressel et al., 1979; Moore & James, 1988; Hsiao et al., 1996). Increase in the serum lipase and amylase activities reported by us clearly indicates that DM results in a state of pancreatic damage. Increased serum lipase activity has also been reported after administration of methidathion, an OPI (Mollaoglu et al., 2003). These results agree with earlier reports of acute pancreatitis in humans after accidental cutaneous exposure to DM (Marsh et al., 1988) and increase in amylase activity reported in dogs after diazinon administration (Dressel et al., 1982). Together, these studies clearly indicate that OPI possess propensity to elicit structural and functional alterations in pancreatic milieu that may be associated with disruptions in euglycemic conditions. From these studies, it may be argued that OPI may present a great threat to pancreatic functions in human beings and such threats may have far-reaching consequences on gluco-regulation in human beings.

Table 2. Antioxidant enzyme activities in pancreas of rats administered oral doses of Dimethoate for 30 days.

Group	Enzyme Activity			GR[3]	GST[4]
	SOD[1]	CAT[2]	GPX[3]		
CTR	26.42 ± 2.2	9.38 ± 0.31	27.18 ± 5.24	17.50 ± 1.60	0.03 ± 0.004
DM1	42.72 ± 0.38[a]	10.24 ± 0.32	25.23 ± 3.89	19.72 ± 2.03	0.04 ± 0.003[a]
DM2	56.23 ±1.18[a,b]	15.44 ± 0.51[a,b]	13.85 ± 2.20[a,b]	25.30 ± 1.30[a,b]	0.06 ± 0.003[a,b]

1 units/mg protein; 2μmol/min/mg protein; 3nmol/ min/ mg protein; 4μmol/ min / mg protein Values are mean ± SEM (n=6)

a Comparison of control (CTR) and other groups;

b Comparison of DM1 (DM: 20mg /kg b.w/d) group with DM2 (DM: 40 mg/kg b.w/d) group

3.2 Adrenal involvement

Studies undertaken by several researchers to investigate the mechanisms mediating hyperglycemic effects of OPI have mainly focused on the involvement of cholinergic stress and adrenal functions. We have extensively studied the mechanistic involvement of adrenals in glucotoxicity of OPI in rats mainly under acute and short-term exposure regimes. The rationale for studying adrenal involvement emerged from the typical hyperglycemic behaviour of single dose (oral) of two OPI-acephate and monocrotophos. Single dose of acephate and monocrotophos elicited rapid and transient hyperglycemia after administration. Both OPI were administered to overnight-fasted rats at 1/10 doses of their LD_{50} (LD50; acephate-1400mg/kg b.w., monocrotophos-18mg/kg b.w.). As depicted in **Fig. 4**, both acephate and monocrotophos induced reversible hyperglycemia with peak occurring at 2h after exposure. Acephate induced peak hyperglycemia at

2h (87%), which tended to normalize thereafter and attained near-control values at 8h after administration (Joshi & Rajini, 2009). Similarly, monocrotophos also induced rapid hyperglycemia with peak occurring at 2h (103%). Interestingly, monocrotophos induced hyperglycemia exhibited steep reversibility compared to acephate, with normalization occurring at 6h (Joshi & Rajini, 2010). This trend observed in the present study is consistent with other reports, which demonstrated reversible hyperglycemia in experimental animals following OPI administration. While Malathion has been reported to cause reversible hyperglycemia in rats (Gupta, 1974; Rodrigues et al., 1986; Seifert, 2001; Lasram et al., 2008), acute exposure to diazinon induced reversible hyperglycemia in mice (Seifert, 2001).

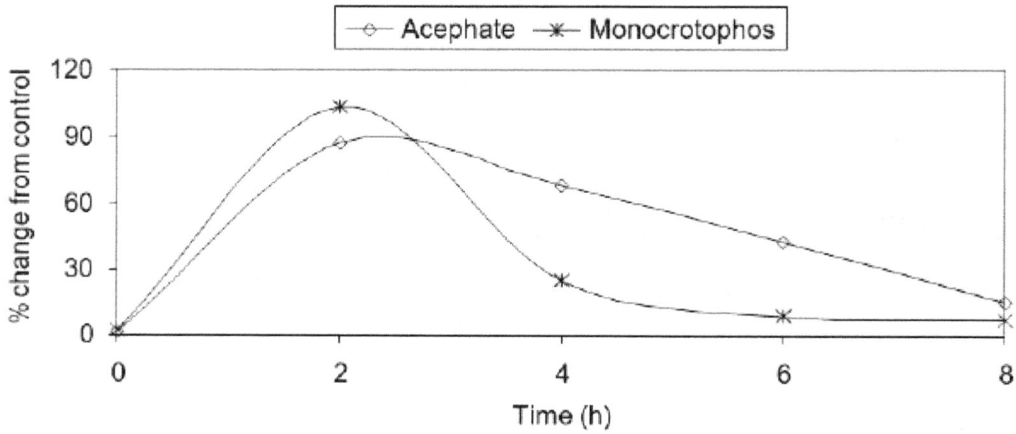

Fig. 4. Time-course of blood glucose levels in rats administered a single oral dose of acephate (140mg/kg b.w.) and monocrotophos (1.8mg/kg b.w.).

Based on the above results, we reasoned that the reversible hyperglycemia could be triggered by transient changes in the hormonal milieu of glucose homeostasis. Adrenals are an important part of the endocrine system and play a key role in glucose homeostasis by secreting glucocorticoid and amine hormones. Glucocorticoid hormones (GCs) (mainly cortisol in man and corticosterone in rodents) are secreted by the adrenal cortex under the control of hypothalamic-pituitary-adrenal axis. Glucocorticoid hormones, along with other key hormones, act to maintain blood glucose levels within narrow limits (Andrews & Walker, 1999). GCs, glucagon and epinephrine raise blood glucose by inhibiting glucose.uptake in the periphery and stimulating hepatic glucose release. Hepatic gluconeogenesis serves as the main source of hepatic glucose production during state of prolonged fasting and contributes significantly to development of diabetes mellitus (Pilkis & Granner, 1992). GCs facilitate gluconeogenesis as they exert permissive effect on the process by transcriptional activation of key enzymes of gluconeogenesis viz., glucose-6-phosphatase (G6Pase) (Argaud et al., 1996), phosphoenolpyruvate carboxykinase (PEPCK) (O'Brien et al., 1990) and tyrosine aminotransferase (TAT)

(Ganss et al., 1994). Increased glycogenolysis and gluconeogenesis appear to be the two chief mechanisms underlying OPI-induced hyperglycemia. Fenitrothion-induced increase in blood glucose in *S. mossambicus* was associated with decreased hepatic glycogen (Koundinya & Ramamuthi, 1979) and sub chronic exposure of rats to acephate, which caused slight increase in blood glucose also caused depletion of liver glycogen in rats (Deotare & Chakrabarthi, 1981). Abdollahi et al. (2004a) reported increased activity of GP and phosphoenolpyruvate carboxykinase (PEPCK) following sub chronic exposure to Malathion. Acute exposure to diazinon has been shown to cause depletion of liver glycogen with increased activity of glycogen phosphorylase, and also increased activities of gluconeogenesis enzymes in liver (Matin et al., 1989). Valexon is reported to have increased the activity of G6Pase in liver of rats (Kuz'minskaia et al., 1978). OPI and other AChE inhibiting organophosphate compounds exert strong influences on functioning of hypothalamic-pituitary-adrenal (HPA) axis, leading to increased circulating levels of corticosteroid hormones in vivo. This is particularly true in the case of acute exposure to AChE inhibiting compounds. Studies have shown elevated corticosteroid hormones levels in response to AChE-inhibiting compounds and role of AChE inhibition in the phenomenon. Single dose of Chlorfenvinphos, acephate and methamidophos have been demonstrated to elevate circulating levels of corticosterone and aldostserone after administration of a single dose (Osicka-Kaprowska et al., 1984; Spassova et al., 2000). Soman has been reported to increase plasma corticosterone levels in rodent models (Hudon & Clement, 1986; Fletcher et al., 1998). More importantly, the stressogenic potential (hypercorticosteronemia and induction of liver tyrosine aminotransferase activity) of soman was effectively abrogated by reactivators of inhibited acetylcholinesterase (Kassa, 1995 & 1997). Similarly, stressogenic potential of Cyclohexylmethyl phosphonofluoridate (AChE inhibitor) has been reported to be eliminated by HI-6 (AChE reactivator) (Kassa & Bajgar, 1995). Thus, it is clearly evident that AChE-inhibiting OPI elicit hyper stimulation of adrenal functions, leading to induction of gluconeogenesis enzymes in liver.

Based on the time-course of reversible hyperglycemia induced by acephate and monocrotophos, further experiments were carried out to investigate the adrenal effects of OPI and its role in the ensuing hyperglycemia. We assessed the effects of 2 or 6h exposure to either acephate (oral) or 2 or 4h exposure to monocrotophos (oral) on plasma corticosterone, adrenal cholesterol, blood glucose, key liver gluconeogenesis enzymes (G6Pase and TAT) and hepatic glycogen content in rats. Interestingly, we observed that both acephate and monocrotophos induced strong hypercorticosteronemia with concomitant hyperglycemia and induction of liver gluconeogenesis enzyme activities. Further, hypercorticosteronemia was associated with decrease in adrenal cholesterol pools (effect of monocrotophos on adrenal pools described in the section on 'comparison between single and repeated dose effects'), which is the precursor for corticosterone synthesis (**Table 3 & 4**). Depletion in adrenal cholesterol pools may therefore be attributable to increased synthesis and secretion of corticosterone. Interestingly, both OPI did not cause depletion in hepatic glycogen content. At time points that represented normalization of blood glucose levels, there was phenomenal increase in liver glycogen levels. The data presented above clearly demonstrates co-existence of hypercorticosteronemia and induction of liver gluconeogenesis enzyme activities with hyperglycemia in OPI treated rats, indicating that OPI may trigger induction of liver gluconeogenesis machinery as result of hypercorticosteronemia, leading to hyperglycemia.

Table 3. Biochemical effects of acephate (140mg/kg b.w.) in rats

	At time interval after administration		
	0h	2h	6h
Plasma corticosterone *	30.9±3.4a	55.0±2.5b	44.0±2.7b
Adrenal cholesterol**	26.5±1.4a	15.6±0.56b	12.5±0.47b
Blood glucose ***	101.6±4.6a	182.4±5.2b	142.7±5.2c
Liver G6Pase#	90.14±4.38a	171.93±5.61b	112.84±4.18c
Liver TAT ##	14.28±1.34a	26.31±0.87b	23.7±0.48b
Hepatic glycogen$	316.2±34.90a	325.3±29.12a	1145.0±27.92b
(Joshi and Rajini, 2009)			

Data analyzed by ANOVA followed by Tukey Test (n=6)

** μg/dl; ** mg/g tissue; *** mg/dl*

 # glucose-6-phosphatase (nmol/min/ mg protein);

tyrosine aminotranferase (nmol/min/mg protein); $ μg/g tissue

Table 4. Biochemical effects of monocrotophos (1.8mg/kg b.w.) in rats

	At time interval after administration		
	0h	2h	4h
Plasma corticosterone *	36.62±1.2a	73.82±3.8b	45.65±1.8a
Blood glucose **	95.2±1.8a	194.8±3.7b	121.3±1.9c
Liver TAT #	15.86±0.8a	32.27±1.2b	26.87±1.8c
Hepatic glycogen##	213.8±49.2a	216.4±21.1a	925.7±27.6b
(Joshi and Rajini, 2010)			

Data analyzed by ANOVA followed by Tukey Test (n=6)

** μg/dl; ** mg/g tissue ; # tyrosine aminotranferase (nmol/min/mg protein); ## μg/g tissue*

Indeed, role of adrenals in glucotoxicity of OPI has been explored earlier. Matin et al., (1989) earlier demonstrated that single dose diazinon (OPI) caused hyperglycemia and induction of liver gluconeogenesis enzymes in normal rats while these changes did not manifest in adrenalectomized rats, indicating the involvement of adrenals in the glucotoxicity of diazinon. Our attempts to study the adrenal and glycemic effects of acephate and monocrotophos revealed that two compounds, which exhibit anticholinesterase property, elicited similar effects. Thus, the effects raised question whether the adrenal and glycemic effects are mediated through the anticholinesterase property of OPI. To address the question, we studied the extent of AChE inhibition elicited by monocrotophos at 2 and 4h after administration. Influence of cholinergic antagonists was investigated at 2h after administration on stressogenic (hypercorticosteronemia and induction of liver TAT activity) and hyperglycemic potential of monocrotophos. For the purpose of mechanistic investigations, we employed two muscarinic cholinergic antagonists-atropine sulphate, a general ACh receptor antagonist that can pass through blood brain barrier (BBB) (Guarini et al., 2004) and methyl atropine nitrate, which is a peripherally active antagonist that does not pass through blood BBB (Pavlov et al., 2006). Both antagonists were administered at 30μmols/kg b.w 3-5 min before monocrotophos (1/10 LD50).

Monocrotophos elicited significant inhibition of AChE activity (>50%) in brain, adrenals and liver at both 2 and 4 h after exposure (**Fig. 5A**). Of the organs studied, maximum inhibition of AChE activity was evident in brain (84 and 78% at 2 and 4 h respectively) while the enzyme activity in adrenals was inhibited to 32 and 34% of control activity at 2 and 4 h after exposure respectively. Similarly, monocrotophos administration reduced liver AChE activities to 47 and 46% of control at 2 and 4 h after exposure respectively. More importantly, we did not observe any spontaneous reactivation of inhibited AChE activity at 4h after administration, which is an important feature of the enzymes' behavior (Reiner and Aldridge, 1967; Reiner, 1971). This indicates that, while hyperglycemic potential of monocrotophos in rats may be a result of its anticholinesterase potency, the reversibility of hyperglycemia is not a consequence of spontaneous reactivation of the enzyme. Reversibility of hyperglycemia may hence be a consequence of counter-regulatory mechanism as reflected by glycogen deposition at 4h after administration. Increase in glycogen content upon 4h exposure is a clear indication of mobilization of glucose into glycogen synthesis pathway as a measure to overcome hyperglycemia.

We also observed that both cholinergic antagonists were potent in offering protection against stressogenic and hyperglycemic potential of monocrotophos. Administration of monocrotophos elicited significant hyperglycemia (103%) **(Fig. 5B)**. Pre- treatment of rats with atropine sulfate (106.04 ±1.83 compared to 191.82 ±7.59 mg/dl of monocrotophos alone) and atropine methyl nitrate (123.49 ±4.12 compared to 191.82 ±7.59 mg/dl of monocrotophos alone) offered significant protection against hyperglycemia induced by monocrotophos. It has been earlier demonstrated that diazinon-induced hyperglycemia was mediated by AChE inhibition, as revealed by protective effects of pralidoxime (AChE reactivator) (Seifert, 2001). Monocrotophos-induced hypercorticosteronemia (112%) was effectively prevented by cholinergic antagonists (**Fig. 5C**). Pre-treatment of rats with atropine sulfate (33.98 ±2.89 compared to 76.63 ±1.76 µg/dl of monocrotophos alone) and atropine methyl nitrate (44.67 ±1.64 compared to 76.63 ±1.76 µg/dl of monocrotophos alone) offered significant protection against hypercorticosteronemia induced by monocrotophos. Monocrotophos induced a marked increase in the TAT activity in liver (107%) (**Fig. 5D**). Pre-treatment of rats with atropine sulfate (20.42 ±1.70 compared to 33.38 ±1.09 nmol/min/mg protein) and atropine methyl nitrate (22.39 ±0.79 compared to 33.38 ±1.09 nmol/min/mg protein) offered significant protection against induction of TAT activity. These results clearly indicated that both physiological stress (hypercorticosteronemia and induction of liver TAT activity) and hyperglycemia manifest as a consequence of peripheral muscarinic cholinergic stimulation. Corticosterone exerts hyperglycemic action by up-regulation of gluconeogenesis machinery. Hence, hypercorticosteronemia and induction of liver TAT (gluconeogenesis enzyme) activity accompanying hyperglycemia raises a question whether hypercorticosteronemia is responsible hyperglycemia in monocrotophos-treated rats.

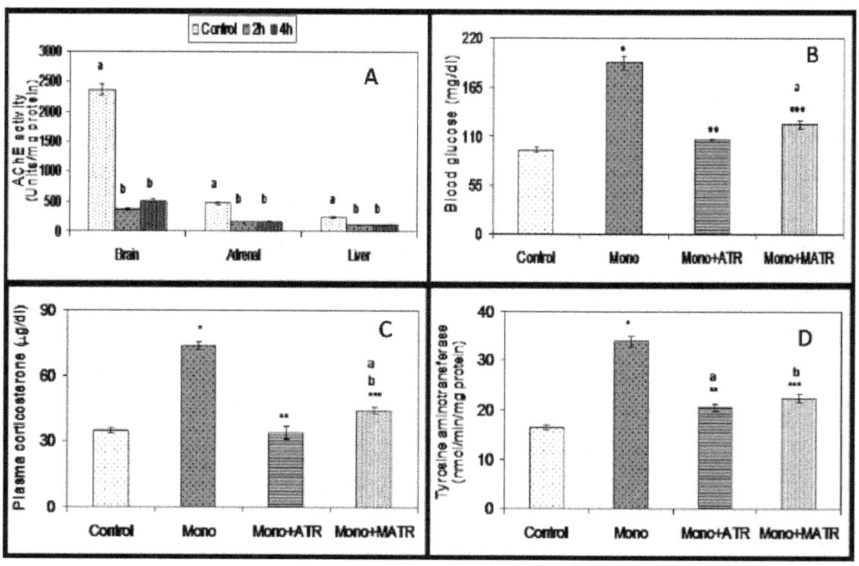

Fig. 5. Protective effects of atropine (ATR) and methyl atropine (MATR) against stressogenic and hyperglycemic potential of monocrotophos (Mono) (Joshi and Rajini, 2010).

Acetylcholine exerts strong influence on functioning of hypothalamus-pituitary-adrenal (HPA) axis. Acetylcholine has been found to increase corticotrophin releasing hormone (CRH) activity of hypothalamus *in vitro* as measured by effect on corticosteroidogenesis, an effect that was antagonized by atropine (Bradbury et al., 1974). ACh has also been shown to increase secretion of immunoreactive CRH from hypothalamus *in vitro* (Calogero et al., 1988), an effect that was antagonized by ACh receptor antagonists, atropine (muscarinic) and hexamethonium (nicotinic). Given the importance of ACh in excitation of HPA axis, assessment of cholinergic stress in activation of HPA axis in monocrotophos treated rats becomes important. The importance of ACh in functioning of HPA axis is further exemplified by the fact that muscarinic receptor agonists such as carbachol (Bugajski et al., 2002) and arecoline (Calogero et al., 1989) were found to increase ACTH and corticosterone *in vivo*. More importantly, the agonist induced increase in ACTH and corticosterone was antagonized by atropine (Bugajski et al., 2002), suggesting role for muscarinic ACh receptor in regulation of HPA axis. Role of anticholinesterase properties of organophosphate compounds in activation of HPA axis is demonstrated by studies showing elimination of stressogenic activity of cyclohexyl methyl phosphonofluoridate (as measured by plasma corticosterone and liver tyrosine aminotransferase activity) by HI-6, a cholinesterase reactivator that sufficiently reactivated inhibited AChE in brain and diaphragm (Kassa & Bajgar, 1995) and protection offered by atropine against diisopropylfluorophosphate induced increase in corticosterone levels (Smallridge et al., 1991). These studies clearly show the influence of ACh and involvement of muscarinic receptors in functioning of HPA axis.

From our data on influence of cholinergic antagonists on stressogenic and hyperglycemic potential of monocrotophos, it could be hypothesized that muscarinic cholinergic stress triggers hypercorticosteronemia, which may lead to induction of liver gluconeogenesis and hyperglycemia. However, experiments conducted with glucocorticoid receptor and adrenergic receptor antagonists revealed that hyperglycemia in mediated by adrenergic mechanisms while hypercorticosteronemia leads to only induction of liver TAT activity (data not shown). Further, we observed that monocrotophos-induced hyperglycemia was completely abolished by a gluconeogenesis inhibitor (data not shown). This establishes that physiological stress and hyperglycemia manifest in monocrotophos treated rats as independent consequence of peripheral cholinergic stress.

We further compared the effects of monocrotophos on adrenal functions and glycemic control in rats following single and repeated doses. Comparison was made between the effects of a single dose (measured 2h after administration) and that of 5 or 10 doses (one dose per day, measured 2h after last dose). In both cases, the oral dose of 1.8mg/kg b.w. was employed for the purpose of comparison. Interestingly, we observed that effects single dose of monocrotophos on adrenal functions and glycemic control was more severe than that of repeated doses. Single dose of monocrotophos elicited hypercorticosteronemia (114%) with concomitant decrease in adrenal cholesterol (33%). These adrenal effects of single dose were accompanied with hyperglycemia (109%) and induction of liver tyrosine aminotransferase activity (113%). However, repeated administration of monocrotophos for 5 or 10days resulted in blunting of responses. In case of repeated exposure, increase in corticosterone was 76 and 67% respectively in 5 and 10d exposure groups with 18 and 13% decrease in adrenal cholesterol. Similarly repeated administration elicited marginal increase in blood glucose (39 and 32%) and induction of liver TAT activity (56 and 61%) (**Fig. 6**).

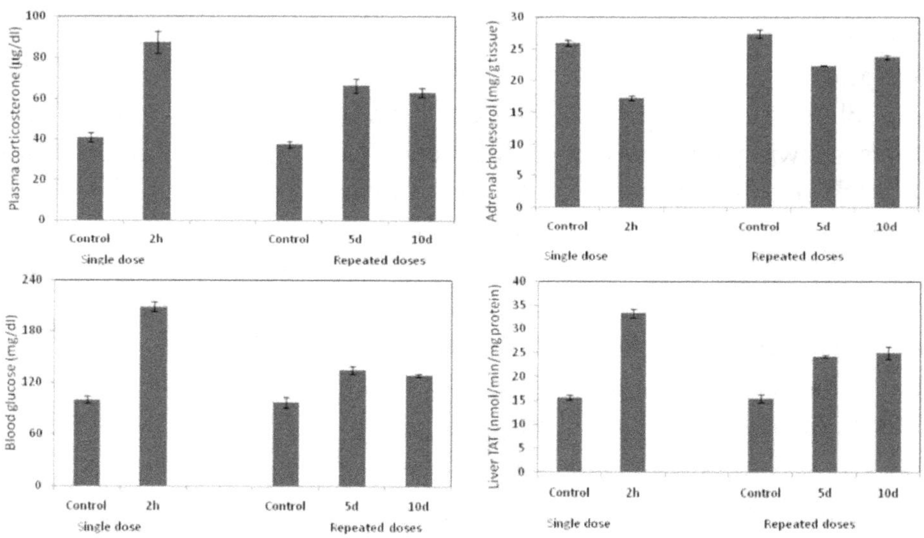

Fig. 6. Adrenal and glycemic effects of monocrotophos.

The above data clearly shows that repeated administration results in blunting of responses. This indicates that multiple administrations are associated with onset of some sort of resistance to the action of OPI. Development of tolerance to cholinesterase inhibitors during multiple administrations is a well documented phenomenon (Brodeur and DuBois, 1964; McPhillips, 1969; Sterri et al., 1980). Tolerance to the elevation of plasma corticosterone by DFP was reported to develop during repeated administration (Kokka et al., 1985). Several studies suggest that cholinergic receptors could be involved in onset of tolerance to OPI, which may be mediated by events such as down regulation of muscarinic receptors (Costa et al., 1982a&b). Tolerance to the toxic effects of dilsulfoton during multiple exposures has been attributed to reduced muscarinic receptor binding in tissues of rats tolerant to the insecticide (Schwab et al., 1981). Blunted responses observed by us in case of repeated administration of monocrotophos may be attributed to tolerance mechanisms such as down regulation of muscarinic receptors. One mechanism that may be responsible for development of resistance is increased blood insulin levels. Comparison of effects of acute and repeated doses of monocrotophos on plasma insulin levels, however, needs to be done. Such a hyperinsulinemic response has been reported in case of exposure to malathion. While malathion caused increase in blood glucose and insulin levels after single exposure and continued dietary administration for 4 weeks, the degree of hyperinsulinemia was markedly greater in dietary group (Panahi et al., 2006). Thus, repeated administration of organophosphorus insecticides leads to blunting of responses. Although blunted, these responses still represent a great threat to euglycemic balance. This is particularly true in the case of constant state of hypercorticosteronemia. This has propensity to affect skeletal muscle glucose metabolism and long-term impairments in such mechanisms may lead to long lasting dysregulation in glucose homeostasis.

3.3 OPI act as pre-disposing factors for onset of diabetes?

Based on our comprehensive studies described above, we have proposed a scheme on the mechanism/s through which OPI might regulate/ disrupt glucose homeostasis (**Fig. 7**). Oxidative stress in pancreatic milieu and glucose intolerance, up regulated gluconeogenesis machinery and hyperglycemia are critical factors in diabetes etiology. With the ability to induce the above-mentioned dysregulations, OPI may have far reaching consequences on diabetic outcomes. This may be a more pertinent issue in the present times since diabetes is fast emerging as a major wide spread disorder that threatens human life. With this realization, our laboratory has also committed to investigate if OPI act as predisposing/aggravating factors for onset or progression of diabetic condition.

We observed that dichlorvos (DDVP) treated rats showed higher (22%) levels of blood glucose compared to normal control rats while as expected, rats injected with the diabetogenic agent, Streptozotocin (STZ) alone also showed elevated (37%) level of blood glucose. However, blood glucose levels of DDVP pre-treated rats administered STZ showed relatively higher blood glucose level compared to all the groups. Liver glycogen levels were significantly lower in rats administered either DDVP (18%) or STZ (19%) alone while, rats administered DDVP followed by STZ revealed further lower levels of glycogen (46 %) **(Table 5)**. Further, we also observed that DDVP pre-treatment resulted in more sever oxidative stress in STZ treated rats. ROS levels were significantly elevated in STZ (40%) and DDVP (55%) groups compared to 'untreated control' group. However, ROS levels were markedly higher (81.23 ± 6.52 pmole DCF/min/mg protein) in 'DDVP+STZ' group of rats. Pancreas of rats administered with either DDVP or STZ alone showed marginally higher levels of the lipid peroxides compared to that in 'untreated controls' while, the levels of lipid peroxides generated in pancreas of 'DDVP+STZ' rats showed significant increase (110%) compared to all other groups. Pancreatic reduced glutathione level in 'DDVP+STZ' rats was significantly lower (37%) while, rats administered with DDVP or STZ alone also had significantly lower levels of GSH, although to a lesser extent (**Table 6**). These results clearly demonstrate that OPI act as pre-disposing factor for diabetes as reflected by higher degree of glucotoxicity of STZ (subdiabetogenic dose) in DDVP treated rats.

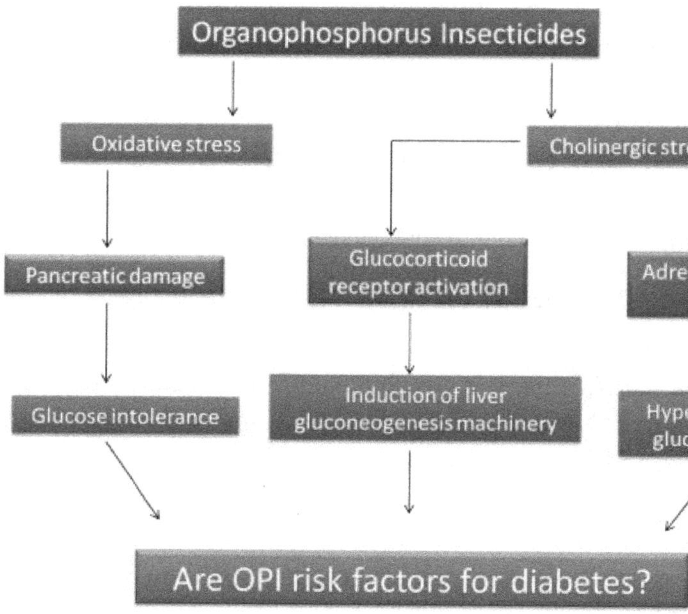

Fig. 7. Proposed scheme for OPI-induced alterations in glucose homeostasis.

Generally, an acute intraperitoneal dose of 40–60 mg/kg b.w is employed to induce significant hyperglycemia in rats. For the present study, we employed a lower dose of 25 mg/kg b.w ('sub-diabetogenic dose') in order to examine if pre-treatment with DDVP renders these rats more susceptible to hyperglycemia. Experimental regime began with two groups with 12 rats each-control and DDVP-treated group. The DDVP-treated group animals were orally administered daily DDVP at 20mg/kg b.w/d (corresponding to 1/5 of LD50 value: 100 mg/kg b.w, determined in a preliminary study) for 10 d. After 10 days, rats of the control group were further divided into two sub groups of six animals each ; the first sub group served as control ('untreated control'), while the second sub group of rats was intraperitoneally injected streptozotocin (STZ, 25 mg/kg b.w.) ('STZ'). The group of rats administered with DDVP was also divided into two sub groups; the first sub group of rats served as DDVP control ('DDVP'), while the second sub group of rats was injected with streptozotocin (i.p, 25mg/kg b.w.) ('DDVP+STZ').

Table 5. Blood glucose and liver glycogen levels in rats treated with DDVP ± STZ (i.p , 25 mg/kg b.w).

Group	Blood glucose[1]	Liver glycogen[2]
CONTROL	113.53[a] ± 2.31	41.55c ± 2.01
DDVP	138.37b ± 4.17	34.20b ± 1.42
STZ	155.03c ± 5.09	33.34b ± 2.23
DDVP+STZ	188.99d ± 4.44	22.62[a] ± 3.52

1 mg/dl; 2mg/g tissue; Values are mean ± SEM (n=6); Mean in the same column with different superscript differ significantly (p<0.05)

Table 6. Oxidative stress parameters in pancreas of rats treated with DDVP ± STZ (i.p , 25 mg/kg b.w).

Group	ROS[1]	TBARS[2]	GSH[3]
Untreated control	38.76a ± 4.04	242.76a ± 19.18	1.07c ± 0.03
DDVP	60.18b ± 4.59	294.94a ± 10.65	0.83b ± 0.01
STZ	54.27b ± 2.89	283.63a ± 7.27	0.78b ± 0.02
DDVP+STZ	81.23c ± 6.52	389.38b ± 38.47	0.67a ± 0.02

1 pmol DCF/min/mg protein; 2nmol/g tissue; 3mg/g tissue

Values are mean ± SEM (n=6); Mean in the same column with different superscript differ significantly (p<0.05)

3.4 Do OPI aggravate diabetic outcomes?

To address this question, we investigated the diabetic outcomes in rats experimentally rendered diabetic and post-treated with monocrotophos. Rats were rendered diabetic with an acute dose (60 mg/kg b.w, i.p) of streptozotocin. Monocrotophos was orally administered at a sublethal dose (1/20 LD_{50}, 0.9 mg /kg b.w./d, 5 days) to both normal and diabetic rats. We observed that monocrotophos *per se* moderately increased (25%) the blood glucose levels in normal rats, but significantly aggravated the hyperglycemic outcome in diabetic rats (56% above diabetic rats). We observed the typical lipid profile alterations among diabetic rats characterized by increase in total cholesterol and triglycerides (TG) in serum. While monocrotophos did not impact lipid profile *per se*, its interaction with diabetic component resulted in severe alterations lipid profile that reflected in phenomenal increase in serum triglyceride. Such augmented impairments may have high bearing on cardiovascular health since the cardiovascular risk index was alarmingly high among diabetic rats treated with monocrotophos (**Fig. 8**). Further, monocrotophos resulted in higher degree of hepatic and renal toxicity as reflected by alterations in serum transaminase activities and blood urea nitrogen values respectively (**Table 7**).

Further, it is now established that oxidative stress in an important consequence of diabetic condition and plays as an important pathophysiological factor in progression of diabetic complications (Maritim, 2003). We observed that STZ-induced diabetes was associated with increased lipid peroxidation (122%), depletion of reduced glutathione (10%) and alterations in activities of two important antioxidant enzymes (superoxide dismutase and catalase) in kidney. Monocrotophos further deteriorated oxidative impairments in kidney as evidenced by further increase in lipid peroxidation (170%) and depletion of reduced glutathione (18%) content. Diabetes was associated with marginal decrease in superoxide dismutase activity in kidney, which was further reduced by monocrotophos treatment (35%) (**Fig. 9**).

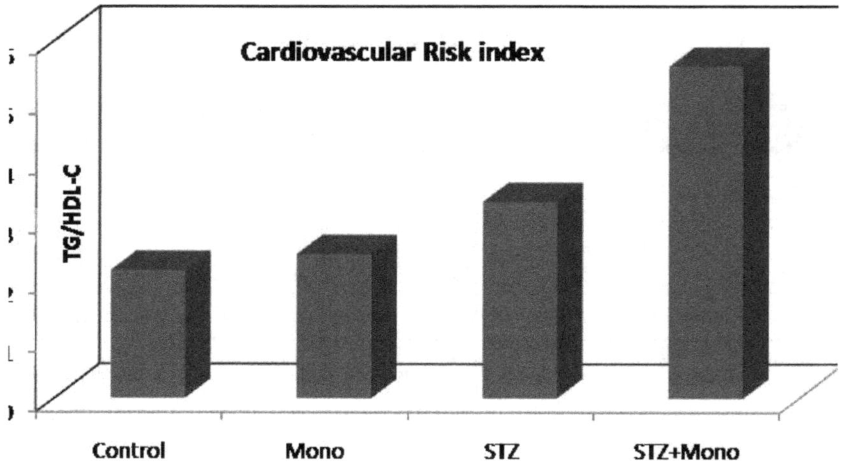

Fig. 8. Effect of repeated oral doses of monocrotophos at 1/20 LD_{50} (0.9 mg/kg b.w) on cardiovascular index in control and diabetic rats (Begum and Rajini, 2011a).

Table 7. Effect of repeated oral doses of monocrotophos at 1/20 LD_{50} (0.9 mg/kg b.w) on blood glucose, lipid profile and hepatic and renal damage markers in serum in control and diabetic rat.

Control		Mono	STZ	STZ+Mono
Blood glucose	101.58 ±	126.91±	382.71 ±	597.94 ±
(mg/dl)	1.4	8.9	14.0ab	12.5cde
TC	38.00 ±	41.92 ±	50.45 ±	50.42 ±
(mg/dl)	2.1	1.9	1.6a	1.2a
HDL-C	31.09 ±	32.68 ±	37.84 ±	35.48 ±
(mg/dl)	1.2	1.7	1.4	1.3
TG	66.74 ±	78.12 ±	125.44 ±	193.52 ±
(mg/dl)	3.5	6.9	9.2a	19.4bcd
BUN	33.08 ±	51.88 ±	71.18 ±	78.05 ±

Control		Mono	STZ	STZ+Mono
(mg/dl)	5.6	8.2	10.1a	5.2b
SC	0.63 ±	0.72 ±	0.81 ±	0.84 ±
(mg/dl)	0.06	0.1	0.1	0.04
Serum ALT	66.19 ±	72.24 ±	80.03 ±	120.77 ±
(U/L)	2.0	0.7	2.4	9.2abc
Serum AST	126.57 ±	156.00 ±	199.78 ±	341.55 ±
(U/L)	0.2	18.7	14.2de	8.5abc
(Begum and Rajini, 2011a)				

Data analyzed by Tukey's HSD test; Mean ± SEM (n=4)

TC: Total cholesterol; HDL-C: High-density lipoprotein ; TG: Triglyceride; BUN: Blood Urea Nitrogen; SC: Serum creatinine

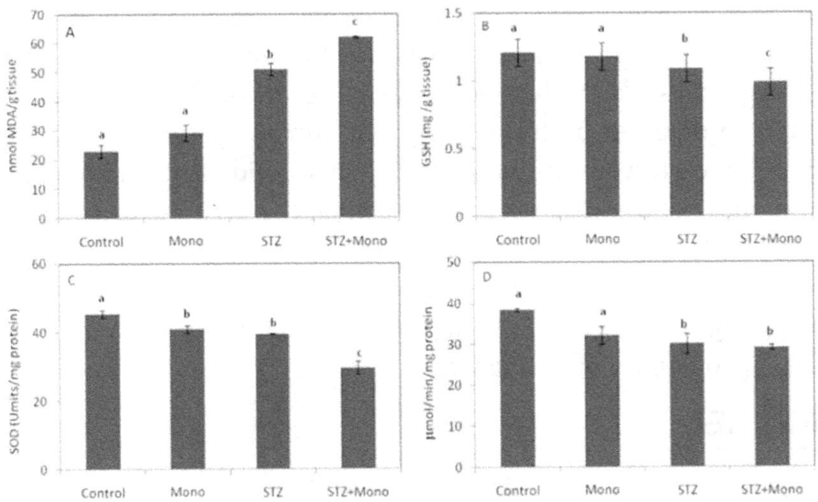

Fig. 9. Effect of repeated oral doses of monocrotophos at 1/20 LD_{50} (0.9 mg/kg b.w) on oxidative balance in kidney of control and diabetic rats (Begum and Rajini, 2011b).

Our work on interaction of OPI with diabetic component clearly shows that OPI can act as both predisposing and aggravating factors for diabetes. The inference becomes an important consideration to be made as the modern world is facing an escalating situation of alarming increase in the incidence of diabetes. Our study employed a low dose of monocrotophos, which *per se* did not interfere with lipid profile in rats, yet causing augmentation of alteration in lipid profile in diabetic rats. However, other studies have clearly demonstrated that several OPI cause alterations in lipid profile, particularly hypertriglyceridemia (Ryhänen et al, 1984; Ibrahim & El-Gamal, 2003; Rezg et al., 2010). Dyslipidemia or lipid abnormalities play an important role in the progression of diabetes (Goldberg, 2001) and these are characterized by lipid derangements including hypertriglyceridemia, low high-density cholesterol (HDL-C), and a high concentration of small dense low-density lipoprotein (LDL) particles. Further, a state of elevated hypertriglyceridemia is commonly associated with insulin resistance and represents a valuable clinical marker of the metabolic syndrome (Grundy et al., 1999). Propensity of OPI to induce hypertriglyceridemia coupled with their permissive effects of gluconeogenesis in liver creates a serious threat to glucose homeostasis.

4. CONCISE SUMMARY

It is almost certain that OPI will interact with etiological factors of diabetes at toxicologically significant levels, given the status of OPI as an environmental pollutant with residues being detected in the biosphere surrounding it and now being demonstrated to make it into the human body. The interaction between the living system and OPI may have a severe two-way impact on glycemic control. OPI are the most probable agents to induce hyperglycaemic responses in humans during exposure, as they are documented facilitators of hepatic glucose output through glycogenolysis and gluconeogenesis. Additionally, OPI may result in a predisposition to diabetes by influencing the human system's responsiveness to insulin through a variety of mechanisms. It is also evident from our research that OPI may also improve diabetic outcomes. Unhealthy diet patterns, obesity, and a lack of physical activity are all risk factors for diabetes in the majority of societies. The burden of constant exposure to OPI (as environmental pollutants) could be a silent aggravating factor that is causing an increase in the incidence of diabetes, given the high probability of risk factors.

5. REFERENCES

Abdollahi, M., Donyavi, M., Pournourmohammadi, S., & Saadat M. (2004a). Hyperglycemia associated with increased hepatic glycogen phosphorylase and phosphoenol pyruvate carboxykinase in rats following sub-chronic exposure to malathion. *Comparative Biochemistry & Physiology (C)*, 137, 343-347.

Abdollahi, M., Ranjbar, A., Shadnia, S., et al. (2004b). Pesticides and oxidative stress; review. *Medical Science Monitor*, 10, 141-147.

Abou-Donia, M. (2003). Organophosphorus Ester-Induced Chronic Neurotoxicity. *Archives of Environmental Health*, 58, 484-497.

Akhgari, M., Abdollahi, M., Kebryaeezadeh, A., et al. (2003). Biochemical evidence for free radical-induced lipid peroxidation as a mechanism for sub chronic toxicity of malathion in blood and liver of rats. *Human & Experimental Toxicology*, 22, 205-211.

Andrews, RC., & Walker, BR. (1999). Glucocorticoids and insulin resistance: old hormones, new targets. *Clinical Science,* 96, 513-523.

Argaud. D., Zhang, Q., Malatra, S., et al. (1996). Regulation of rat liver glucose-6-phosphatase gene expression in different nutritional and hormonal states. *Diabetes*, 45, 1563-1571.

Banerjee, BD., Seth, V., & Ahmed, RS. (2001). Pesticide-induced oxidative stress: perspective and trends. *Reviews on Environmental Health*, 16, 1-40.

Begum, K, & Rajini, PS. (2011a). Monocrotophos augments the early alterations in lipid profile and organ toxicity associated with experimental diabetes in rats. Pesticide Biochemistry & Physiology, 99, 33-38.

Begum, K & Rajini, PS. (2011b). Augmentation of hepatic and renal oxidative stress and disrupted glucose homeostasis by monocrotophos in streptozotocin-induced diabetic rats. *Chemico Biological Interactions*, (In press).

Bradbury, MWB., Burden, J., Hillhouse, EW., & Jones, MT. (1974). Stimulation electrically and by acetylcholine of the rat hypothalamus in vitro. *Journal of Physiology*, 239, 269–83.

Brodeur, J., & DuBois, KP. (1964). Studies on the mechanism of acquired tolerance by rats O, O-diethyl S-2- (ehylthio) ethyl phosphorodithioate (Di-syston). *Archives of International Pharmacodynamics*, 149, 560-570.

Bugajski, J., Gadek-Michalska, A., & Bugajski, AJ. (2002). Effect of constitutive- and inducible-cyclooxygenase in the carbachol-induced pituitary-adrenocortical response during social stress. *Journal of Physiology & Pharmacology,* 53, 453-462.

Calogero AE, Kamilaris TC, Gomez MT, et al. (1989). The muscarinic cholinergic agonist arecoline stimulates the rat hypothalamic-pituitary-adrenal axis through a centrally-mediated corticotropin-releasing hormone-dependent mechanism. *Endocrinology*, 125, 2445-2453.

Calogero, AE., Gallucci, WT., Bernardini, R., et al. (1988). Effect of cholinergic agonists and antagonists on rat hypothalamic corticotropin-releasing hormone secretion in vitro. *Neuroendocrinology*. 47, 303-308.

Costa, LG., Schwab, BW., & Murphy, SD. (1982a). Differential alterations of cholinergic muscarinic receptors during chronic and acute tolerance to organophosphorus insecticides. *Biochemical Pharmaco*logy, 31, 3407-3413.

Costa, LG., Schwab, BW., & Murphy, SD. (1982b). Tolerance to anticholinesterase compounds in mammals. *Toxicology*, 25, 79-97.

Deotare, ST., & Chakrabarti, CH. 1981. Effect of acephate (orthene) on tissue levels of thiamine, pyruvic acid, lactic acid, glycogen and blood sugar. *Indian Journal of Physiology & Pharmacology,* 25, 259-264.

Dressel, TD., Goodale, RL Jr, Zweber B, & Borner JW. 1982. The effect of atropine and duct decompression on evolution of diazinon-induced canine pancreatitis. *Annals of Surgery*, 195: 424-434.

Dressel, TD., Goodale, RL Jr., Ameson, MA., & Borner, JW. (1979). Pancreatitis as a complication of anticholinesterase insecticide intoxication. *Annals of Surgery*, 189, 199-204.

Eto, M. (1974). *Organophosphorus Pesticides: Organic and Biological Chemistry*, 387 pp, CRC Press, Cleveland, OH.

Fletcher, HP, Akbar, WJ, Peoples, RW, & Spratto, GR. (1998). Effect of acute soman on selected endocrine parameters and blood glucose in rats. *Fundamental & Applied Toxicology*, 11, 580-586.

Frick, TW, Dalo, S, O'Leary, JF, et al. (1987). Effects of insecticide, diazinon, on pancreas of dog, cat and guinea pig. *Journal of Environmental Pathology, Toxicology and Oncology*, 7, 1-11.

Fukuto, TR. (1990). Mechanism of action of organophosphorus and carbamate insecticides. *Environmental Health Perspectives*, 87, 245-254.

Galloway, T, & Handy R. (2003). Immunotoxicity of organophosphorus pesticides. *Ecotoxicology*, 12, 345-363.

Ganss, R, Weih, F, & Schutz G. (1994). The Cyclic Adenosine 3'5'-Monophosphate and the glucocorticoids-dependent enhancers are targets for insulin repression of tyrosine aminotransferase gene transcription. *Molecular Endocrinology*, 8, 895-903.

Gokalp, O, Buyukvanh, B, & Cicek E, et al. (2005). The effects of diazinon on pancreatic damage and ameliorating role of vitamin E and vitamin C. *Pesticide Biochemistry & Physiology*, 81, 123-128.

Goldberg, JI. (2001). Diabetic dyslipidemia: causes and consequences. *The Journal of Clinical Endocrinology & Metabolism*, 86, 965–971.

Grundy, SM. (1999). Hypertriglyceridemia, insulin resistance, and the metabolic syndrome. *American Journal of Cardiology*, 83, 25F-29F.

Guarini, S, Cainizzo, MM, Giuliani, D, et al. (2004). Adrenocorticotropin reverses hemorrhagic shock in anesthetized rats through the rapid activation of a vagal anti- inflammatory pathway. *Cardiovascular Research*, 63, 357-365.

Gupta, PK. (1974). Malathion induced biochemical changes in rats. *Acta Pharmacology et. Toxicology*, 35, 191-194.

Hsiao, CT, Yang, CC, Deng, JF, et al. (1996). Acute pancreatitis following organophosphate intoxication. *Journal of Toxicology Clinical Toxicology*, 34, 343-347.

Hudon, M, & Clement, JG. (1986). Effect of soman (pinacolyl methylphosphonofluoridate) on the blood levels of corticosterone and adrenocorticotropin in mice. *Candian Journal of Physiology & Pharmacology,* 64, 1339-1342.

Ibrahim, NA, El-Gamal, BA. (2003). Effects of diazinon, an organophosphate insecticide, on plasma lipid consitutents in experimental animals. *Journal of Biochemistry and Molecular Biology,* 36, 499–504.

Joshi AK, Rajini PS. 2010. Hyperglycemic and stressogenic effects of monocrotophos in rats: Evidence for the involvement of acetylcholinesterase inhibition. *Experimental & Toxicologic Pathology,* (In press).

Joshi, AK, & Rajini, PS. (2009). Reversible hyperglycemia in rats following acute exposure to acephate, an organophosphorus insecticide: role of gluconeogenesis. *Toxicology,* 257, 40-45

Kamath, V, Joshi, AKR, & Rajini, PS. (2008). Dimethoate induced biochemical perturbations in rat pancreas and its attenuation by cashew nut skin extract. *Pesticide Biochemistry & Physiology,* 90, 58-65.

Kamath, V, Rajini, PS. (2007). Altered glucose homeostasis and oxidative impairment in pancreas of rats subjected to dimethoate intoxication. *Toxicology.* 231, 137-146.

Kassa, J, & Bajgar, J. (1995). Comparison of the efficacy of HI-6 and obidoxime against cyclohexyl methyl phosphonofluoridate (GF) in rats. *Human & Experimental Toxicology,* 11, 923-928.

Kassa, J. (1995). Comparison of efficacy of two oximes (HI-6 and obidoxime) in soman poisoning in rats. *Toxicology,* 101, 167-174.

Kassa, J. (1997). Importance of cholinolytic drug selection for the efficacy of HI-6 against soman in rats. *Toxicology,* 116, 147-152.

Kokka, N, Lee, R, & Lomax P. (1985). Effects of organophosphae cholinesterase inhibitors on pituitary adrenal activity and thermoregulation. *In Fifth Annual Chemical Defense Bioscience Review.* 60.

Koundinya, PR, Ramamurthi, R. (1979). Effect of organophosphate pesticide (fenitrothion) on some aspects of carbohydrate metabolism in a freshwater fish, Saotherodon (Tilapia) mossamicus (Peters). *Experientia,* 35, 1632-1633.

Kuz'minskaia, UA, Bersan, LV, & Veremenko, LM. (1978). Activity of the indicator enzymes of liver subcellular structures with the prolonged administration of Valexon. *Vopr. Pitan,* 5, 48-51.

Lasram, MM, Annabi, AB, Rezg R, et al. (2008). Effect of short-time Malathion administration on glucose homeostasis in Wistar rat. *Pesticide Biochemistry & Physiology,* 92, 114-119.

Maritim, AC, Sanders, RA, Watkins (III), JB. (2003). Diabetes, Oxidative Stress, and Antioxidants: A Review. *Journal of Biochemical & Molecular Toxicology*, 17, 24-38.

Marsh, WH, Viekov, GA, & Conradi, EC. (1988). Acute pancreatitis after cutaneous exposure to an organophosphate insecticide. *American Journal of Gastroenterology*, 83, 1158- 1160.

Mathur HB, Johnson S, Mishra R., et al. (2003). Analysis of pesticide residues in bottled water (Delhi Region). CSE Report. http://www.cseindia.org/userfiles/Delhi_uploadfinal_sn.pdf

Mathur, HB, Agarwal, HC, Johnson, S, & Saikia, N. (2005). Analysis Of Pesticide Residues In Blood Samples From Villages Of Punjab. CSE Report. http://www.cseindia.org/userfiles/Punjab_blood_report.pdf

Matin, MA, Khan, SN, Hussain K, & Sattar S. 1989. Effects of adrenalectomy on diazinon- induced changes in carbohydrate metabolism. *Archives of Toxicology*, 63, 376-380.

McPhillips, JJ. (1969). Altered sensitivity to drugs following repeated injections of a cholinesterase inhibitor to rats. *Toxicology and Applied Pharmacology*, 14, 67-73.

Meller, D, Fraser, I, & Kryger, M. (1981). Hyperglycemia in anticholinesterase poisoning. *Candian Medical Association Journal*, 124, 745-748.

Mollaoglu, H, Yilmaz, HR, Gokalp, O, & Altuntas I. (2003). Methidathion un pankreas uzerine erkileri: Vitamin E ve C nin rolu. *Van Tip Dergisi*, 10, 98-100.

Moore, PG, & James, OF. (1988). Acute pancreatitis induced by acute organophosphate poisoning. *Postgraduate* Medical Journal, 57, 660-662.

Namba, T. (1971). Cholinesterase inhibition by organophosphorus compounds and its clinical effects. *Bulletin - World Health Organization*, 44, 289-307.

O'Brien, RM, Lucas, PC, Forest CD, et al. (1990). Identification of a sequence in the PEPCK gene that mediates a negative effect of insulin on transcription. *Science*. 249, 533- 537.

Osicka-Kaprowska, A, Lipska M, Wysocka-Paruzewska, B. (1984). Effect of chlorfenvinphos on plasma corticosterone and aldosterone levels in rats. *Archives of Toxicolicology*, 55, 68-69.

Panahi,P, Vosough-Ghanbari, S, Pournourmohammadi, S, et al. (2006). Stimulatory Effects of Malathion on the Key Enzymes Activities of Insulin Secretion in Langerhans Islets, Glutamate Dehydrogenase and Glucokinase. *Toxicology Mechanisms & Methods*, 16, 161–167.

Pavlov, VA, Ochani M, Gallowitsch-Puetra, M, et al. (2006). Central muscarinic cholinergic regulation of the systemic inflammatory response during endotoxemia. *Proceedings of the National Academy of Sciences, USA*, 103, 5219-5223.

Pilkis, SJ, & Granner, DK. (1992). Molecular physiology of the regulation of hepatic gluconeogenesis and glycolysis. *Annual Review of Physiology,* 54, 885-909.

Rahimi, R, & Abdollahi, M. (2007). A review on the mechanisms involved in hyperglycemia induced by organophosphorus pesticides. *Pesticide Biochemistry & Physiology,* , 115- 121.

Ranjbar, A, Pasalar, P, & Abdollahi, M. (2002). Induction of oxidative stress and acetylcholinesterase inhibition in organophosphorous pesticide manufacturing workers. *Human & Experimental Toxicology,* 21, 179-182.

Reiner, E, & Aldridge, WN. (1967). Effect of pH on inhibition and spontaneous reactivation of acetylcholinesterase treated with esters of phosphorus acids and carbamic acids. *Biochemical Journal,* 105: 171-179.

Reiner, E. (1971). Spontaneous reactivation of phosphorylated and carbamylated cholinesterase. *Bulletin of World Health Organization,* 44, 109-112.

Rezg, R, Mornagui, B, Benahmed, M, etal. (2010). Malathion exposure modulates hypothalamic gene expression and induces dyslipedemia in Wistar rats. *Food & Chemical Toxicology.* 48, 1473-1477.

Rodrigues, MA, Puga, FR, Chenker, E, & Mazanti, MT. (1986). Short-term effect of malathion on rats' blood glucose and on glucose utilization by mammalian cells in vitro. *Ecotoxicology & Environnemental Safety,* 12 , 110-113.

Ryhänen, R, Herranen, J, Korhonen, K, et al. (1984). Relationship between serum lipids, lipoproteins and pseudocholinesterase during organophosphate poisoning in rabbits. *International Journal Biochemistry.* 16: 687-690.

Sanghi, R, Pillai, MK, Jayalekshmi, TR, & Nair A. (2003). Organochlorine and organophosphorus pesticide residues in breast milk from Bhopal, Madhya Pradesh, India. *Human & Experimental Toxicology.* 22, 73-6.

Schwab, BW, Hand, H, Costa, LG, & Murphy, SD. (1981). Reduced muscarinic receptor binding in tissues of rats tolerant to the insecticide disulfoton. *Neurotoxicology.* 2, 635-47.

Seifert, J. (2001). Toxicological significance of the hyperglycemia caused by organophosphorous insecticides. *Bulletin of Environmental Contamination & Toxicology,* 67, 463-469.

Sevillano, S, de la Mano, AM, Manso, MA, et al. (2003). N-acetylcysteine prevents intra- acinar oxygen free radical production in pancreatic duct obstruction-induced acute pancreatitis. *Biochimica et Biophysica Acta,* 20, 177-184.

Smallridge, RC, Carrb, FE, & Feina, HG. (1991). Diisopropylfluorophosphate (DFP) reduces serum prolactin, thyrotropin, luteinizing hormone, and growth hormone and increases adrenocorticotropin and corticosterone in rats:

Involvement of dopaminergic and somatostatinergic as well as cholinergic pathways. *Toxicology & Applied Pharmacology,* 180, 284-295.

Sogorb, MA, & Vilanova, E. (2002). Enzymes involved in detoxification of organophosphorus, carbamate and pyrethroid insecticides through hydrolysis. *Toxicology Letters,* 128, 215-228.

Soltaninejad, K, & Abdollahi, M. (2009). Current opinion on the science of organophosphate pesticide and toxic stress: A systematic review. *Medical Science Monitor,* 15, RA75- RA90.

Spassova, D, White, T, & Singh, AK. (2000). Accute effects of acephate and methamidophos on acetylcholinesterase activity, endocrine system, and amino acid concentrations in rats. *Comparative Biochemistry & Physiology Part (C),* 12, 79-89.

Sterri, SH, Lyngaas, S, & Fonnum, F. (1980). Toxicity of soman after repetitive injection of sublethal d

CHAPTER 4

The Toxicity of Permethrin and Fenitrothion

1. INTRODUCTION: AN OVERVIEW

Fenitrothion: A yellow-brown liquid with an unpleasant odor at room temperature, fenitrothion (O,O-dimethyl O-4-nitro-m-tolyl phosphorothioate; CAS No. 122-14-5) is an organophosphorus insecticide that was introduced in 1959 by both Sumitomo Chemical Company and Bayer Leverkusen. It was subsequently introduced by American Cyanamid Company (Hayes, 1982; Hayes and Laws, 1990; Worthing and Walker, 1987).

The emergence of organophosphorus insecticides coincided with the significant expansion of agriculture and agribusiness following World War II. At that time, parathion, a renowned organophosphorus insecticide, was employed in significant proportions to prevent the rice-stem borer on a global scale. Nevertheless, parathion was considered an exceedingly hazardous substance due to its high acute toxicity. An oral dose of 3-5 mg/kg is typically fatal in humans. The subsequent case report also encountered the high toxicity and persistence in humans (Clifford and Nies, 1989). The groin and scrotal areas of a 25-year-old employee at a pesticide-formulating facility were contaminated as a result of an accidental spill of a 76% parathion solution. Despite being able to shower and change his clothes promptly, the subsequent nausea and diarrhea necessitated that he consult a doctor two days later. The parathion-saturated uniform was deposited in a plastic bag for burning by the worker. However, the contaminated apparel was laundered and subsequently was worn by a second worker in succession until he experienced symptoms similar to those of the first worker. The coveralls were once more laundered and utilized by a third intoxicated laborer. In total, three employees experienced toxic reactions to parathion. This case demonstrates the toxic nature of parathion and its ability to remain on apparel despite multiple launderings. In May and June 1986, 49 individuals in Sierra Leone were acutely poisoned by parathion, and 14 of them subsequently passed away, according to Etzel et al. (1987). In order to investigate the factors that were linked to the development of symptoms such as excess salivation, excess weeping, increased urination, diarrhea, convulsions, and loss of consciousness, a case-control study was conducted on the 21 cases and 22 household controls. Each case and control were questioned regarding the foods and beverages they had consumed during the four hours preceding their illness (cases) or on the day of the case's illness (controls). The odds ratio of cases (12.7; 95% confidence interval (CI), 2.4-83.8) for consuming bread was substantially elevated, indicating that cases were more likely than controls to have consumed bread within the four hours prior to falling ill.. In addition, when stratified by age, the odds ratio was far higher in children under 18 years (odds ratio, 21.7; 95% CI, 2.4-264.6) than adults (odds ratio, 2.3; 95% CI, 0.02-195.9). This may be due to the higher consumption of parathion based on body weight or higher susceptibility to the insecticide in the former than the latter. Parathion was detected from residue floor on the truck that had brought the wheat flour from the milling factory to the general store where the baker purchased it, suggesting that the flour had been contaminated during transport. The authors estimated that 10-15 ml of parathion may have spilled onto a 22.5 kg bag of flour in the truck. Besides these, many cases of parathion intoxication have been reported to date (Aardema et al., 2008; Eyer et al., 2003; Hoffmann and Papendorf, 2006; Laynez et al., 1997).

In light of this background, fenitrothion was developed in place of parathion for its highly selective toxicity to insects over humans and animals. While the structure of fenitrothion is similar to that of parathion, its residual effects and acute toxicity are lower than parathion (Miyamoto, 1969). An oral LD50 of parathion is approximately 6 mg/kg for rats, against 330 mg/kg for fenitrothion, which is more rapidly broken down and does not persist in areas where they are used. Fenitrothion is effective against a wide range of pests on rice, cereals, fruits, vegetables, stored grains, cotton and forests, and also in public health programs for control of flies, mosquitoes and cockroaches. Fenitrothion is produced at the rate of 15,000 to 20,000 tons per year worldwide, and is available in emulsifiable concentrates, ultra-low- volume concentrates, powders, granules, dustable powders, oil-based sprays and in combination with other pesticides.

Permethrin: Permethrin (3-phenoxybenzyl (1*RS*,3*RS*;1*RS*,3*SR*)-3-(2,2-dichlorovinyl) (-2,2- dimethylcyclopropanecarboxylate; CAS No. 52645-53-1) was first synthesized in 1973 and marketed in 1977 as a photostable pyrethroid (Elliott et al., 1973). Approximately 600 tons per year is at present used worldwide not only in agriculture but also in forestry, household settings, and public health programs.

Pyrethroids were a significant advancement in terms of their insecticidal potential; however, they demonstrated a relatively low potential for mammals. Their development was particularly opportune in light of the identification of issues associated with the use of DDT. Pyrethroids are initially composed of the identification of pyrethrum components, which were derived from East African chrysanthemum flowers. These flowers have been recognized for their insecticidal properties for a long time. The environment benefits from the low mammalian toxicity and negligible persistence of pyrethrum, which promptly eliminates flying insects. However, it is of poor efficacy when used in the field. Bioallethrin, tetramethrin, resmethrin, and bioresmethrin were among the first-generation pyrethroids that were developed in the 1960s. They are more active than natural pyrethrum, but they are unstable in sunlight. Then, a second generation of more persistent compounds, including permethrin, cypermethrin, and deltamethrin, was discovered. They are significantly more resistant to degradation by light and air, rendering them suitable for agricultural applications.

The control of body lice and household noxious insects, as well as the protection of stored cereals, cotton, and other crops, is extremely effective with permethrin. Technical products are a mixture of cis and trans isomers in the ratio of 40:60 or 25:75, and they are available in emulsifiable concentrates, ultra-low-volume concentrates, wettable powders, and dustable powders. They are a brown or yellowish-brown liquid.

2. ABSORPTION, METABOLISM AND EXCRETION IN LABORATORY ANIMALS AND HUMANS

In mammals, fenitrothion and permethrin are absorbed via gastrointestinal or respiratory tract and skin, and are rapidly metabolized and excreted. *Fenitrothion*: After uptake into the body, fenitrothion is metabolized by hepatic cytochrome P450 (CYP) to form fenitrooxon, which is thought to have a higher potential acute neurotoxicity than the parent compound. Fenitrooxon is further metabolized to dimethylphosphate and 3-methyl-4-nitrophenol (MNP) by paraoxonase 1 (PON1). MNP and methylphosphate are also produced by glutathione-S-aryltransferase (GST) and PON1. In another pathway, fenitrothion is directly metabolized to MNP and dimethylthiophosphate by PON1 or MNP and methylthiophosphate by GST and PON1 (Figure 1). Interestingly, its major metabolic route differs between mammals and birds as mentioned later. Most of the metabolites are excreted in urine within 24 hours in humans (Nosal and Hladka, 1968), and within 2-4 days in the rat, guinea-pig, mouse, and dog (Miyamoto et al., 1963; Miyamoto, 1964). Species and sexes differences are observed in the composition of the metabolites. MNP, which is also contained in diesel exhaust emissions, has potential adverse effects on the reproductive systems in mammals and birds. Fenitrothion at doses of 0.18 and 0.36 mg/kg per day was administered to 12 human volunteers for 4 days (Meaklim et al., 2003). Pharmacokinetic parameters could only be determined at the high dosage, because the blood levels of fenitrothion at the low dosage were below the detectable level. Fenitrothion concentrations showed a wide range of interindividual variability, with peak blood levels achieved 1-4 hours after dosing, and the half-life ranged from 0.8 to 4.5 hours. Serum concentrations of fenitrothion were measured in 15 patients after acute fenitrothion intoxication, who admitted to the hospital 0.5-12 hours after the ingestion of 5-50 g fenitrothion (Koyama et al., 2006). The serum fenitrothion concentrations ranged from undetectable (< 0.01 μg/ml) to 9.73 μg/ml. Serum fenitrothion concentrations were less than 7 μg/ml in the patients with mild intoxication, while in the severe cases, the levels were more than 7 μg/ml. The elimination half-lives in the mild cases were 9.9 ± 7.7 hours (mean ± SD), and the serum fenitrothion concentrations declined below the detectable level in 48 hours. The elimination half-lives relating to two severe cases were 5.3 and 6.7 hours in the alpha phase (under direct hemoperfusion), and 35 and 52 hours in the beta phase, respectively. The serum fenitrothion concentrations fell below the detectable level in 300 hours.

Permethrin: Regarding its metabolism, permethrin is converted to 2,2-dichlorovinyl-2,2-dimethylcyclopropane-1-carboxylic acid and 3-phenoxybenzyl alcohol (3PBAlc) by carboxyl- esterase. The latter metabolite is followed by oxidation to form 3-phenoxybenzaldehyde, and finally 3-phenoxybenzoic acid (3PBA) (Figure 2). The metabolites are reported to be endocrine- disrupting agents, but most studies mention permethrin toxicity is derived from itself (Yuan et al., 2010). In general, *trans* isomer is more rapidly metabolized than *cis* isomer, which is related to the lower susceptibility of *cis* isomer to enzymatic hydrolysis of the ester linkage (Soderlund and Casida, 1977; Zhang et al., 2008). Besides hydrolytic pathway by carboxylesterase, oxidative metabolic pathway of both *cis*-and *trans*-permethrin in rat and human hepatic microsomes was recently reported (Scollon et al., 2009). The toxicokinetics of permethrin (with a *cis:trans* ratio of 25:75) was investigated after single oral doses to rats (Anadon et al., 1991). The plasma level of permethrin was maximal within 4 hours after dosing, and then was slowly eliminated from plasma with a half-life of 12.4 hours. The bioavailability of permethrin was found to be 60.7%. The maximum permethrin concentrations in the central and peripheral nervous system were higher than plasma concentrations, and declined with half-life similar to those of plasma. Clearance of *trans*- and *cis*-permethrin from the blood was also investigated in a man who drank an emulsifiable concentrate formulation of permethrin (consisting of 43.5% *cis* and 56.5% *trans*) (Gotoh et al., 1998). The serum concentrations of *cis*- and *trans*-permethrin peaked 3-4 hours after ingestion and then declined, with *trans*-permethrin cleared from the blood more quickly than *cis*-permethrin. Levels of the *trans* isomer were below the detectability threshold within 25 hours after exposure, whereas *cis* isomer was still detectable 10 days after exposure. The present study indicated that the differential persistence of *cis* and *trans* isomers in human is consistent with a difference in the metabolic rate of *cis*- and *trans*- permethrin in animal studies (Anadon et al., 1991).

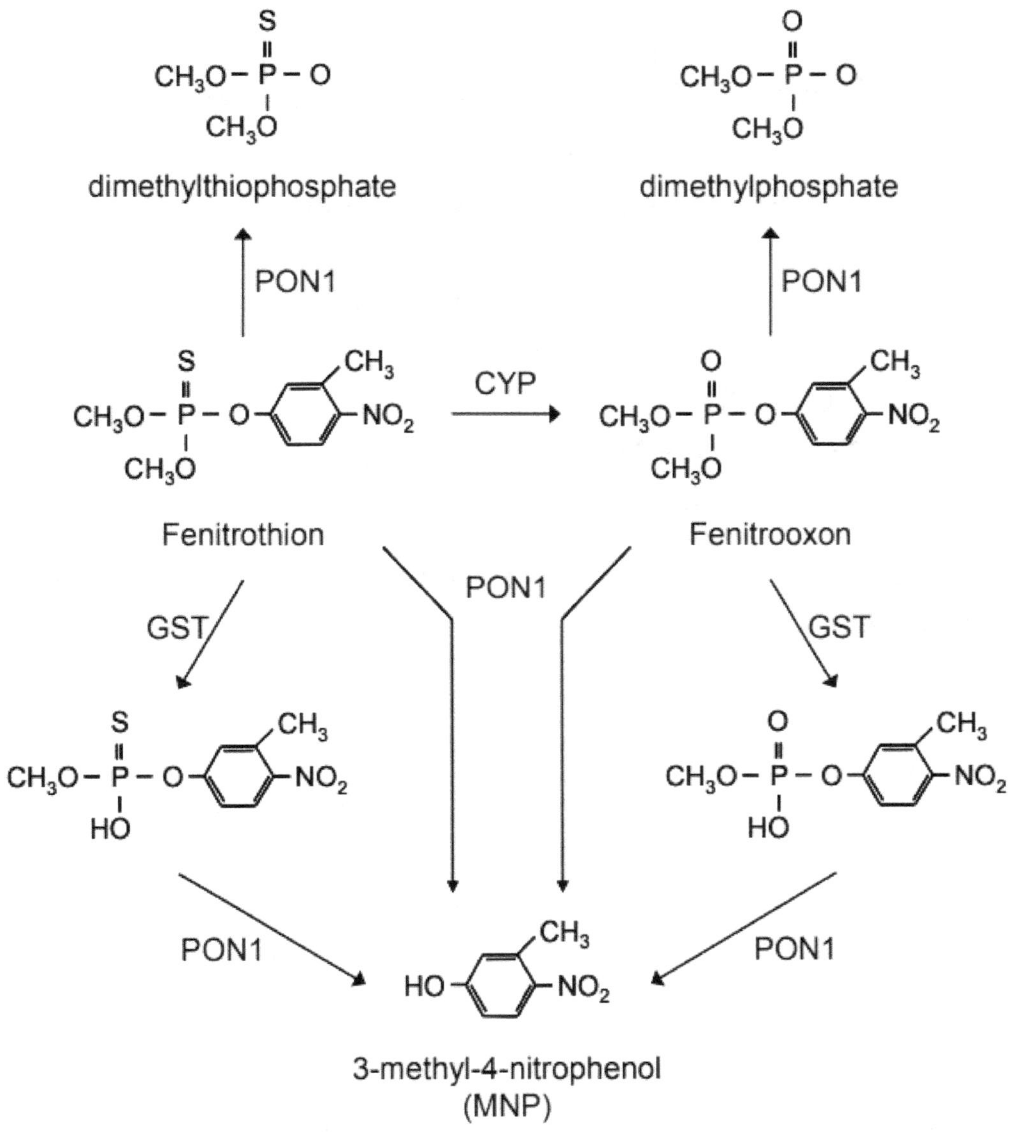

Fig. 1. Metabolic pathway of fenitrothion in mammals or birds. This figure was adopted from WHO (1992) with slight modifications. In mammals, dimethylthiophosphate, dimethylphosphate and MNP were the major urinary metabolites of fenitrothion and fenitrooxon. In birds, MNP was the major urinary metabolite of fenitrothion and fenitrooxon. GST activity was found to be lower than in mammals.

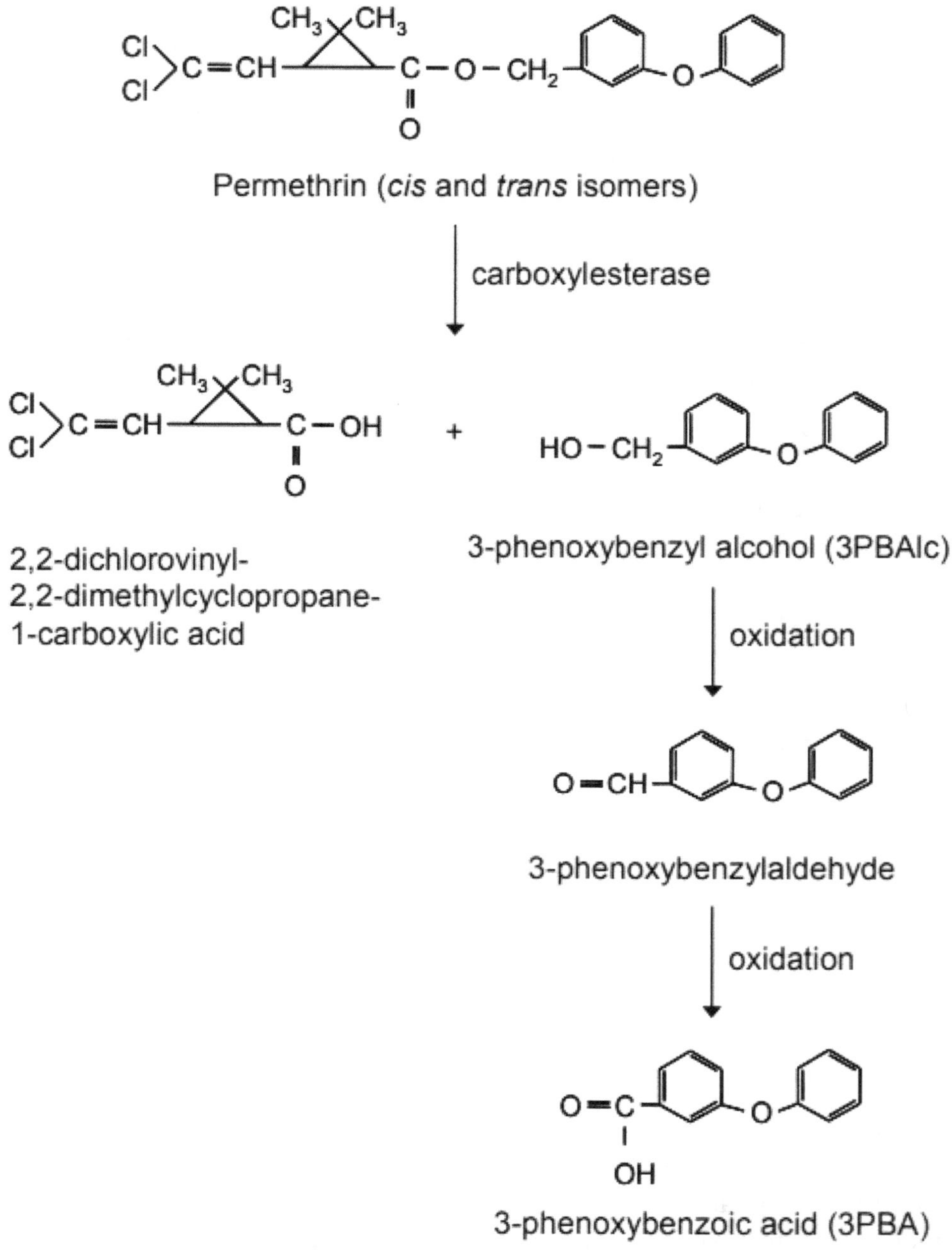

Fig. 2. Metabolic pathway of permethrin in mammals. This figure was adopted from WHO (1990) with slight modifications. Carboxylesterase plays an important role in the hydrolytic biotransformation of permethrin. The expression of carboxylesterase is ubiquitous with high levels in various tissues. Amon various animal tissues, the highest hydrolase activity is typically found in the liver and other tissues, such as testis, kidney and plasma. Unlike mouse, rat, rabbit, horse and cat, human plasma contains no carboxylesterase.

3. TOXICITY FOR EXPERIMENTAL ANIMALS AND HUMANS

Fenitrothion

Since it is a cholinesterase (ChE) inhibitor, exposure to fenitrothion causes ChE activity depression in plasma, red blood cells, brain, and liver tissues. The acute toxicity of fenitrothion is considered to be low in mammals, because of the high metabolic rate (Hayes, 1982; Spencer, 1981).

Animal: The no-observed-adverse-effect levels (NOAEL), based on brain-ChE activity, were 10 mg/kg diet in both short- and long-term studies on rats, in long-term studies on mice, and were 50 mg/kg diet in short-term studies on dogs, respectively (WHO, 1992). Fenitrothion was given to female rats by gavage every other day from gestational day 6-15 at doses 3, 15, 30 and 45 mg/kg (Berlińska and Sitarek 1997). The maternal death rates were 39% and 88% at doses of 30 and 45 mg/kg, respectively. At 30 mg/kg, fenitrothion caused a significant decrease in maternal body weight gain, food consumption, hemoglobin and hematocrit values, and absolute weights of liver and kidney, but an increase in relative weights of adrenal and ovary. At 15 mg/kg, fenitrothion significantly decreased maternal relative liver weight. Although fenitrothion at doses of 3-30 mg/kg did not induced teratogenic effects, at 30 mg/kg it showed embryotoxicity, such as a significant increase in the frequency of early resorption per litter, postimplantation loss, and fetuses and litters with dilation of the cerebral ventricles. Furthermore, fenitrothion produced delayed ossification of sternum and cranium, and decreased fetal body weight and length. The frequency of fetuses and litters with dilation of the cerebral ventricles was increased at a dose of 15 mg/kg. Thus, Berlińska and Sitarek concluded that the NOAEL for developmental toxicity in rats was 3 mg/kg per day, and the lowest-observed-adverse-effect level was 15 mg/kg per day.

Recent studies showed the endocrine-disrupting effect of fenitrothion. Berger and Sultatos (1997) demonstrated that fenitrothion treatment caused a dose-dependent decrease in 2- hydroxyestradiol and 4-hydroxyestradiol production in mouse hepatic microsomes even at a dosage as low as 7 mg/kg, and an increase in 16 alpha-hydroxyestrone and estriol production. In another study, 7-week-old castrated Sprague-Dawley rats were subcutaneously treated with testosterone propionate (50 µg/day in 0.2 ml corn oil) and orally with corn oil vehicle or fenitrothion (15 or 30 mg/kg per day) once a day for 7 days (Tamura et al. 2001). Both fenitrothion doses caused significant decreases in the weights of ventral prostate, seminal vesicle, and levator ani plus bulbocavernosus muscles. In contrast, blood acetylcholinesterase activity was only inhibited at the higher dose (30 mg/kg). Tamura et al. also demonstrated in an *in vitro* experiment that fenitrothion blocked dihydrotestosterone-dependent androgen receptor (AR) activity in a concentration- dependent and competitive manner in HepG2 human hepatoma liver cells, which were transiently transfected with human AR and an AR-dependent luciferase reporter gene, suggesting that fenitrothion may be a competitive androgen receptor antagonist.

On the other hand, Okahashi et al. (2005) suggested that lower-dose fenitrothion did not cause disruption of endocrine systems in animals. They administered fenitrothion to Crj:CD(SD)IGS parental rats at concentrations of 10, 20, and 60 (3.81 mg/kg per day) ppm in the diet for 10 weeks prior to mating, and throughout mating, gestation and lactation. Their offspring were exposed from weaning until maturation at the age of 10 weeks. In the parental animals, brain cholinesterase activity was remarkably reduced in males exposed to 60 ppm fenitrothion and in females exposed to 20 and 60 ppm fenitrothion. Reproductive performance, organ weights, histopathology, and sperm analytical parameters were not influenced. In the offspring, no effects on anogenital distance, retention of areolae/nipples, onset of puberty, organ weights, histopathological findings, and sperm parameters were observed. In conclusion, fenitrothion had no effects on the reproductive or endocrine systems of the parental animals and their offspring, even at a toxic dose suppressing brain cholinesterase activity in parental animals. The concentration of 60 ppm (3.81 mg/kg per day) is 750 times higher than the acceptable daily intake (ADI) of fenitrothion (0.005 mg/kg body weight). Therefore, any potential risk of exposure may be negligible, and fenitrothion at in-use levels in the environment may be unlikely to cause disruption of human endocrine systems.

Human: WHO (1990) classified technical grade fenitrothion as "moderately hazardous" (Class II). ADI of fenitrothion was established as 0.005 mg/kg body weight by the Joint FAO/WHO Expert Committee on Pesticide Residues in 2000. Nosal and Hladka (1968) reported that administration of fenitrothion at a single oral dose of 0.042-0.33 mg/kg body weight and repeated administration of 0.04-0.08 mg/kg body weight to human volunteers did not cause inhibition in plasma and erythrocyte ChE, and the urinary MNP was completely excreted within 24 hours. Chronic symptoms of exposure to fenitrothion in humans include general malaise, fatigue, headache, loss of memory and ability to concentrate, anorexia, nausea, thirst, loss of weight, cramps, muscular weakness and tremors.

Permethrin

Permethrin acts on the axons in the peripheral and central nervous systems, causing prolonged opening of sodium channels. The acute toxicity of permethrin in mammalians is relatively low, though the LD50 value varies considerably according to the vehicle used and the *cis:trans* isomeric ratio (FAO, 1999; U.S. EPA, 2007).

Animal: NOAEL is assigned at 5 mg/kg body weight per day for permethrin with an isomer ratio of *cis:trans* 40:60 from the viewpoint of the effects on liver weight in 2-year and 26- week studies in rats, and a 3-month study in dogs. NOAEL is not available for respective *cis* and *trans* isomers (WHO, 1990). The rat appeared to be the most sensitive species with an oral LD50 of 400 mg/kg body weight for *cis:trans* 40:60 permethrin administered in corn oil, against 650 mg/kg body weight in mice. The neurotoxicity of intravenous- or orally- administered *cis*-permethrin is over 10-fold greater than that of *trans* isomer. Neonatal rats are more sensitive than adult rats to the acute toxic effects of permethrin, which are thought to be related to differences in permethrin metabolism.

In their acute neurotoxicity study, Freeman (1993a) performed a functional observational battery (FOB) approximately 12 hours following administration of 10, 150 or 300 mg/kg of technical grade permethrin (mixture of *cis* and *trans*) in corn oil to male and female Sprague- Dawley rats. At doses of 150 and 300 mg/kg, permethrin caused salivation, tremor, splayed hindlimbs, abnormal posture, staggered gait, decreased grip strength, exaggerated reaction to sound, exaggerated hindlimb flexion, convulsions, and mortality. No treatment-related effects were observed at the lowest dose of 10 mg/kg. In a behavioral neurotoxicity study (McDaniel and Moser, 1993), technical grade permethrin was administered by gavage in corn oil to Long-Evans hooded rats, at doses of 25, 75 and 150 mg/kg, and the FOB were evaluated at 2 and 4 hours following treatment. Results of the present study are consistent with the acute regulatory study (Freeman, 1993a) including tremor, chromodacryorrhea, decreased grip strength and an exaggerated startle response. However, the absence of salivation, splayed hindlimbs and convulsions and the presence of aggressive sparring in the latter study (McDaniel and Moser, 1993) were inconsistent with the findings of the regulatory acute neurotoxicity study (Freeman, 1993a).

In a subchronic neurotoxicity study (Freeman, 1993b), technical grade permethrin was administered through the diet to male and female Sprague-Dawley rats, at concentrations of 250, 1500 and 2500 ppm (18, 101 and 170 mg/kg per day, respectively). At the 1500 and 2500 ppm dietary levels, permethrin produced tremor, splayed hindlimbs, abnormal posture, a staggered gait, and decreased grip strength. No such effects were observed at the lowest dose of 250 ppm.

Effects of permethrin on endocrine or reproductive function are investigated, but the report is very limited. Castrated rats (5-week-old) were pretreated with testosterone propionate and orally given permethrin (mixture of *cis* and *trans*, 24.8% and 71.8%) at doses of 10, 50 and 100 mg/kg per day for 10 days. A mixture of *cis*- and *trans*-permethrin showed anti-androgen-like effects on male rats such as significant reductions in androgen-dependent sex accessory tissue (ventral prostate, seminal vesicles, levator ani and bulbocavernosus muscles, Cowper's gland and glans penis) weights (Kim et al., 2005). *cis*-Permethrin at 0, 35 and 70 mg/kg was orally administered to IRC mice for 6 weeks, and male reproductive toxicity was investigated. This chemical dose-dependently decreased testicular and plasma testosterone levels, along with a dose-dependent increase in circulating LH and declines in epididymal sperm count and sperm motility (Zhang et al., 2007). Testicular residue concentrations of *cis*-permethrin from the individual animals were also strongly inversely correlated with testicular testosterone levels. The exposure-related reductions in mRNA and protein expression levels of peripheral benzodiazepine receptor, steroidogenic acute regulatory protein and cytochrome P450 side- chain cleavage, which are involved in testosterone synthesis in testis, were observed, as well as structural changes in Leydig cell mitochondria, suggesting that the mitochondrial damage caused by permethrin exposure may result in a reduction of testosterone synthesizing elements and thereby decrease testosterone levels. In a follow-up study in mice, *cis*-permethrin induces reproductive toxicity whereas at the same dose *trans*-permethrin does not because of a faster metabolic rate than *cis* isomer (Zhang et al., 2008). Zhang et al. also reported that *cis*-permethrin caused structural abnormalities in the seminiferous tubules. However, it must be noted that these studies to date have used dose levels much higher than encountered by non- occupationally exposed humans.

Human: Permethrin is a moderately to practically non-toxic pesticide in EPA toxicity class II or III, depending on the formulation. Formulations in the case of possible eye and skin irritation are grouped into class II. Permethrin belongs to the type I group of pyrethroids because it lacks a cyano group, and typically causes tremor (T-syndrome), incoordination, hyperactivity, prostration, and paralysis. An ADI of 0.05 mg/kg body weight for *cis:trans* 40:60 or 25:75 permethrin was established in 1987. Rishikesh et al. (1978) evaluated staff involved with bagging, mixing, or spraying a 5% preparation of permethrin (cis/trans ratio, 25:75) in Nigeria by a questionnaire and urinalysis. Regardless of the protective equipment worn by the sprayers, only 2 mg of permethrin was absorbed after exposure to 6 kg of permethrin, which was excreted in 24 hours.

4. TOXICITY FOR ECOSYSTEM

Fenitrothion

In the environment, fenitrothion is degraded by photolysis and hydrolysis. In the presence of ultraviolet radiation or sunlight, the half-life of fenitrothion in water is less than 24 hours.

The presence of micro-flora may accelerate degradation. Miyamoto et al. (1966) studied the degradation of fenitrothion by *B. subtilis*. The major metabolite was aminofenitrothion, and other minor metabolites detected were dimethyl thiophosphoric acid and desmethyl fenitrothion. In the bacteria, aminofenitrothion is further degraded to desmethyl aminofenitrothion, but the rate is slower than the parent compound. No reduction of desmethyl fenitrothion to desmethyl aminofenitrothion was detected, and dimethyl phosphoric acid was not formed from aminofenitrothion (Figure 3). Thus, the degradation of fenitrothine in *B. subtilis*, may be quite different from the metabolic route of experimental animals and humans. In the absence of sunlight or microbial contamination, fenitrothion is stable in water. In soil, biodegradation is the primary route, though photolysis may also play a role. Airborne concentrations of fenitrothion and its levels in water may decrease rapidly by photolysis and hydrolysis. The concentrations of fenitrothion that are likely to be found in the environment do not have any effects on microorganisms in soil or water. In laboratory studies, fenitrothion is highly toxic for aquatic invertebrates in both freshwater and seawater, while fish are less sensitive to fenitrothion than invertebrates, and the most sensitive life stage is early larva. Fenitrothion is highly toxic to bees (LD50, 0.03-0.04 μg/bee) when bees are exposed to direct treatment or to dried residues on foliage (U.S. EPA, 1987).

Furthermore, fenitrothion was found to be highly toxic to upland game birds, but not so toxic to waterfowl. Indeed, the acute oral LD50 values were determined to be 23.6 and 1190 mg/kg body weight for bobwhite quail and mallards, respectively. Even in reproduction studies, NOEL was 10 mg/kg body weight for the quail and 100 mg/kg body weight for the mallard, respectively. There are quantitative differences in the composition of metabolites of fenitrothion between mammalian and avian species. For example, in rats, mice and dogs, demethylated products at the *P-O*-methyl linkage by GST accounted for 30 to 60% of the total urinary metabolites (Hollingworth et al., 1967; Miyamoto et al., 1976), whereas in the birds only 10 to 15%. This may be due to lower GST activity in avian species compared with that in mammalians. Mihara et al. (1979) also revealed that oxidative activities of the *m*-methyl group of fenitrothion and fenitrooxon in livers from hen, quail, pheasant and duck were higher than those of mammalian liver, while *O*-demethylate activity for fenitrothion or fenitrooxon was lower in these birds. In birds, MNP is the major metabolite of fenitrothion by hydrolysis, though a pathway exists with oxidation of the *m*-methyl group of fenitrothion or fenitrooxon. The metabolite MNP is then conjugated with uridine diphosphate glucuronic acid or 3'-phosphoadenosine-5'-phosphosulfate by catalytic action of uridine diphosphate glucuronosyltranferase (UGT) and sulfotransferase (SULT), respectively (Mackenzie et al., 1997). Hepatic UGT and SULT activities investigated in vitro for MNP in Japanese quail, mice and rats revealed lower UGT activity for MNP in quail than rats and mice, but no significant difference in SULT activity (Lee et al., 2007). In addition, the SULT activity was only one-tenth of the UGT activity, suggesting that the latter enzyme plays an important role in MNP elimination in vivo. Li et al. (2008) reported that the birds treated with 100 mg/kg of MNP induced acute toxicological responses such as dyspnea and tremor, and finally death. MNP may cause acute toxicity and death, possibly by a rapid decrease in blood pressure followed by ischemic shock, because the potential vasodilatory action of MNP had been reported in rats (Mori et al., 2003; Taneda et al., 2004). However, none of the rats died after treatment with 100 mg/kg of MNP (Li et al., 2007), suggesting that the sensitivity to MNP is higher in quail than in rats. For these reasons, fenitrothion causes higher toxicity in birds than in mammals.

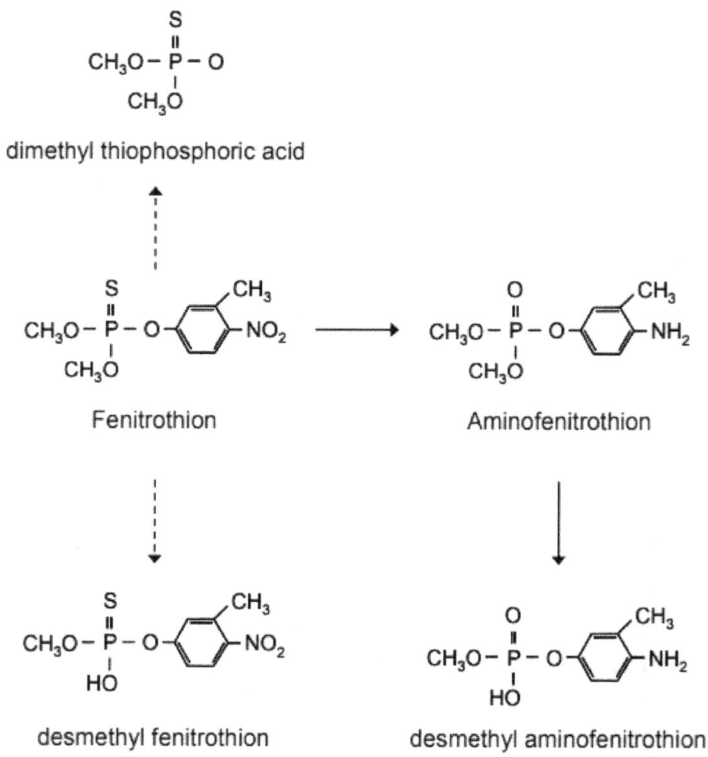

Fig. 3. Pathway of degradation of fenitrothion by *B. subtilis* in environment. This figure was adopted from Miyamoto et al. (1966) with slight modifications. Aminofenitrothion was the major metabolite of fenitrothion in *B. subtilis.*

Permethrin

Permethrin is photodegraded by sunlight in water and on soil surfaces. Under aerobic conditions in soil, permethrin degrades with a half-life of 28 days. Permethrin deposited on plants degrades with a half-life of approximately 10 days. Thus, in the environment, permethrin is hydrolyzed, and the resultant acid and alcohol are conjugated. However, permethrin itself evidences very little movement within the environment, because it binds very strongly to soil particles and is nearly insoluble in water and not expected to leach or contaminate groundwater. Permethrin has been shown to be highly toxic for aquatic arthropods and fish, because they have lower levels of carboxylesterase activity than.mammals. However, the extreme susceptibility to permethrin may be ascribed to its high sensitivity to sodium channels rather than low carboxylesterase activity. Permethrin is also highly toxic to honey bees (LD50, 0.11 μg/bee), yet exhibits very low toxicity to birds when given orally or fed in the diet. LD50 is >3000 mg/kg body weight for a single oral dosage and >5000 mg/kg diet for dietary exposure, respectively. One of the reasons for the different toxicity of permethrin among species is negatively correlated its toxic action to their body temperature, thus generally showing more acute effects on cold-blooded animals (insects, fish, etc) than warm-blooded animals (mammals and birds).

5. INTERACTIVE TOXICITY OF INSECTICIDE MIXTURE

Organophosphorus insecticides are being increasingly used in combination with pyrethroid insecticides. Fenitrothion is used in combination with other pesticides to enhance ChE inhibition by nature. However, fenitrothion inhibits not only ChE but also other esterase activity such as carboxylesterase. Trottier et al. (1980) reported that the oral administration of fenitrothion to male CD rats at a dose of 0, 2.5, 5, 10, or 20 mg/kg per day for 30 consecutive days significantly decreased liver carboxylesterase activity (by 50-80%) on days 8-30 at doses more than 2.5 mg/kg per day but had returned to control values by day 45 (15 days after termination of treatment) at all doses except 20 mg/kg per day, at which a decrease of 25% was still observed. At this dose, the values had returned to normal by day 87 (57 days after termination of treatment). A significant decrease in renal carboxylesterase activity (by 20-70%) was also observed on days 8-30 at doses over 5 mg/kg per day. Recovery of the activity was rapid, and the values were comparable to those of controls by day 38 (8 days after the end of treatment).

As described at the metabolism of permethrin, carboxylesterase plays an important role in detoxication of permethrin. Ortiz et al. (1995) examined the interactions between a commercial formulation of methyl parathion and a commercially formulated product of permethrin in male rats. When rats were treated with the mixture, 380 mg/kg of methyl parathion reduced the LD50 of permethrin by only 9.0%, whereas when rats received methyl parathion at 464 mg/kg, the LD50 of permethrin was reduced by 37% ($P < 0.001$). Results indicated that methyl parathion modified the acute toxicity of permethrin. Another study examined the effect of organophosphorus insecticide dichlorvos on excretion levels of urinary *cis*-permethrin-derived 3PBA in rats (Hirosawa et al., 2011). After *cis*-permethrin injection (20 mg/kg) via the tail vein of rats pretreated intraperitoneally with dichlorvos (low dose, 0.3 mg/kg; high dose, 1.5 mg/kg), the amounts of urinary 3PBA excretion over 48 hours were decreased to 81.1% and 70.3% of dichlorvos non-treated rats in the low- and high-dose dichlorvos groups, respectively. The plasma concentration of *cis*-permethrin- derived 3PBAlc in high-dose dichlorvos group was significantly lower than that in the dichlorvos non-treated group one hour after *cis*-permethrin injection. In contrast, no differences were observed in the excretion levels of urinary 3PBA after injection of 3PBAlc between the dichlorvos non-treated group and the high-dose dichlorvos group. These results suggested that dichlorvos may have inhibited the metabolism of the co-exposed *cis*- permethrin and thereby decreased the amount of urinary 3PBA excretion. In our recent study, we evaluated male reproductive toxicity after co-exposure to diazinon (3 mg/kg) and *cis*-permethrin (35 mg/kg) in mice. Exposure to diazinon alone and the mixture with *cis*- permethrin inhibited plasma and liver carboxylesterase activities. In the co-exposed mice, the urinary *cis*-permethrin metabolite decreased compared to that in mice exposed to *cis*- permethrin alone. The co-exposure significantly decreased plasma testosterone levels and increased the number of degenerated germ cells within the seminiferous tubule, whereas exposure to each chemical did not. We concluded that diazinon inhibited the plasma and liver carboxylesterase activities and the metabolic rate of co-exposed *cis*-permethrin, which resulted in accentuating the reproductive toxicity of *cis*-permethrin (Wang et al., unpublished data submitted to the Journal).

Recently, we were unable to locate any research on the interaction between fenitrothion and permethrin. Consequently, we hypothesized that the toxicity of permethrin may be increased by fenitrothion through a reduction in carboxylesterase activity. Fenitrothion and permethrin may be combined, even if not intentionally in a mélange, due to their use in the same location. The risk assessments of combined toxicity to mammals are currently insufficient, and additional detailed studies are necessary.

6. REFERENCES

Aardema H, Meertens JH, Ligtenberg JJ: Organophosphorus pesticide poisoning: cases and developments. *Neth J Med* 66:149-153, 2008.

Anadon A, Martinez-Larranaga MR, Diaz MJ, et al: Toxicokinetics of permethrin in the rat. *Toxicol Appl Pharmacol* 110:1-8, 1991.

Berger CW Jr, Sultatos LG: The effects of the phosphorothioate insecticide fenitrothion on mammalian cytochrome P450-dependent metabolism of estradiol. *Fundam Appl Toxicol* 37:150-157, 1997.

Berlińska B, Sitarek K: Disturbances of prenatal development in rats exposed to fenitrothion. *Rocz Panstw Zakl Hig* 48:217-228, 1997.

Clifford NJ, Nies AS: Organophosphate poisoning from wearing a laundered uniform previously contaminated with parathion. *JAMA* 262:3035-3036, 1989.

Elliott M, Farnham AW, Janes NF, et al: A photostable pyrethroid. *Nature* 246:169-170, 1973 Etzel RA, Forthal DN, Hill RH Jr, et al: Fatal parathion poisoning in Sierra Leone. *Bull World Health Organ* 65:645-649, 1987.

Eyer F, Meischner V, Kiderlen D, et al: Human parathion poisoning. A toxicokinetic analysis. *Toxicol Rev* 22:143-163, 2003.

FAO: Pesticide Residues in Food, Toxicological Evaluations; Food and Agriculture Organization of the United Nations and World Health Organization: Rome, 1999.

Freeman C: Permethrin technical: acute neurotoxicity screen in rats. *FMC Corporation* Study No. A92-3646, 1993a.

Freeman C: Permethrin technical: subchronic neurotoxicity screen in rats. *FMC Corporation* Study No. A92-3647, 1993b.

Gotoh Y, Kawakami M, Matsumoto N, et al: Permethrin emulsion ingestion: clinical manifestations and clearance of isomers. *J Toxicol Clin Toxicol* 36:57-61, 1998.

Hayes WJ: Pesticides Studied in Man. *Williams and Wilkins.* Baltimore, London. 1982.

Hayes WJ, Laws ER: Handbook of Pesticide Toxicology, Volume 2: Classes of Pesticides. *Academic Press Inc. NY*, 1990.

Hirosawa N, Ueyama J, Kondo T, et al: Effect of DDVP on urinary excretion levels of pyrethroid metabolite 3-phenoxybenzoic acid in rats. *Toxicol Lett* 203:28-32, 2011.

Hoilingworth RM, Metcalf RL, Fukuto TR: The selectivity of sumithion compared with methylparathion: Metabolism in the white mouse. *J Agric Food Chem* 15:242-249, 1967.

Hoffmann U, Papendorf T: Organophosphate poisonings with parathion and dimethoate. *Intensive Care Med* 32:464-468, 2006.

Kim SS, Lee RD, Lim KJ, et al: Potential estrogenic and antiandrogenic effects of permethrin in rats. *J Reprod Dev* 51:201-210, 2005.

Koyama K, Suzuki R, Kikuno T: Serum fenitrothion concentration and toxic symptom in acute intoxication patients. *Chudoku Kenkyu* 19:41-47, 2006.

Laynez BF, Martínez GL, Tortosa FI, et al: Fatal food poisoning by parathion. *Med Clin (Barc)* 108:224-225, 1997.

Lee CH, Kamijima M, Li C, et al: 3-Methyl-4-nitrophenol metabolism by uridine diphosphate glucuronosyltransferase and sulfotransferase in liver microsomes of mice, rats, and Japanese quail (Coturnix japonica). *Environ Toxicol Chem* 26:1873- 1878, 2007.

Li C, Suzuki AK, Takahashi S, et al: Effects of 3-methyl-4-nitrophenol on the reproductive toxicity in female Japanese quail (Coturnix japonica). *Biol Pharm Bull* 31:2158-2161, 2008.

Li C, Taneda S, Suzuki AK, et al: Effects of 3-methyl-4-nitrophenol on the suppression of adrenocortical function in immature male rats. *Biol Pharm Bull* 30:2376-2380, 2007.

Mackenzie PI, Owens IS, Burchell B, et al: The UDP glycosyltransferase gene superfamily: recommended nomenclature update based on evolutionary divergence. *Pharmacogenetics* 7:255-269, 1997.

McDaniel KL, Moser VC: Utility of a neurobehavioral screening battery for differentiating the effects of two pyrethroids, permethrin and cypermethrin. *Neurotoxicol Teratol* 15:71-83, 1993.

Meaklim J, Yang J, Drummer OH: Fenitrothion: toxicokinetics and toxicologic evaluation in human volunteers. *Environ Health Perspect* 111:305-308, 2003.

Mihara K, Misaki Y, Miyamoto J: Metabolism of fenitrothion in birds. *J Pesticide Sci* 4:175- 185, 1979.

Miyamoto J, Kitagawa K, Sato Y: Metabolism of organophosphorus insecticides by *Bacillus subtilis*, with special emphasis on Sumithion. *Jpn J Exp Med* 36:211-225, 1966.

Miyamoto J: Mechanism of low toxicity of Sumithion toward mammals. *Residue Rev* 25:251- 264, 1969.

Miyamoto J, Mihara K, Hosokawa S: Comparative metabolism of m-methyl-14C-sumithion in several species of mammals *in vivo*. *J Pest Sci* 1:9-21, 1976.

Miyamoto J, Sato Y, Kadota T, et al: Studies on the mode of action of organophosphorus compounds. Part I. Metabolic fate of 32P-labelled Sumithion and methylparathion in guinea-pigs and white rats. *Agric biol Chem* 27:381-389, 1963.

Miyamoto J: Studies on the mode of action of organophosphorus compounds. Part III. Activation and degradation of Sumithion and methylparathion in vivo. Agric biol Chem 28:411-421, 1964.

Mori Y, Kamata K, Toda N, et al: Isolation of nitrophenols from diesel exhaust particles (DEP) as vasodilatation compounds. *Biol Pharm Bull* 26:394-395, 2003.

Nosál M, Hladká A: Determination of the exposure to fenitrothion (O,O-dimethyl-O-3-methyl-4-nitrophenyl-thiophosphate) on the basis of the excretion of p-nitro-m- cresol by the urine of the persons tested. *Int Arch Arbeitsmed* 25:28-38, 1968.

Okahashi N, Sano M, Miyata K, et al: Lack of evidence for endocrine disrupting effects in rats exposed to fenitrothion in utero and from weaning to maturation. *Toxicology* 206:17-31, 2005.

Ortiz D, Yáñez L, Gómez H, et al: Acute toxicological effects in rats treated with a mixture of commercially formulated products containing methyl parathion and permethrin. *Ecotoxicol Environ Saf* 32:154-158, 1995.

Rishikesh N, Clarke JL, Martin HL, et al: Evaluation of Decamethrins and Permethrin Against *Anopheles gambiae* and *Anopheles funestus* in a Village Trial in Nigeria. World Health Organization, Division of Vector Biological Control; 1978.

Scollon EJ, Starr JM, Godin SJ, et al: In vitro metabolism of pyrethroid pesticides by rat and human hepatic microsomes and cytochrome P450 isoforms. *Drug Metab Dispos* 37:221-228, 2009.

Soderlund DM, Casida JE: Effects of pyrethroid structure on rates of hydrolysis and oxidation by mouse liver microsomal enzymes. *Pestic Biochem Physiol* 7:391-401, 1977.

Spencer EY: Guide to the Chemicals Used in Crop Protection. 7th edition. Publication 1093. *Research Branch. Agriculture Canada.* 1981.

Tamura H, Maness SC, Reischmann K, et al: Androgen receptor antagonism by the organophosphate insecticide fenitrothion. *Toxicol Sci* 60:56-62, 2001.

Taneda S, Kamata K, Hayashi H, et al: Investigation of vasodilatory substances in diesel exhaust particles (DEP): isolation and identification of nitrophenol derivatives. *J Health Sci* 50:133-141, 2004.

Trottier B, Fraser AR, Planet G, et al: Subacute toxicity of technical fenitrothion in male rats. *Toxicology* 17:29-38, 1980.

U.S. Environmental Protection Agency: Guidance for the Reregistration of Pesticide Products Containing Fenitrothion. US EPA, Office of Pesticide Programs, Registration Div., Washington, DC. 132 pp. July, 1987.

U.S. Environmental Protection Agency: Reregistration Eligibility Decision (RED) for Permethrin. US EPA, Office of Prevention, Pesticides and Toxic Substance, Office of Pesticide Programs, U.S. Government Printing Office: Washington, DC, 2007.

WHO: Environmental Health Criteria 133: Fenitrothion. World Health Organization, International Programme on Chemical Safety, Geneva, Switzerland, 1992.

WHO: Environmental Health Criteria 94: Permethrin. World Health Organization, International Programme on Chemical Safety, Geneva, Switzerland, 1990.

WHO: The WHO recommended classification of pesticides by hazard and Guidelines to classification 1990-1991, Geneva, World Health Organization (WHO/PCS/90.1), 1990.

Worthing CR, Walker SB: The Pesticide Manual: A World Compendium. 8th edition. Published by The British Crop Protection Council, 1987.

Yuan C, Wang C, Gao SQ, et al: Effects of permethrin, cypermethrin and 3-phenoxybenzoic acid on rat sperm motility in vitro evaluated with computer-assisted sperm analysis. *Toxicol In Vitro* 24:382-386, 2010.

Zhang SY, Ito Y, Yamanoshita O, et al: Permethrin may disrupt testosterone biosynthesis via mitochondrial membrane damage of Leydig cells in adult male mouse. *Endocrinology* 148:3941-3949, 2007.

Zhang SY, Ueyama J, Ito Y, et al: Permethrin may induce adult male mouse reproductive toxicity due to *cis* isomer not *trans* isomer. *Toxicology* 248:136-141, 2008.

CHAPTER

DDT and Its Metabolites in Mexico 5

1. INTRODUCTION: AN OVERVIEW

DDT (dichlorodiphenyltrichloroethane) was initially produced in 1874, with its insecticidal capabilities identified in 1939 by Paul Hermann Müller (Stapleton 1998). In 1944, the U.S. military commenced considerable utilization of DDT for mosquito control, especially in the Pacific, where significant World War II operations occurred in areas with high malaria prevalence (Stapleton 1998). In 1955, the World Health Organization (WHO) initiated a worldwide malaria control program utilizing DDT; by 1958, 75 nations had participated, and at the campaign's zenith, 69,500 tons of pesticides, predominantly DDT [1,1,1-trichloro-2,2-bis(p-chlorophenyl)ethane], were administered to 100 million residences annually (Wijeyaratne, 1993). To combat malaria, residences were treated biannually with DDT wettable powder to eliminate resting adult Anopheles mosquitoes. The Stockholm Convention on Persistent Organic Pollutants, effective from 17 May 2004, prohibited the use of 12 chemicals, including DDT (UNEP, 2004). Nevertheless, one exemption clause permits malaria-endemic countries to utilize DDT only for the control of disease vectors. The United Nations Environment Program predicts that approximately 25 nations will utilize DDT under exemptions from the pesticide ban on DDT. (POPs, 2009). Thus, in regard to presence of DDT around the world can be divided into three scenarios: Sites where DDT is still in use; sites where the presence is due to DDT sprayed several years ago, and sites where the presence of DDT is the result of a long-range transport of the insecticide to areas where it was never used like the Antarctic. In Mesoamerica (Mexico, Costa Rica, El Salvador, Guatemala, Honduras, Mexico, Nicaragua and Panama) DDT was used until the year 2000, Mexico and Nicaragua being the last nations that applied the insecticide in agriculture and for the control of malaria. Table 1 lists the period and the total amount of DDT used in each Mesoamerican country by the malaria control programs. The amount used (approximately 85,000 tons between 1946 and 1999) together with the high environmental persistence of DDT and its metabolites, provide the necessary conditions for DDT to become a contaminant of concern for this region of the world (ISAT, 2002). Taking into account the environmental persistence and the toxicity of DDT, a program for the control of malaria without using insecticides in Mesoamerica was developed between 2004 and 2007, with assistance from the Pan American Health Organization [PAHO; (Chanon et al., 2003; PAHO, 2008)]. The discontinuation of DDT in Costa Rica, El Salvador, Guatemala, Honduras, Mexico, Nicaragua, and Panama was part of a regional initiative endorsed by the Global Environment Facility (GEF) and the United Nations Environmental Program, with involvement from the North American Commission for Environmental Cooperation (CEC).DDT in Mexico

In 1944, and for the first time, houses were sprayed with DDT in Temixco, Morelos, Mexico (Stapleton 1998). The spray was applied to the walls and ceilings of residences. Studies done two months after the spraying, showed that there was a 99% reduction in the incidence of Anopheles (Stapleton 1998). In 1947–48, the spraying of DDT began in other Mexican regions, such as Veracruz, Mexico City and Baja California (Stapleton 1998). By 1948, the first clear evidence of malaria control appeared in the areas first sprayed with DDT; the overall parasite rate in the state of Morelos was found to be 10%, and the rate in the sprayed towns was found to be 1% (Stapleton 1998). In 1936 it was estimated that half of the Mexican population lived in endemic regions and was subject to a malaria mortality rate of 0.5%, or about 36,000 deaths per year (Stapleton 1998).During the 1930s and 1940s, malaria became the third cause of death in the country. However, the antimalaria campaign was not generalized until 1956 (CCE 1998). The success of DDT was outstanding, malaria cases decreased from 41,000 in 1955 to 4,000 in 1960 (Fernández de Castro 1998); in 1970 the campaign was relaxed and the cases increased to 57,000 (Fernández de Castro 1998). However, this was also the time in which DDT production peaked in Mexico, with more than 80 thousand tonnes produced annually (CCE 1998). In recent years, the incidences of malaria have declined significantly, to less than 5,000 cases. Since 1982 there have been no deaths from this disease. As, showed, Malaria is a long- standing public health problem that has inhibited development in large areas of the country. Approximately 60% of the Mexican territory, representing an area inhabited by close to 45 million people, provides an environment suitable for malaria transmission. This includes the Pacific coast, the Gulf of Mexico slopes, the Yucatan peninsula and interior basins of the high plateau. (CCE 1998). In actuality, Mexico operates a malaria control program that has substantially reduced the incidence of this disease. In 1995, Mexico initiated an integrated pest management approach for malaria to reduce the heavy dependence on pesticides. Much of the success of Mexico's malaria control program (there have been no recorded deaths from malaria since 1982) is due to improvements in sanitation, increased disease surveillance, and integrated pest management schemes that focus pesticide applications on critical habitats and stages in the mosquito's life cycle (Government of Mexico 1998). Since 1998, DDT was substituted with pyrethroids in the malaria control program. In other hand, In the area of agriculture, as much as 1,000 tonnes per year were used (CCE 1998).Application rates in the north of Mexico, were among the highest in the world (CCE 1998). However, the growing concern about DDT persistence has had a significant impact on agricultural practices in Mexico. During the early 1970s the US Food and Drug Administration (USFDA) began rejecting the importation of commodities due to high residue levels, especially of DDT (CCE 1998). Therefore, some agricultural areas changed to newer pesticides in order to comply with the USFDA regulations. By 1990, DDT was limited to campaigns addressing public sanitation (CCE 1998). In recognition of DDT's environmental and human health effects, Mexico shifted the emphasis of its anti-malarial campaigns away from DDT beginning in the 1980s and 1990s, and the use of the pesticide was gradually reduced. In 1997, the Intergovernmental Forum on Chemical Safety agreed there was sufficient evidence to take international action to restrict and reduce the use of DDT.

Table 1. History of DDT use in Mesoamerica countries (ISAT, 2002).

Country	Period of use	Total tons
México	1957-2000	69,545.00
Nicaragua	1959-1991	2,172.00
Costa Rica	1957-1985	1,387.00
Guatemala	1958-1979	4,790.00
Honduras	1950-1978	2,640.00
El Salvador	1946-1973	4,271.00
Panamá	1967-1971	189.00

2. ENVIRONMENTAL PATHWAYS OF EXPOSURE TO DDT

The physicochemical properties of DDTs (Table 2) show the extent of their volatility and the high KOW/KOA value shows that they are more likely to partition into environmental sectors which exhibit greater organic phases (biota, soil, and sediment). The concentration of DDTs in the water samples may be limited due to characteristically low water solubility. In other hand, the exposure pathways are the processes by which DDT may be transported from the pollution source to living organisms. In the malaria areas, the source of DDT was the household-spraying of the insecticide. Since the beginning of the control program of malaria, DDT was sprayed on the ceilings and walls, both indoors and outdoors. Therefore, after spraying, indoor dust (or indoor soil in some cases), and the external surface soil in those areas next to the dwellings, were the media first to become contaminated with DDT. From these points, the insecticide could be transported from one medium to another by different processes.

Table 2. Physicochemical properties of DDT and its metabolites at 250 C. (Sahsuvar et al. 2003; Shen and Wania 2005).

Compound	Molecular weight	Vapor pressure (Pa)	Aqueous solubility (mol/m3)	Henry´s law constant (Pa m3/mol)	Log Kow*	Log Koa
p´p-DDT	354.5	0.00048	0.00042	1.1	6.39	9.73
p´p-DDE	319.0	0.00340	0.00079	4.2	6.93	9.70
p´p-DDD	321.0	0.00120	0.00230	0.5	6.33	10.03

Soil and Dust

Several studies have identified indoor house dust as an important pathway of toxicant exposure. Often levels of pollutants found in house dust, including compounds banned long ago such as DDT, are significant sources of exposure for the general population, especially children (Butte and Heinzow 2002; Hwang et al. 2008; Rudel et al. 2003). Moreover, analyses of compounds in house dust are a measure of indoor contamination, but may also provide valuable information on the assessment of human indoor exposure (Butte and Heinzow 2002). Also, outdoor soil is considered an important exposure pathway for the general population and children to compounds banned long ago (Herrera-Portugal et al. 2005). However, it is important to note that longer residence times and elevated contaminant concentrations in the indoor environment may increase the chance of exposure to these contaminants by 1,000-fold compared to outdoor exposure (Hwang et al. 2008).

Tables 3 and 4 show DDT levels in outdoor and indoor surface soils, respectively. Taking into account the guideline for DDT in residential soil: 0.7 mg/kg from Canada (Environment Canada, 2007) different scenarios have been observed in Mexico. Regarding outdoor levels, in general lower levels were found in household outdoor samples (Table 3). With exception of levels found in Chiapas and Oaxaca, that have levels lower than Canadian guide (Table 3). In other hand, high levels are recorded in indoor levels in different regions of Mexico, generally higher than Canadian guideline (Table 4). Also, we can note that the higher levels of DDT in those environment media were found in indoor dust samples, generally with levels above the Canadian guideline (Table 5). Moreover, the data in Tables 3, 4 and 5 indicate high levels of total DDT in soil and dust in all regions studied in Mexico when compared with studies around the world.

Water

DDT, DDD and DDE (DDTs) are only slightly soluble in water, with solubilities of 3.4 ppb, 160 ppb and 120 ppb, respectively (ATSDR 2010). In this regard, sedimentation is the most important factor for the disappearance of DDT from water. However, it has also been suggested that contaminated sediments are a main source of DDT inputs to the water column (Zeng 1999). In order to study the degree of pollution in water bodies located in tropical areas, DDTs were quantified in a relatively small stream. The levels of total DDT found in the tropical area was 280 pg/L (Carvalho et al. 2009). In other hand, levels of total DDT found in the tropical area in United states of America was 10300 pg/L (California 1999)

Sediments

As stated above, sediments act as the primary reservoir for excess quantities of DDT. Therefore, it is very important to analyze the concentrations in this medium. In Table 6 it is shown that DDT concentrations in Mexican samples are lower than those detected in other countries, where DDT was used either for the control of malaria or for agricultural practices. Whether this difference can be explained by an increased degradation or by a DDT mass reduction caused by water currents carrying suspended DDTs out of the contaminated area, are issues that deserve further research. However, we cannot exclude another explanation. The Mexican studies, results of which are shown in Table 6, were not designed to assess the amount of DDT in sediments due to spraying. In fact, a sediment sample collected in a river near an area where the insecticide was used intensively for vector control, had DDT concentrations of up to 70.0 mg/kg (Gonzalez-Mille et al. 2010). Discvutir disminución.

Table 3. Total DDT levels in outdoor surface soil (mg/Kg) in different Mexican Regions.

Location	Total DDT (mg/Kg)	Region	Reference
Chiapas	0.95	Southeastern	Martínez-Salinas et al. 2011
Chiapas	8.20	Southeastern	Yañez et al. 2002
Oaxaca	0.90	Southeastern	Yañez et al. 2002
Tabasco	0.04	Southeastern	Torres-Dosal et al. 2011
Chihuahua	0.45	North	Díaz-Barriga et al. 2011
Veracruz	0.01	Southeastern	Espinosa-Reyes et al. 2010
Puebla and Mexico	0.07	Central	Waliszewski et al. 2008

Table 4. Total DDT levels in indoor soil (mg/Kg) in different Mexican Regions.

Mexican state	Total DDT (mg/Kg)	Region	Reference
Chiapas	6.8	Southeastern	Martínez-Salinas et al. 2011
Chiapas	7.1	Southeastern	Yañez et al. 2002
Oaxaca	0.15	Southeastern	Yañez et al. 2002
Chihuahua	0.95	North	Díaz-Barriga et al. 2011

Food and Biota

Due to their lipophilic attributes and high persistence, the DDTs may bioaccumulate significantly in animal species (Fisher 1995). Furthermore, biomagnification has been observed; for example, DDT concentration increased with each successive trophic level in a food chain (Fisher 1995).Taking into account these properties, food ingestion can be considered a pathway of exposure. In Mexico, studies have been done in different food items, such as fish, hen's egg, butter and cow's milk and muscle. In Table 7, total DDT levels in different food items are presented. Considering fish, the concentrations of DDT, in organisms collected in Mexico, are above normal values. As is shown in Table 8, where DDTs levels in Fish are depicted for different countries. We can note that the food item with high levels of DDT are food rich in fat as butter and cow´s milk (Table 7).

Table 5. Total DDT levels in dust (mg/Kg) in different Mexican Regions.

Location	Total DDT (mg/Kg)	Region	Reference
Chiapas	6.9	Southeastern	Martínez-Salinas et al. 2011
Chihuahua	1.0	North	Díaz-Barriga et al. 2011

Table 6. Total DDT levels in sediment (mg/Kg) in different Mexican Regions.

Location	Total DDT (μg/Kg)	Region	Reference
Estado, Mexico (bay)	1.5	Southeastern	Noreña-Barroso et al. 1998
Mexico (bay)	0.6	Southeastern	Noreña-Barroso et al. 2007
Mexico (river)	74.0	Southeastern	Gonzalez-Mille et al. al. 2007
Mexico (lagoon)	4.9	Southeastern	Botello et al. 2000
China (Bay)	7.8		Liu et al. 2011

Location	Total DDT (μg/Kg)	Region	Reference
China (river)	3.8		Tan et al. 2009
Korea (bay)	3.4		Khim et al. 2001
Japan (bay)	1.2		Kim et al. 2007

Air

Because DDTs have a Henry's Law constant value of 10–4/10–5 atm m3 mol, they are considered moderate volatile compounds [5]. Therefore, these compounds can be transported by air, either in the gaseous phase or adsorbed to atmospheric particles [5].Photodegradation of DDT occurs slowly; thus, residues of these pesticides are ubiquitous in the atmosphere, although at lower concentrations. Information on the atmospheric levels of OCs in Mexico is scarce. Previous studies in southern Mexico found that DDT and toxaphene concentrations in air were 1-2 orders of magnitude above levels in the Laurentian Great Lakes and arctic regions (24-26). Atmospheric levels in southern Mexico were generally higher than those in central Mexico (27), Costa Rica (28, 29) and Cuba (30), and comparable to those in Belize (31).

Recently, two important studies regarding DDT and its metabolites in air has been developed. Passive air samplers (PAS) were deployed at four sampling sites at the southern Mexico in 2002-2004 and eleven sampling sites across Mexico during 2005-2006 (referencia). The total DDT levels ranged from 239 to 2360 pg/m3 in 2002-2004 (referencia) and from 15 to 1975 pg/m3 in 2005-2006 (refernecia). Table ? shows the Total DDT air levels in the sampling sites in both studies. Total DDT tended to be higher in the south (poner siglas) and some central sites (sigles). While, the other central and northern sites had lower total DDT levels (poner siglas). It is important to note that the higher levels were found in tropical sites where DDT was used for health campagnes or for agriculture as MT, CEL and VC.

Table 7. Total DDT levels in food items (mg/Kg) in different Mexican Regions.

Location	Food	Total DDT (mg/Kg)	Region	Reference
Mexico country	Butter	88.0		Waliszewski et al. 2003
Veracruz	Cow's milk	39.0	Southeastern	Pardio et al. 2003
Campeche	Oysters	5.9	Southeastern	Carvalho et al. 2009
Campeche	Oysters	1.5	Southeastern	Gold-Bouchot et al. 1995
Tabasco	Oysters	6.2	Southeastern	Botello et al. 1994

Location	Food	Total DDT (mg/ Kg)	Region	Reference
Campeche	Shrimps	0.25	Southeastern	Gold-Bouchot et al. 1995
Campeche	Mussels	1.44	Southeastern	Gold-Bouchot et al. 1995
Baja California	Mussels	9.16	North	Gutierrez-Galindo et al. 1988

Table 8. Total DDT levels in fish (mg/Kg) in different Mexican Regions.

Location	Total DDT (ng/g lipid)	Region	Reference
Veracruz, Mexico	25.0	Southeastern	Gonzalez-Mille et al. 2010
Chiapas, Mexico	4.7	Southeastern	Pérez-Maldonado et al. 2010
Hidalgo, Mexico	7.3	Central	Fernandez-Bringas et al. 2008
Costa Rica	0.6		Pérez-Maldonado et al. 2010
Honduras	< LOD (below detection limit)		Pérez-Maldonado et al. 2010
Nicaragua	3.9		Pérez-Maldonado et al. 2010
El Salvador	3.8		Pérez-Maldonado et al. 2010
Guatemala	< LOD		Pérez-Maldonado et al. 2010

Table 9. Total DDT levels in air (pg/m3) in different Mexican Regions.

Community	Total DDT (pg/m3)	Region	Reference
Baja California	338	North	Wong et al. 2009
Chihuahua	34	North	Wong et al. 2009
Yucatan	1975	Southeastern	Wong et al. 2009
Colima	750	North	Wong et al. 2009
Veracruz	129	Southeastern	Wong et al. 2009
Morelos	500	Central	Wong et al. 2009
Sinaloa	76	North	Wong et al. 2009
Mexico DF	55	Central	Wong et al. 2009
Nuevo Leon	15	North	Wong et al. 2009
San Luis Potosi	21	Central	Wong et al. 2009
Veracruz	50	Southeastern	Wong et al. 2009
Tabasco	239	Southeastern	Alegría et al. 2008

Community	Total DDT (pg/m3)	Region	Reference
Chiapas	2360	Southeastern	Alegría et al. 2008
Chiapas	547	Southeastern	Alegría et al. 2008
Veracruz	1200	Southeastern	Alegría et al. 2008

3. HUMAN EXPOSURE TO DDT

Biomonitoring studies are a useful instrument in formulating environmental health policies. For example, in Mexico during the last decade, studies in children led to the reduction or elimination of different chemicals such as lead, lindane and inclusive DDT. Furthermore, the biomonitoring of susceptible populations is a valuable method for the identification of critical contaminants, as has been shown in the United States with the National Health and Nutritional Examination Survey (NHANES III; Needham et al., 2005). Information about human exposure to chemicals is very limited; besides, in relation to children, the information is even more scarce. In this regard, in this text, we shall present data for breast milk, blood serum and adipose tisssue. Results have been done in adults and children.

In other hand, a biomarker of exposure is a xenobiotic substance or its metabolite (s), that is measured within a compartment of an organism. The preferred biomarkers of exposure are generally the substance itself or substance-specific metabolites in readily obtainable body fluid(s) or excreta. DDT and its metabolites DDD,DDE, DDA, and MeSO2-DDE (3- methylsulphonyl-DDE), can be measured in adipose tissue, blood serum, urine, feces, semen, or breast milk .

Breast Milk

Psychological and medical studies have underlined the benefits of nursing which raises immunological defenses and provides a healthier development of the baby. Parallel findings have increased concern about the excretion of drugs and environmental contaminants contained in breast milk, since it is considered the main route for eliminating deposited organochlorine pesticides from a mother's body (Jensen and Slorach 1991; Sonawane 1995; Cupul-Uicab et al. 2008).

Because of their lipophilic nature and high persistence, DDT and its metabolites accumulate in lipophilic human body parts, particularly in lipid-rich tissues such as adipose tissue and subsequently translocated and excreted through milk fat. A major concern is that milk is the first (and in some areas the only) food for the newborn child.

Concentrations of DDTs (DDT, DDD and DDE) in human milk have been shown to be higher in communities exposed to this insecticide, than in non-exposed populations (Table 10). For example, levels from a cotton area where DDT was used for agricultural purposes (Coahuila) were higher or similar to those obtained in samples collected in a malarious area (Veracruz, Yucatan) where DDT was extensively used (Table 10). And both were higher than the concentrations quantified in urban areas (Mexico DF), where DDT has never been used.

The World Health Organization's Acceptable Daily Intake (ADI) for DDTs is 20 mg/kg/day (Lu 1995). In this regard, considering a body mass of 5 kg, a milk intake of 0.85 kg/day, a proportion of fat in milk of 0.035, and a DDT concentration in milk of 10.4 mg/kg (total DDT concentration in samples collected during 1996 1997 in a suburban malarious area; Albert et al. 1980), the estimated daily intake is three times higher than the ADI. Furthermore, if the maximum range concentration found in some studies (36.5 mg/kg; Albert et al. 1980) is taken into account, the ADI is surpassed 11 times. However, when calculated the ADI with DDT concentration in milk of 2.4 mg/kg (Waliszewski et al. 2009), the ADI calculated is lower than 20 mg/kg/day. It is important to remark that the chronological levels obtained by Waliszewski et al. (1996, 1999, 2001, 2002, 2009) have a decreasing tendency. That result coincides with the restriction and prohibition for DDT use in Mexico.

Serum

In this document, we presented data regarding DDT and its metabolites levels in children (Table 11) and adults (Table 12).

Children appear to be particularly suitable for a monitoring program, as they are not directly exposed to occupational pollution; thus, children normally reflect present trends of environmental exposure more accurately than do adults (Link et al., 2005). Moreover, it is well established that children are potentially at a higher risk than adults for adverse health effects from exposure to many environmental chemicals (Guzelian et al., 1992; Bearer, 1995; Carlson, 1998; Galson et al., 1998; Aprea et al., 2000; Needham and Sexton, 2000; Adgate and Sexton, 2001; Brent et al., 2004; IPCS, 2006).

Table 10. Total DDT levels in human milk (ng/g lipid) in different Mexican Regions.

Location	Total DDT	Region	Reference
Coahuila	10400.0	North	Albert et al. 1980
Mexico DF	900.0	Central	Torres-Arreola et al.1999
Veracruz	7815.0	Southeastern	Pardio et al. 1998
Veracruz	6280.0	Southeastern	Waliszewski et al.1996
Veracruz	4700.0	Southeastern	Waliszewski et al.1999
Veracruz	4700.0	Southeastern	Waliszewski et al.2001

Location	Total DDT	Region	Reference
Veracruz	3740.0	Southeastern	Waliszewski et al.2002
Morelos	4320.0	Central	Lara et al. 2000
Yucatan	3065.0	Southeastern	Rodas-.Ortíz et al.2008
Veracruz	2335.0	Southeastern	Waliszewski et al. 2009

Adipose tissue

Adipose tissue biopsy has been used in epidemiological studies to assess chronic exposure to DDT. This is a logical choice because the DDTs are accumulated in adipose tissue due to its lipid solubility. The half-life of DDT in human adipose tissue is approximately seven years (Woodruff et al. 1994).

As in serum, DDTs in adipose tissue are a good biomarker of exposure for communities exposed to DDT. When compared to an urban non-exposed community [46], the levels of DDTs (especially those of DDE), were higher in the exposed population (Table 9). In the same table it can be observed that the concentrations of DDT in adipose tissue from workers of the malaria program were higher than the levels found in people living in an agricultural area or in malarious areas. In the workers, a linear model that included an index of chronic exposure, the use of protective gear,and recent weight loss explained 55% of the variation of p,p'-DDE conncentrations in adipose tissue. The index of chronic exposure was constructed according to worker position and based on the historical duration and intensity of DDT application [48].

When the concentrations of DDTs in adipose tissue were expressed by age group, two groups were identified as the most exposed. Those groups were children and elderly people [49]. The levels in elderly people can be explained by the accumulation of DDT in a chronic exposure scenario,whereas the concentration in children may be the result of an exposure to multiple pathways (soil, household dust, air, water, food, etc.). It is interesting that the group less exposed to DDT was the 0–2 years, a group that may be exposed to DDT through lactation [49].

Table 11. Total DDT levels in serum (ng/g lipid) of children living in different Mexican Regions.

Location	Total DDT (ng/g lipid)	Region	Reference
Chiapas	22280.0	Southeastern	Trejo-Acevedo et al. 2009
Oaxaca	7500.00	Southeastern	Perez-Maldonado et al. 2006
Quintana Roo	11300.0	Southeastern	Perez-Maldonado et al. 2006
Chihuahua	35000.00	North	Díaz-Barriga et al. 2011

Location	Total DDT (ng/g lipid)	Region	Reference
Queretaro	2170.0	Central	Trejo-Acevedo et al. 2009
Durango	2270.0	North	Trejo-Acevedo et al. 2009
San Luis Potosí	1990.0	Central	Trejo-Acevedo et al. 2009
Guanajuato	940.0	Central	Trejo-Acevedo et al. 2009
Veracruz	1910.0	Southeastern	Trejo-Acevedo et al. 2009
Michoacan	550.0	Central	Trejo-Acevedo et al. 2009
Zacatecas	700.0	Central	Trejo-Acevedo et al. 2009

A monitoring program of DDTs in adipose tissue is needed in order to assess the body burden, now that in Mexico this insecticide has been eliminated from the malaria program. However, due to ethical constraints, it is not always possible to obtain adipose tissue samples form healthy individuals. Therefore, alternative matrices are needed; for example, a good correlation between adipose tissue concentration and levels in human milk [50] or human serum [51] has been reported .When the geometric DDE levels in lipid bases are used for the estimation of the adipose tissue/serum DDE ratio, a value near unity is obtained [51].

4. HEALTH EFFECTS

DDT and its metabolites have been associated with neurological effects (Dorner and Plagemann 2002; Fenster et al. 2007; Torres-Sánchez et al. 2007; Rocha-Amador et al. 2009), asthma (Sunyer et al. 2006), immunodeficiency (Dewailly et al. 2000; Vine et al. 2000; Vine et al. 2001; Belles-Isles et al. 2002; Bilrha et al. 2003; Cooper et al. 2004; Dallaireet al. 2004), apoptosis (Pérez-Maldonado et al. 2004) and DNA damage in immune cells in children (Yáñez et al. 2004; Herrera-Portugal et al. 2005b).

Table 12. Total DDT levels in serum (ng/g lipid) of adults living in different Mexican Regions.

Location	Total DDT (ng/g lipid)	Region	Reference
Chiapas	12750.0	Southeastern	Yañez et al. 2002
Oaxaca	8050.0	Southeastern	Yañez et al. 2002
Tabasco	8700.0	Southeastern	Torres-Dosal et al.2011
Mexico, DF	20.0	Central	Lopez-Carrillo et al.1997
Veracruz	4500.00	Southeastern	Waliszewski et al.2000
Morelos	20.0	Central	Lopez-Carrillo et al.2001
San Luis Potosí	1715.0	Central	Yañez et al. 2004

Table 13. Total DDT levels in Adipose Tissue (ng/g lipid) of adults living in different Mexican Regions.

Location	Total DDT	Region	Reference
Coahuila	18400.0	Central	Albert et al. 1980
Mexico DF	6100.0	Central	Albert et al. 1980
Puebla	2700.0	Central	Albert et al. 1980
Veracruz	10000.0	Southeastern	Waliszewski et al.1996
Veracruz	61000.0	Southeastern	Rivero-Rodriguez etal. 1997
Veracruz	5700.0	Southeastern	Waliszewski et al.2001
Veracruz	2600.0	Southeastern	Waliszewski et al.2010
Puebla	800.0	Central	Waliszewski et al.2010
Veracruz	1900.0	Southeastern	Waliszewski et al.2011
Veracruz	1400.0	Southeastern	Herrero-Mercado et al. 2010
Veracruz	900.0	Southeastern	Herrero-Mercado et al. 2011

4.1 Cancer

Although it has been suggested that the estrogenic activity of DDE may be a contributing factor for development of breast cancer in women, levels of these compounds are not consistently elevated in breast cancer patients. It was initially reported that levels of *p,p*2- DDE were elevated in breast cancer patients (serum or tissue) versus controls [52].More recent studies and analysis of organochlorine levels in breast cancer patients versus controls show that these contaminants are not elevated in the latter group [53–56]. The study of occupationally exposed workers has not found clear increased risks for other cancers [57].

Two case-control studies of breast cancer have been carried out in Mexico City, with conflicting results. The first study, conducted by Lopez-Carrillo et al. [58] in Mexico City, compared 141 cases of breast cancer with 141 age-matched controls. All subjects were identified at three referral hospitals between March 1994 and April 1996. The arithmetic mean of serum DDE in lipid basis was 562 ppb±676 for the cases and 505 ppb±567 for the controls.The age-adjusted odds ratios for breast cancer regarding the serum level of DDE were 0.69 (95% confidence interval, 0.38–1.24) and 0.97 (CI, 0.55–1.70) for the contrasts between tertile 1 (lowest level) and tertiles 2 and 3, respectively. These estimates were unaffected by adjustment for body mass, accumulated time of breast-feeding and menopause, and other breast cancer risk factors. These results do not lend support to the hypothesis that DDT is causally related to breast cancer. The second study conducted by Romieu et al. [59] compared 120 cases and 126 controls, selected from six hospitals in Mexico

City, from 1989 to 1995. Serum DDE levels in lipid basis were higher among cases (mean=3840 ppb±5980) than among controls (mean=2510 ppb±1970).After adjusting for age, age at menarche,duration of lactation, Quetelet index, and serum DDT levels, serum DDE levels were positively r elated to t he risk for b reast cancer (adjusted O RQ1–Q2=1.24), 382 F. Díaz-Barriga et al. (CI, 0.50–3.06; ORQ1 –Q3=2.31, 95 percent, CI, 0.92–5.86; ORQ1 –Q4=3.81, CI, 1.14–12.80). The increased risk associated with higher serum DDE levels was more apparent among postmenopausal women (ORQ1 –Q4=5.26, 95%, CI, 0.80–34.30). Serum DDT level was not related to the risk for breast cancer. In addition to the differences in the comparison of cases and controls, the difference in the serum DDE levels among the women studied is remarkable. Participants from both studies came from similar hospitals, and there were no apparent differences between case and control selection that could explain this divergence. Differences in laboratory procedures is the most feasible explanation.

4.2 Endocrine disruption

DDT is known to have adverse effects on wildlife via endocrine disruption.Clear effects include thinning of the eggshell, feminization, reproduction impairment and development effects [60]. In Mexico two studies in humans have reported findings in this area. Gladen and Rogan [61] found that DDE might affect women's ability to lactate in a study conducted in an agricultural town in northern Mexico. Two hundred and twenty-nine women were followed from childbirth until weaning or until the child reached 18 months of age.DDE was measured in breast milk samples taken at birth, and women were followed to see how long they lactated.Median duration was 7.5 months in the lowest DDE group and 3 months in the highest. The effect was confined to those who had lactated previously - but not for first pregnancies - and it persisted after statistical adjustment for other factors. Rodriguez et al. [62] conducted a study aimed at determining the capability of long-term exposure to DDT of altering the normal endocrine function of the hypothalamus-hypophysis-gonads axis in humans. This included 70 workers dedicated to control malaria in the State of Guerrero,Mexico. The main activities of these workers were the application of pesticides,detection of malaria cases and promotion of preventive measures for control vectors. The average time of exposure to technical grade DDT was 25 years (range: 4–35), their last exposure being 5 months before sampling. An interview gathered information on the occupational history, reproductive performance, life styles and other relevant factors. Blood and urine samples were collected to measure serum levels of DDT and metabolites as well as levels of LH, FSH, prolactine, and testosterone. Participants ranged in age from 22 to 69 years, and had been employed in the sanitation campaign from 4 to 37 years. Ninety- seven percent of the participants were sprayers of DDT at some time in their occupational history, and 15% are current sprayers. Average levels of DDT and metabolites expressed as µg/g of extractable lipids were: total DDT, 60.1; *p,p*2-DDE, 37.41; *p,p*2-DDT, 21.52; *p,p*2- DDD, 1.07 and *o*2*p*-DDT 0.11. Results show a positive association of LH and FSH with DDT metabolites. An increase of 10 *m*g/g of *p,p*2-DDE was associated with an increase of 1.95 UI/L in LH (p=0.01); and 1.10 UI/L per each 10 *m*g/g of *p,p*2-DDT (p=0.02). FSH increased 1.09 UI/L per each 10 *m*g/g of *p,p*2- DDT (p=0.03). There was a negative association of DDE with testosterone, DDT in Mexico 383 especially for those participants under 55 years of age. These associations suggest direct toxicity to the testicles, especially the Leydig cells, as observed with antineoplasic drugs.

4.3 Genotoxicity

Some studies have reported genotoxic effects in humans heavily exposed to DDT [8]. Therefore, this area has been studied in Mexico. Studies were done in workers from the control program of malaria, and in women living in malarious areas. Herrera et al. [63] evaluated chromosomal translocations in a sample of the above-mentioned workers.Nineteen male sprayers (median age 46 years; range: 22–64),working in campaigns to control malaria vectors in the state of Guerrero, Mexico were included in this study.DDT data obtained in a previous study from eleven individuals, 5 women and 6 men, living in Mexico City, and occupationally unexposed to DDT were used as reference group. Chromosomal aberrations in lymphocytes were analyzed with a chromosome painting technique, a highsensitivity technique for detecting complex chromosomal translocations. *o,p*2- DDT was the only isomer significantly associated with the frequency of chromosomal translocations (p=0.003). Individuals presenting the higher levels of *o,p*2-DDT in serum (>0.79 *m*g/g fat; n=4) had a mean frequency of chromosomal translocations (5.1¥1000 metaphases), two times higher than that observed in workers occupationally exposed to 0.5 Gy of radiation.A positive relationship between the duration of exposure to DDT, measured as years working for the vector control program, and chromosomal translocations was observed (Fig.2). These results suggest an increased risk for diseases with a genetic component, such as cancer.

Yañez et al. [64] evaluated the association of blood DDT levels and DNA damage using the single cell gel electrophoresis assay. A group of 53 postpartum women were selected from two different areas in San Luis Potosí to assure different exposure levels, one with antecedents of malaria and DDT spraying and the other without malaria. Mean and range levels of DDT, DDE and DDD in whole blood were 5.57 ppm (0.02–20.69), 6.24 ppm (0.04– 39.16) and 1.16 ppm (0.01–5.63), respectively. The significant correlation of DNA damage, measured as DNA migration with the logarithm of DDT,DDE and DDD was 0.60, 0.62 and 0.43, respectively. Fig. 3 shows the shape of the association of DNA migration with DDE concentration in whole blood, as obtained by the regression: DNA migration= 71.58+7.62(logDDE), this association was not modified by age, smoking habits, nutrition or occupation. This observational finding in the epidemiological study with postpartum women was reevaluated in an in vitro study.Human blood cells were exposed to three doses of DDT, DDD and DDE.DNA damage was assessed by two different techniques: single-cell electrophoresis and flow cytometry. Results obtained by either technique showed that DNA damage was induced by the three organochlorides and a dose-response was observed with DDT. The data indicate DDT-induced DNA fragmentation, a phenomenon previously found with DDE and DDD.

5. REFERENCES

1. Stapleton DH (1998) Parassitologia 40 :149
2. North American Commission for Environmental Cooperation (1998) North American Regional Action Plan on DDT. In: The Sound Management of Chemicals Initiative. Regional Commitments and Action Plans.Montreal, Canada, p 67
3. Fernández de Castro J (1988) Panorama histórico y epidemiológico del paludismo en México. Secretaría de Salud,México
4. Implementation Plan to Control Malaria with Alternatives Methods to DDT in Mexico 1999–2001.Government of Mexico, January 1999
5. ISAT, 2002. Diagnóstico situacional del uso de DDT y el control de la malaria. Informe regional para México y Centroamérica. Instituto Salud Ambiente y Trabajo, México. <http://www.cec.org/files/PDF/POLLUTANTS/InfRegDDTb_ES_EN.pdf> (accessed 07.07.09.).
6. ATSDR (2010) Toxicological Profile for DDT,DDE, and DDD.Agency for Toxic Substances and Disease Registry,Atlanta,Georgia
7. Zeng EY,Yu CC, Tran K (1999) Environ Sci Technol 33:39
8. Fisher SW (1995) Rev Environ Contam Toxicol 142:87
9. Albert LA (1996) Rev Environ Contam Toxicol 147:1
10. Lu FC (1995) Reg Toxicol Pharmacol 21 :352
11. Woodruff T,Wolff MS, Davis DL,Hayward D (1994) Environ Res 65:132

CHAPTER

6

Detection of Dichloro dipheny ltrichloroethane (DDT) in Croatia and Assessment of Its Genotoxicity

1. INTRODUCTION: AN OVERVIEW

Pesticide refers to any agent or combination of compounds employed to prevent, eliminate, or manage pests. The Food and Agriculture Organization (FAO) defines a pest as an organism that: (1) serves as a vector for human and animal diseases; (2) disrupts or inflicts damage upon any phase of the production, transportation, or storage of food, wood, and agricultural products; (3) diminishes the quality of animal feed or pharmaceuticals. The comprehensive definition of pesticide encompasses all substances that: (1) function as plant growth regulators, defoliants, desiccants, or agents to prevent premature fruit drop; (2) are applied to crops to safeguard them from deterioration during storage and transportation (Food and Agriculture Organization [FAO], 2005). Each pesticide comprises an active chemical designed to hurt the pest and an inert carrier substance that enhances storage, handling, application, efficiency, and safety of the pesticide. (World Health Organization [WHO], 1990). For easier dealing with large numbers of pesticides, there are several classifications which divide pesticides in some classes. According to the chemical structure pesticides can be classified as: pyrethroids, carbamates, thiocarbamates, dithiocarbamates, organophosphates, organochlorides, phenoxy and benzoic acid herbicides, triazines, triazoles, ureas etc. (Kamir, 2000). Another classification is done considering the target organism, so pesticides can be classified as: insecticides, fungicides, herbicides, virucides, avicides, molluscicides, nematocides, rodenticides etc. (WHO, 1990). It is important to notice that pesticide's mode of action is not restricted just to the target organism; it also has negative effects on other organisms. That is why the World Health Organization (WHO) recommends the classification of pesticides by hazard. The base of this classification is determining the lethal dose

(LD_{50}) for 50% of rat population after acute oral and dermal pesticide exposure. Depending of the LD_{50} value, as presented in Table 1, pesticides are classified as extremely, highly, moderately or slightly hazardous (WHO, 2009a).

It can be assumed that fighting against the pests started when human civilization was introduced to agriculture which ensured enough food for further development. The first known historical use of pesticide was observed in Mesopotamia 2500 BC. In ancient Sumer, individuals utilized sulfur for agricultural treatment, effectively exterminating pests (Miller, 2004). Circa 500 BC, the ancient Greek historian Herodotus documented the utilization of the castor-oil plant as a mosquito repellant. Inhabitants of Egypt residing in marshy regions utilized castor-oil lamps that emitted an awful odor. They also secured fishing nets around their beds prior to sleeping. This information indicates that ancient Egyptians employed comparable methods to manage mosquitoes as contemporary insecticide-treated nets (Charlwood, 2003; WHO, 2007). Following an extended era of utilizing natural pesticides (such as arsenic, mercury, nicotine sulfate, and pyrethrum), the 20th century witnessed the advent of manufactured pesticides.

Table 1. The WHO pesticide classification based on LD_{50} (adapted from WHO, 2005).

Class	LD_{50} for rats (mg/kg of body mass)				
		Oral		Dermal	
		Solids	Liquids	Solids	Liquids
I a	Extremely hazardous	5 or less	20 or less	10 or less	40 or less
I b	Highly hazardous	5 – 50	20 – 200	10 – 100	40 – 400
II	Moderately hazardous	50 – 500	200 – 2000	100 – 1000	400 – 4000
III	Slightly hazardous	over 500	over 2000	over 1000	over 4000

DDT (1,1,1-trichloro-2,2-di(4-chlorophenyl)ethane) was one of the most used pesticides in the mid 20th century. This organochloride was synthesized in 1874 by Othmar Zeidler and its pesticides properties were discovered in 1939 by Paul Hermann Müller who was awarded Nobel Prize in Physiology or Medicine for this discovery (Agency for Toxic Substances and Disease Service [ATSDR], 2002; Turusov et al., 2002). Commercially used DDT is a mixture of several isomeric forms: *p,p'*-DDT (85%), *o,p'*-DDT (15%), *o,o'*-DDT (trace amounts) and its breakdown products DDE (1,1-dichloro-2,2-bis(*p*-chlorophenyl)ethylene) and DDD (1,1-dichloro-2,2-bis(*p*-chlorophenyl)ethane). The basic chemical properties of *p,p'*- DDT, *p,p'*-DDE and *p,p'*-DDD are shown in Table 2. According to WHO, DDT is moderately hazardous pesticide with LD_{50} of 113 mg/kg (WHO, 2009a).

In the period of almost 35 years, 2 million tons of DDT was used to control malaria and typhus, thus contaminating water, soil and air. When DDT and its metabolites enter the environment they have the potential to stay adsorbed in the sediment for more than 100 years. The best example of DDT's persistency is that it has been found in the areas 1000 km off the spraying spot and even in the Arctic animals and ice where it has never been used (ATSDR, 2002). One of the most popular acts that contributed global DDT ban in 1973, was Rachel Carson's book *Silent Spring* written in 1962 where she emphasized the problem of the DDT use (Carson, 2002). Today DDT is still legally produced (China, India, and North Korea) and used in the countries of the third world (e.g. India, Ethiopia, South Africa etc.) for malaria control (WHO, 2009b).

There are four possible ways for DDT to enter organisms: (1) ingestion, (2) inhalation, (3) dermal exposure, and (4) placental transport. Once in the body DDT is metabolized mainly in the liver and partly in the kidneys to its most familiar metabolites DDE and DDD. Detailed metabolic pathways are shown in Figure 1. Conjugated forms of 2,2-bis(4- chlorophernyl)-acetic acid (DDA) are then excreted through urine and faeces. The most dangerous property of DDT is that it possesses high bioaccumulation and biomagnification potential. As the trophic level rises, the concentration of DDT and/or its metabolites increases. All three compounds are reported to be harmful to either human or animals (Brooks, 1986; Gold & Brunk, 1982, 1983; Morgan & Roan, 1974; Peterson & Robinson, 1964). As mentioned in the book *Silent spring*, DDT (especially metabolite DDE) has negative effects on bird's reproductive system by reducing Ca2+ transport which results with eggshell thinning thus increasing lethality (Carson, 2002).

Table 2. Chemicals properties of *p,p'*-DDT, *p,p'*-DDE, and *p,p'*-DDD (adapted from ATSDR, 2002), a room temperature, b at 25°C, *NA – not available.

	p,p'-DDT	*p,p'*-DDE	*p,p'*-DDD
Chemical formula	$C_{14}H_9Cl_5$	$C_{14}H_8Cl_4$	C14H10Cl4
Chemical structure	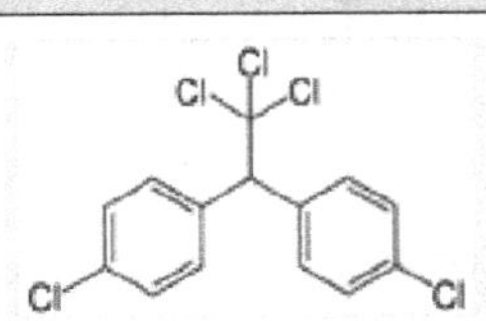	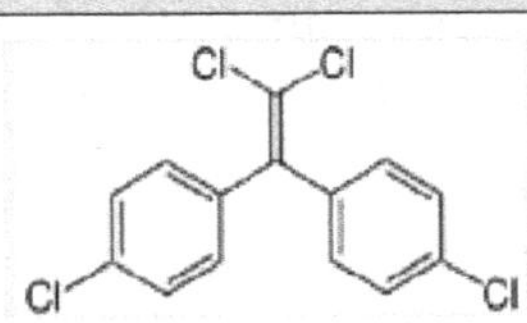	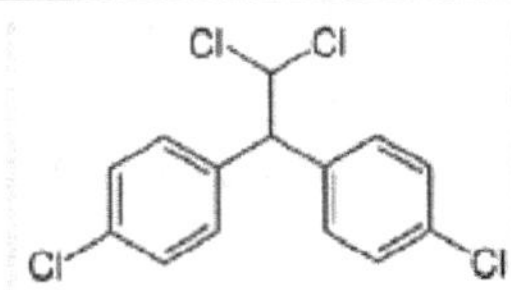
Molecular mass	354.49	318.03	320.05
Physical state	Solida	Cristalline solida	Solida
Colour	Colourless crystals, white powder	White	Colourless crystals, white powder

	p,p'-DDT	*p,p'*-DDE	*p,p'*-DDD
Melting point	109 °C	89 °C	109 – 110 °C
Boiling point	Decomposes	336 °C	350 °C
Solubility in water	0.025 mg/Lb	0.12 mg/Lb	0.09 mg/Lb
Solubility in organic solvents	Slightly soluble in ethanol, very soluble in ethyl ether and acetone	Lipids and most organic solvents	NA*

Not only birds are affected by this persistent pollutant, animals with higher amounts of fatty tissue have high levels of DDT and its metabolites (e.g. polar bears, orcas, belugas etc. (Crinnion, 2009; Galassi et al., 2008; Glynn et al., 2011; Okonkwo et al., 2008)). Although the acute DDT poisoning is rare, there are reports of negative effects on human health when exposed to low concentrations of DDT and its metabolites for longer periods. As shown in animals, the major damage was done on reproductive system where DDT can interfere with reproductive hormones. Also, there have been some reports regarding neurotoxicity, hepatotoxicity, imunotoxicity and genotoxicity (ATSDR, 2002). According to International Agency for Research on Cancer (IARC) DDT is classified as possible carcinogen to humans (IARC, 2009) and the increase in frequency of some cancers was detected (e.g. breast, testicular cancer etc.; Aubé et al., 2008; McGlynn et al., 2008).

As mentioned above, DDT introduced revolution in control of vector-borne diseases. Today, there are reports that some *Anopheles* species are resistant to DDT (Morgan et al., 2010) which implies that new methods of fighting vectors should be introduced. According to WHO one can observe the decrease in both number of countries using pesticides to control vectors (from 72 countries in 2006 to 61 countries in 2007) and in total amount of used DDT for vector control (from 5.1 million kg in 2006 to 3.7 million kg in 2007). Also WHO stimulates the import of new technologies to improve efficiency of insecticide-treated nets and long-lasting insecticidal nets (WHO, 2007). The importance of introducing new technologies for non harmful and efficient global pest fighting can be seen in estimation that only in United States, environmental and social costs of pesticide use are 9645 million dollars per year (Pimentel, 2005).

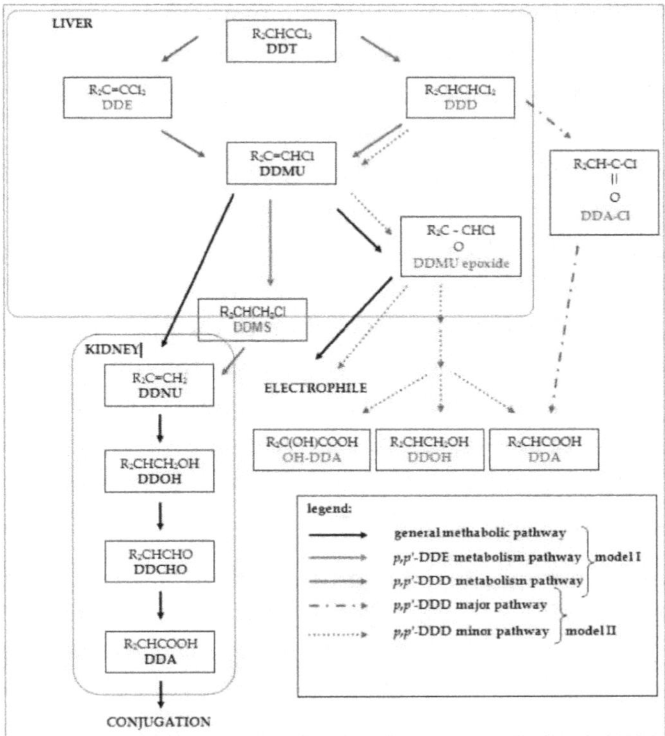

Fig. 1. Models of DDT metabolic pathways (adapted from Gold & Brunk, 1982 and Peterson & Robinson, 1964).

2. PRESENCE OF DDT IN CROATIA

2.1 Production and use in Croatia

Croatia, as a Mediterranean country, has been affected by malaria since the ancient times, especially in the coastal area. Only areas along the Velebit and Biokovo and the districts Hvar and Supetar have been spared from the disease. Most exposed to malaria have been the cities Nin, Benkovac, Obrovac, Skradin, Knin, Drniš, Šibenik, Imotski, Vrgorac and Metković area, the valley of the Neretva River, where malaria was called *Morbus naronianus*Neretva disease. In 1820 French pharmacists Pierre Joseph Pelletier & Jean Biename Caventou isolated the alkaloid quinine, while doctor Lujo Adam from the island Lošinj was probably the first in the world who injected quinine sulphate into the veins of people sick from malaria. In 1902 doctor Rudolf Battara conducted the first controlled study of drug quinine prophylaxis on the overall population in Nin. The success of the study encouraged implementation of this prophylaxis in other places in Dalmatia in the upcoming years (Dugački, 2005).

DDT was first used in Croatia in 1941, when spraying was carried out against the lice of people as part of educational campaigns in the School of Public Health. The first mass use of DDT was carried out on the front of Srijem in 1945 by spraying soldiers as prevention of typhoid fever with the preparation "neocid", which was donated by the International Red Cross (Bakić, 2011). Companies Chromos from Zagreb and Zorka from Šabac, with their preparations Pantakan and Pepein, were the first producers of DDT in Croatia and the former Yugoslavia. Since 1946 DDT was produced in the form of dust, and since 1949 in the form of concentrated emulsions (Table 3). Production of pesticides in Croatia was small, but large quantities of pesticides were imported. Data on quantities of imported pesticides does not exist, except for the year 1957, when active substances for plant protection products in quantities of 1435 tons were imported. The reason for the continuing production decline of DDT was the lack of foreign exchange quotas for the purchase of organic substances necessary for the production of pesticides. Production and use of DDT in Croatia was until 1972, when its use was banned in agriculture, while in forestry DDT was still used until 1984 (Table 4; National Implementation Plan [NIP], 2004).

Table 3. The planned and produced quantities of DDT, according to the available data for Yugoslavia in some years (adapted from Hamel, 2003).

Year and purpose	Type of pesticide	Amount (t)	Note
1958. production	DDT	2150	1600 t sprayer concentrate
1959. plan	DDT	3000	550 t emulsifiable concentrate
1959. consumption	DDT	3122	
	DDT + Lindane	687	
1963. production	DDT	2327	
	DDT + Lindane	1854	4.6-6.6 % DDT + 0.3-0.7 lindane content of active substance in preparation

Table 4. Consumption of DDT in kg per year in agriculture and forestry in Republic of Croatia (adapted from NIP, 2004).

Year	Consumption of DDT in forestry (kg/year)	Consumption of DDT in agriculture (kg/year)
1963.	2312	-
1964.	-	1784
1965.	53428	1196
1966.	280	14051
1967.	-	16323
1968.	600	4183

1969.	-	6051
1970.	-	5450
1971.	2363	4296
1972.	4912	1078
1973.	884	0
1974.	8437	0
1975.	6907	0
1976.	8437	0
1979. – 1987.	18658	0

Today, list of active substances permitted for use in the Republic of Croatia is synchronized with the official list of active substances permitted in the means of the European Union. The regulations require from pesticide manufacturers to provide data on all possible risks to human health and the environment, as well as data on the effectiveness of pesticides and information on possible contamination in order to obtain licenses for the production and use.

2.2 Methodology

Residues of DDT and its metabolites were analyzed in samples of surface water, soil and food. One of the most sensitive techniques for measuring the rest of DDT and its metabolites in samples from the environment is gas chromatography with electron capture detector (GC/ECD). DDT and its metabolites were determined by applying analytical methods or modifications of the method: International Organization for Standardisation [ISO] 6468 (2002), ISO 10382 (2002), Reference Methods for Marine Pollution (United Nations Environment Programme/International Atomic Energy Agency [UNEP/IAEA], 1982) and EN 1528 1-4 (1996), according to the scheme; extraction, purification of the extracts and quantitative analysis.

2.2.1 Sample extraction and cleanup

Water samples, two to four litres, were extracted with methylene chloride (HPLC grade for spectroscopy, Merck, Darmstadt, Germany) as solvent using an Ultra Turrax system. Extracts were dried over granular anhydrous sodium sulphate and concentrated using rotary evaporator (ISO 6468, 2002). The soil sample was sieved (<2 mm) and stored at room temperature for two days before the experiments. 10 g soil sample (dry matter) was extracted in an ultrasonic bath (ISO 10382, 2002) and the extract was dried by passing them through anhydrous sodium sulphate. Determination of DDT in food samples was based on method EN 1528 1-4 (1996). Approximately 10-20 g of food were homogenized and extracted with hexane, the extract was evaporated in a rotary evaporator and in nitrogen stream to ensure dryness. Contents of fat were determined weighing a dry sample. Milk fat portions were extracted from each individual sample of milk, cheese, cream and yoghurt according to the method by Sannino et al. (1996). Butter does not normally require extraction procedures. An automated gel permeation chromatographic (GPC) procedure was used to determine DDT residues in fatty foods. About three grams of the fat or less was dissolved into methylene chloride and cleaned up by GPC with a Biobeads SX3 column (OI Analytical, College Station, TX, USA) and a methylene chloride as eluant. About 10 g of fish tissue was weighed, homogenized with anhydrous sodium sulphate and extracted with pesticide grade hexane (Merck, UNEP/IAEA, 1982). The extract was condensed in a rotary flask vaporator to a specific aliquot (5 ml). The aliquot was then subjected to acid treatment by adding concentrated sulphuric acid (Merck). All samples were cleaned up with florisil, commercially available cartridges, 3-6 mL, 500-1000 mg (Kodba & Voncina, 2007). Extracts were concentrated using evaporator in nitrogen stream. An aliquot of each extract was transferred to vials for the quantitative analysis.

2.2.2 Gas chromatograph analysis

The samples were analyzed by gas chromatograph (GC) Shimadzu (Models GC 2010 and GC 17A series, Tokyo, Japan) equipped with autosampler and electron capture detector (ECD) on two fused silica capillary columns of different polarity. Nitrogen was used as the carrier gas with a flow rate of about 1.4 mL/min and as the makeup gas. The injection was set at splitless mode. The injection port and detector temperature were 250 and 300 0C, respectively. The compounds *p,p'*-DDE, *p,p'*-DDD, *o,p'*-DDT and *p,p'*-DDT were identified by comparing peak retention times between samples and known standards. The standard samples of the 18 pesticides were obtained from Dr. Ehrenstorfer GmbH (Augsburg, Germany, and AccuStandard Inc., New Haven, CT, USA) with the purities of 97-99%. Calibration standard curves were created and DDT residues were quantitatively determined by comparison of the retention times and peak areas of the sample chromatogram with those of standard solutions run under the same operating conditions. Peak confirmation was done by running the samples and the standard on another column and comparing. The concentrations of DDT residues in each sample were reported as ng/L or µg/kg. Limit of quantification for every matrix is presented in Table 5. The laboratory has participated (annually) in intercalibration study proficiency test.

Table 5. Limit of quantification for *p,p'*-DDE, *p,p'*-DDD, *o,p'*-DDT and *p,p'*-DDT.

Limit of quantification					
		p,p'-DDE	*p,p'*-DDD	*o,p'*-DDT	*p,p'*-DDT
Water	µg/L	0.0005	0.001	0.001	0.001
Soil (dry mater)	µg/g	0.002	0.003	0.004	0.003

Limit of quantification					
		p,p'-DDE	*p,p'*-DDD	*o,p'*-DDT	*p,p'*-DDT
Milk (liquid) and milk products (milk fat)	µg/g	0.0001	0.0004	0.0004	0.0004
Fish and fish products (wet weight)	µg/g	0.0005	0.001	0.0006	0.001
Food (fat)	µg/g	0.001	0.002	0.004	0.002

2.3 Results

During his lifetime a human being has been exposed to DDT and other pesticides through food, water and the environment. A level of these exposures can be established by analysing of samples from the environment. All food samples come from the market in Croatia while samples of surface water and soil were from different areas of Croatia. Food, particularly dairy products, meat and fish, has been identified as the primary immediate intake route of DDT and other organochlorine pesticides for the population (Johansen et al., 2004).

Samples of surface water of the rivers Sava, Krka, Mrežnica, Kupa, Zrmanja and Cetina were analyzed. The concentration of total DDT in surface waters in Croatia for the period 2009-2010 amounted from below detectable limit to 0.021 µg/L, with median value 0.0016 µg/L (Table 6). The average concentrations of total DDT in surface waters were 0.0026 µg/L. These results comply with those reported by Drevenkar et al. (1994), Drevenkar & Fingler (2000), and Fingler et al. (1992) in their papers. DDT may reach surface waters primarily by runoff, atmospheric transport and drift, or by direct application (e.g. to control mosquito- borne malaria). DDT is practically insoluble in water; but some DDT may be adsorbed onto the small amount of particulate matter present in water (ATSDR, 2002).

The concentration of DDT and metabolites in analyzed samples of soil were below 0.5 µg/g of dry matter (legally prescribed limit values). These analyses were conducted on the samples collected over the last two years and only from a few locations and therefore cannot be found sufficient to make any kind of general conclusions about the current situation. There is no systematic monitoring of DDT in soil in Croatia, resulting with little relevant data. Studies such as Picer et al. (2004) are rare examples of organized research in this area in Croatia, conducted as a part of post war damages in areas were there were concerns that soil was contaminated.

Data on levels and distribution of DDT and other persistent organic pollutant (POPs) in surface waters and soils in Croatia is insufficient, despite these hydrophobic substances being extremely important for assessing environmental contamination. DDT compound in sample of fish tissue were present in very low concentrations, although it is well known that DDT bioaccumulates in marine species. The mean value of DDT in fish tissue, analyzed in year 2007, was 3.8 μg/kg wet weights, with median value of 4.7 μg/kg . Similar results are reported by Krautchaker & Reiner (2001), and Bošnir et al. (2007). The five dairy products have been examined for the rest of DDT: milk, butter, cheese, cream and yoghurt. The mean values of the residual concentrations of total DDT in the examined dairy products were 29.6 μg/kg fat, respectively, with median 25.2 μg/kg (Table 6). The presence of DDT in milk and milk products has also been reported by Krautchaker & Reiner (2001) and Bošnir et al. (2010) as well as in other countries (Nevein et al., 2009; Dawood et al., 2004; Bulut et al., 2010). These studies have found that DDT complex were the most frequent contaminants in dairy products. Heck et al. (2006) concluded there is no difference in DDT in raw and pasteurized milk. The concentration of DDT in meat products sampled at food markets was 1.2-740.0 μg/kg fat, with median value 16.2 μg/kg fat (Table 6). These results should be taken with caution since the origin of meat is unknown; whether the meat is from domestic production or imported from other countries of the world. Meat is imported primarily from developing countries where there is limited or no control over the use and/or control of pesticide residues in foods. These results comply with those reported by Krautchaker & Reiner (2001), Covaci et al. (2004), and Tompić et al. (2011).

Table 6. Concentrations of DDT compounds in different samples.

	Value range	Mean	Median
Surface and ground water (ng/L)	0.6 – 20.5	2.6	1.6
Soil (μg/g; dry matter)	0 – 0.005	0.002	0.002
Fish and fish products (μg/kg; wet weight)	0.2 – 31.0	3.8	4.7
Meat and meat products (μg/kg; fat)	1.2 – 74.0	51.7	16.2
Milk and milk products (μg/kg; milk fat)	11.3 – 79.9	29.6	25.2

Lamb meat results (Table 7) demand special attention, since the concentrations of *p,p'*-DDE were generally high. Similar results were found by Tompić et al. (2011) in samples of lamb imported from Bulgaria. *p,p'*-DDE are found in every examined sample of lamb meat, which indicates the need for continuous monitoring of concentrations of this metabolites in samples of lamb meat.

Table 7. Concentrations of DDT compounds in samples of lamb meat.

	p,p'-DDE (μg/kg)	p,p'-DDD (ng/g)	p,p'-DDT (ng/g)	Total DDT (ng/g)	Total DDT (μg/kg)
Lamb meat	739.0	-	0.4	74.3	743.0
Lamb meat	173.0	-	0.5	17.8	178.0

	p,p'-DDE (μg/kg)	p,p'-DDD (ng/g)	p,p'-DDT (ng/g)	Total DDT (ng/g)	Total DDT (μg/kg)
Lamb meat	609.0	0.6	-	1.5	615.0
Lamb meat	964.0	6.1	1.2	103.7	1037.0
Lamb meat	354.0	1.4	0.5	37.3	373.0

Generally, it was observed that the total DDT or its metabolites residues were bellow acceptable and legally prescribed boundaries. These results highlight the need for regular analyzing of a larger number of samples from the environment to DDT residues and other chemicals of POPs, especially in imported food.

3. CYTOGENETIC METHODS FOR DETECTION OF PESTICIDE GENOTOXICITY

Pesticide exposure is ubiquitous, due not only to agricultural pesticide use and contamination of foods, but also to the extensive use of these products in and around residences. Because of their biological activity, the use of pesticides may cause undesired effects to human health. Pesticides tend to be very reactive compounds that can form covalent bonds with various nucleophilic centers of cellular biomolecules, including DNA (Crosby, 1982). For instance, the induction of DNA damage can potentially lead to adverse reproductive outcomes, the induction of cancer and many other chronic diseases (Ribas et al., 1996; Lander et al., 2000; Meinert et al., 2000; Ji et al., 2001). A great variety of tests and test systems based on microbes, plant and animals have been developed in order to asses the genotoxic effects of xenobiotic agents, including pesticides. Biomonitoring studies on human populations exposed to pesticides are employing circulating lymphocytes as biomarkers of exposure (and perhaps of effect). Those studies have essentially focused on cytogenetic end- points such as chromosomal aberrations (CA), sister-chromatid exchanges (SCE) and micronuclei (MN) frequency. Genetic damage at the chromosomal level entails an alternation in either chromosome number or chromosome structure, and such alternations can be measured as CA or MN frequency. The SCE analysis was also adopted as an indicator of genotoxicity, although the exact mechanism that leads to an increased exchange of segments between sister chromatids is not known in detail at present (Palani-Kumur & Panneerselvam, 2008). Recent studies revealed the nucleotide pool imbalance can have severe consequences on DNA metabolism and it is critical in SCE formation. The modulation of SCE by DNA precursors raises the possibility that DNA changes are responsible for the induction of SCE and mutations in mammalian cells (Popescu, 1999; Ashman & Davidson, 1981). While increased levels of CA have been associated with increased cancer risk (Hagmar et al, 1994, 1998), a similar conclusion has not been reached for SCE or MN. However, high levels of SCE and MN frequency have been observed in persons at higher cancer risk due to occupational or environmental exposure to a wide variety of carcinogens (Fučić et al, 2000; Vaglenov et al, 1999; Fenech et al, 1997). Evidence of CA increases, mainly as structural chromosomal aberrations in occupationally exposed populations. The sensitivity of SCE is lower than that of the CA test in detecting genotoxic effects related to pesticide exposure and fewer data are therefore, available for MN than for the other cytogenetic endopoints (Bolognesi, 2003). Exposure to potential mutagens or carcinogens can provide an early detection system for the initiation of cell disregulation. Biomarkers of effect are generally pre-clinical indicators of abnormalities and the most frequently used in genotoxicity assessment are comet assay and cytokinesis-block micronucleus test that are being proposed as a useful biomarkers for early effects. The cytogenetic endpoints can give indication of genetic damage; hence they are used as effective biomarkers of exposure *in vivo* and *in vitro*. In recent years, the comet (single-cell gel) assay has been established as a useful technique for studying DNA damage and repair (Tice, 1995). The comet assay combines the single-cell approach typical of cytogenetic assays with the simplicity of biochemical techniques for detecting DNA single strand breaks. The advantages of the comet assay include its simple and rapid performance, its sensitivity for detecting DNA damage, and the use of extremely small cell samples (Hartmann et al., 1998). The advantage of micronucleus assay is its simplicity and speed over the assay of chromosomal aberration. Both techniques have become an important tool for genotoxicity testing because of their simplicity of scoring and wide applicability in different cell types. These techniques became the methods of choice for studies of environmental and occupational exposure to air pollutants, metals, radiation, pesticides, and other xenobiotics.

3.1 Comet assay

The comet assay, also known as the single-cell gel electrophoresis assay (SCGE), is a method for detecting DNA strand breakage (single-strand DNA breaks, alkali-labile sites, double- strand DNA breaks, incomplete repair sites, and inter-strand cross-links) in virtually any nucleated cell (Collins et al., 2004, 2008; Shaposhnikov et al., 2008).

First quantification of DNA damage in individual cells was done by Rydberg & Johanson (1978). After gamma–irradiation they embedded cells in agarose on microscopic slides and lysed under mild alkali conditions. Upon neutralization, the cells were stained with acridin orange and the extent of DNA damage was measured by the ratio of green (indicating double–stranded DNA) to red (indicating single–strand DNA) fluorescence. To enhance the sensitivity for the DNA damage detection, Östling & Johanson (1984) proposed that strand breaks would enable DNA loops to stretch out upon electrophoresis, so the microelectrophoretic procedure under pH of 9.5 was developed. As reported by Singh et al. (1988) this pH of 9.5 is below the limit for DNA unwinding, and was notify to detect only double strand breaks (DSB), with more strongly alkaline conditions (pH 10 or above) needed for unwinding and detection of single strand breaks (SSB). It has been shown that neutral or mildly alkaline comet assay has the same limit of detection of DNA damage (SSB) as the alkaline comet assay, although the use of neutral pH does effect the comet image obtained (Collins, 2004). The comet tails are less pronounced at netural pH, and this can be an advantage when a less sensitive method is needed, for example when investigating cells that have large amount of background, or induced damage is high (Angelis et al., 1999).

Alkaline version of the comet assay was presented by Singh et al. (1988) in which DNA is allowed to unwind at pH>13. In their paper DNA damage was measured as the migrating distance of DNA from the nucleoid. In 1990, Olive et al. (1990) also under alkaline conditions developed the concept of the tail "moment", a combination of tail length and DNA content, as a measure of DNA damage. Also, in 1990, the name "Comet assay" was introduced and the application of the first image analysis program was described (Olive, 1989; Olive et al., 1990; Sviežená et al., 2004). Image analysis has become essential for objective measurement of low–dose effects, or for distinguishing small differences among sub–populations of cells. Strong alkaline conditions enabled clearer images, and besides SSBs other types of DNA damage could be detected, such as alkaline labile sites (Tice et al., 2000). Olive et al. (1990) revealed that employing milder alkaline (pH 12.3) conditions prevents conversion of alkaline labile sites into breaks. Under the pH>13 alkali labile sites are formed into SSBs, thus revealing otherwise hidden damage. Therefore, by modifying the pH of lysis and/or electrophoresis over the range of 9.5–13.5, one can apply a comet assay of different sensitivity, but of similar limits of detection (Collins et al., 1997; Angelis et al., 1999; Wong et al., 2005).

In its basic form, comet assay gives limited information on the type of DNA damage being measured. Single strand breaks detected by standard alkaline method are not the most interesting of lesions, because they are quickly repaired, and are not regarded as a significant lethal or mutagenic lesion. Many genotoxic agents do not induce strand breaks directly. They may create apurinic/pyrimidinic (AP) sites, which are alkali labile and are probably converted to breaks while DNA is in the electrophoresis solution at high pH. Furthermore, it is not possible to determine whether the high level of breaks in the comet assay is the indicator of high damage or efficient repair, due to temporary presence of breaks in the lesions repair via base excision or nucleotide excision (Collins et al., 1997).

More recently, the assay was modified further to enable the detection of specific kinds of DNA damage by combining the assay with the use of a purified DNA repair enzymes, which recognize the lesions along the DNA and convert them into the breaks expressed as an increase in comet DNA migration. Briefly, the DNA in the gel, following lysis, is digested with a lesion-specific repair endonuclease, which introduces breaks at sites of damage. In principle, any lesion for which a repair endonuclease exists can be detected in this way. To date, endonucleases most commonly used in the modified comet assay are the bacterial enzymes which recognize different types of oxidative damage. The first enzyme to be used was endonuclease III, a glycosylase which recognizes a variety of oxidized pyrimidines in DNA and removes them, leaving an AP site (Doetsch et al., 1987). Formamidopyrimidine DNA glycosylase (FPG) has the ability to convert altered purines, including 8-oxoguanine, into DNA breaks (Collins, 2007; Dušinská & Collins, 1996). When using these enzymes to measure oxidative DNA damage the usual practice is to incubate a slide with buffer alone in parallel with the enzyme slide. Slide with buffer would be a valid control slide, due to small increase in strand breaks on incubation without enzyme (Collins, 2009; Gajski et al., 2008). Recently, a mammalian analogue of FPG, 8-oxoguanidine DNA glycosylase, or OGG1, has been applied in the comet assay (Smith et al., 2006). OGG1 is the major base extension repair enzyme that initiates the repair of oxidative base lesion. It is a bifunctional DNA glycosylase possessing both DNA glycosylase and AP lyase activities (Boiteux & Radicella, 2000). In human it is named hOGG1. hOGG1 recognizes both 8-oxoguanine (8-oxodG) and 8- oxoadenine (8-oxodA) and removes these oxidized bases from double-stranded DNA, initiating the base lesion repair process (Smith et al., 2006). 8-hidroxy-2-deoxyguanosine (8- OHdG) lesion causes G→T and A→C transversions (Moriya, 1993) that have been reported as the sites of spontaneous oncogene expression and ultimately cancer manifestation (Valko et al., 2004; Bartsch, 1996; Shinmura & Yokota, 2001). Deletion of the hOGG1 gene was shown to be associated with accumulation of 8-OHdG lesion and increase in mutational risk (Hansen & Kelley, 2000; El-Zein et al., 2010).

For evaluation of DNA specific damage the comet assay has also been coupled with the method of fluorescent in situ hybridization (Comet-FISH). Since its initial development, Comet-FISH has been used to handle a number of quite different questiones. First, it was used to identify chromosome-specific areas on electrostretched DNA fibres and to determine their special distribution (Santos et al., 1997). Further applications were then to detect region-specific repair activities (Horvathova et al., 2004; McKenna et al., 2003; Mellon et al., 1986), genotoxic effects in total DNA and in telomeres (Arutyunyan et al., 2005), or in tumor relevant genes, like TP53 (Schaeferhenrich et al., 2003). Also specific chromosomal alternations (Harreus et al., 2004) were studied as were genetic instabilities (Tirukalikundram et al., 2005). Comet-FISH has also been used to discriminate between DNA double- and single-strand breaks (Fernandez et al., 2001). Whereas results from the Comet assay alone reflect only the level of overall DNA damage, the combination with the FISH-technique allows the assignment of the probed sequences to the damaged or undamaged part of the comet (tail or head, respectively). If two fluorescence signals are obtained with a probe for a particular gene in the head of a comet, this indicates that the gene is in an undamaged region of DNA, whereas the appearance of a spot or several spots in the tail of a comet indicates that a break or breaks has/have occurred in the proximity of the probed gene.

3.2 Micronucleus assay

Human exposure to environmental mutagens can be monitored using cytokinesis–block micronucleus (CBMN) assay (Natarajan et al., 1996) which is an efficient biomarker for diagnosing genetic damage and/or genome instability at the chromosome/molecular level in animal and/or human cells. It provides a comprehensive measure of chromosome breakage, chromosome loss, chromosome rearrangments, non-disjuction, gene amplification, necrosis and apoptosis (Fenech, 2000, 2006; Kirsch-Volders et al., 2000).

In the classical cytogenetic techniques, chromosomes are studied directly by observing and counting aberrations in methaphases (Natarajan & Obe, 1982). The complexity and laboriousness of enumerating aberrations in methaphase and the confounding effect of artefactual loss of chromosomes from methaphase preparations has stimulated the development of a simpler system of measuring chromosome damage. More than a century ago micronuclei were described in the cytoplasm of erythrocytes and were called "fragment of nuclear material" by Howell or "corpuscules intraglobulaires" in the terminology of Jolly in the late 1800s and early 1900s. To the hematologists these structures are known as "Howell-Jolly bodies". Similar structures were described in mouse and rat embryos and in *Vicia faba* (Thoday, 1951) and called "fragment nuclei" or "micronuclei". In the early 1970s the term micronucleus test was suggested for the first time by Boller & Schmidt (1970) and Heddle (1973) who showed that this assay provided a simple method to detect the genotoxic potential of mutagens after *in vivo* exposure of animals using dividing cell population such as bone marrow erythrocytes. A few years later it was shown by Countryman & Heddle (1976) that peripheral blood lymphocytes could also be used for the *in vivo* micronucleus approach and they recommended using micronuclei as a biomarker in testing schemes. As only dividing cells could express micronuclei, for the *in vitro* micronucleus studies it was necessary to establish cell proliferation and micronucleus induction at the same time. The decisive breakthrough of micronuclei as assay for *in vitro* genotoxicity testing came with work of Fenech & Morley (1986) that developed the CBMN assay. In the CBMN assay, cells that have completed nuclear division are blocked from performing cytokinesis using cytochalasin–B and are consequently readily identified by their binucleated appearance (Fenech & Morley, 1985, 1986). Whereas micronuclei originate from chromosome fragments or whole chromosomes that leg behind at anaphase during nuclear division, their occurrence is proven in binucleated cells. As a consequence, the CBMN assay has been shown to be more accurate and more sensitive than the conventional methods that do not distinguish between dividing and nondividing cells (Fenech & Morley, 1986; Fenech, 1991; Kirsch-Volders & Fenech, 2001).

Baseline or spontaneous micronucleus frequencies in culture human lymphocytes provide an indicator of accumulated genetic damage occurred during the lifespan of circulating lymphocytes. The half-life and mean lifespan of T-lymphocytes has been estimated to be three to four years, respectively (Natarajan & Obe, 1982; Buckton et al., 1967). The observed genetic instability may also reflect accumulated mutations in the stem cell lineage from which the mature lymphocytes originate. The type of mutations that could contribute to spontaneous micronuclei include: 1) mutations to kinetochore proteins, centromeres, and spindle apparatus that could lead to unequal chromosome distribution or whole chromosome loss at anaphase, and 2) unrepaired DNA strand breaks induced endogenously or as a result of environmental mutagens, which may result in acentric chromosome fragments. Studies using kinetochore antibodies to identify whole chromosomes suggest that approximately 50% of spontaneously occurring micronuclei are the consequence of whole chromosome loss and the rest are presumably derived from acentric chromosome fragments (Thompson & Perry, 1988; Fenech & Morley, 1989; Eastmond & Tucker, 1989). The spontaneous micronucleus frequency refers to the incidence of micronucleus observed in the absence of the environmental risk or exposure that is being assessed. The spontaneous micronucleus frequency of a population has to be established to determine acceptable normal values as well as providing baseline data for those situations when spontaneous micronucleus frequencies for individuals is not known before exposure. Micronuclei harbouring whole chromosomes are primarily formed from failure of the mitotic spindle, kinetochore, or other parts of the mitotic apparatus or by damage to chromosomal sub- structures, alterations in cellular physiology, and mechanical disruption. An increased number of micronucleated cells is a biomarker of genotoxic effects and can reflect exposure to agents with clastogenic modes of action (chromosome breaking; DNA as target) or aneugenic ones (aneuplodigenic; effect on chromosome number; mostly non-DNA target) (Albertini, 2000). The advantage of the CBMN assay is its ability to detect both clastogenic and aneugenic events, leading to structural and numerical chromosomal aberrations, respectively (Kirsch-Volders et al., 2002; Mateuca et al., 2006). Micronuclei observed in cultured lymphocytes are believed to arise primarily *in vitro* from: 1) chromatid-type chromosomal aberrations formed during DNA replication on a damaged template, 2) chromosome-type aberrations initiated before the mitosis and duplicated at replication, or 3) disturbances of mitotic apparatus leading to chromosome lagging. Micronuclei arising *in vivo*, inducible by both clastogenic and aneugenic mechanisms, can be scored in exfoliated epithelial cells (Salama et al., 1999) sampled, e.g., from buccal or nasal mucosa or urine, or in peripheral blood mononuclear cells e.g., isolated lymphocytes (Surrallés et al., 1996; Albertini, 2000). The CBMN assay is the preferred method for measuring micronuclei (MNi) in cultured human and/or mammalian cells because scoring is specifically restricted to once-divided cells. For scoring MNi uniform criteria should be used. Only MNi not exceeding 1/3 of the main nucleus diameter, clearly separable from the main nucleus and with distinct borders and of the same color as the nucleus, should be scored. In practice, 1000–2000 cells are often scored per subject in lymphocyte studies utilising the CBMN technique, while more cells (3000–5000 per subject) are evaluated in epithelial cells due to the lower baseline MNi frequency. In the CBMN method, only binucleate cells should be analysed for MNi; further divisions of a binucleate cell, usually resulting in cells with 3–4 nuclei, are highly irregular and show high MNi rates (Fenech et al., 2003).

The discovery of kinetochore-specific antibodies in the serum of scleroderma CREST (*Calcinosis, Raynaud's phenomenon, Esophageal dysmotility, Sclerodactyly and Telangiectasia*) patients (Moroy et al., 1980) has made it possible to determine the contents of micronuclei. Kinetochore immunofluorescence has been rapidly developed for the *in situ* detection of aneuploidy and chromosome breakage in human micronuclei (Degrassi & Tanzarella, 1988; Hennig et al., 1988; Eastmond & Tucker, 1989). Also, the FISH technique using chromosome- specific DNA probes has improved the detection and evaluation of structural chromosomal aberrations. The combination of the micronucleus assay with FISH using a DNA probe specific to the centromeric regions, or with antibodies that specifically stain kinetochore proteins, provides the methodology to distinguish between micronuclei containing either one or several whole chromosomes, which are positively labeled (centromere positive micronucleus), or acentric chromosome fragments, which are unlabeled due to the absence of centromere (centromere negative micronucleus) (Natarajan et al., 1996; Mateuca et al., 2006; Benameur et al., 2011). Except for cytogenetic damage measured by the number and distribution of micronuclei, according to the new criteria for micronuclei scoring, the CBMN assay also detects the nucleoplasmic bridges (NPBs), as well as nuclear buds (NBUDs). Current evidence suggests that NPBs derive from dicentric chromosomes which the centromeres have been pulled to the opposite poles of the cell during the anaphase stage, and are therefore indicative of the DNA mis-repair, chromosome rearrangement or telomere end-fusion. According to the new criteria applicable to the CBMN assay, NBUDs arise from the elimination of the amplified DNA and possibly from the elimination of the DNA-repair complexes, which therefore, may be considered a marker of gene amplification and altered gene dosage (Fenech, 2006; Fenech & Crott, 2002; Fenech et al., 2011; Garaj- Vrhovac et al., 2008; Lindberg et al., 2006; Thomas et al., 2003).

The significance of the CBMN assay lies in the fact that every cell in the system studied is scored cytologically for its viability status (necrosis, apoptosis), its mitotic status (mononucleated, metaphase, anaphase, binucleated, multinucleated) and its chromosomal instability or damage status (presence of MNi, NPBs, NBUDs and number of centromere probe signals amongst nuclei of binucleated cell if such molecular tools are used in combination with the assay). In this respect, the micronucleus assay has evolved into a comprehensive method employed in measuring chromosomal instability of the phenotype and altered cell viability and represents an effective tool to be used in research of cellular and nuclear dysfunctions caused by *in vitro* or *in vivo* exposure to toxic substances (Fenech, 2006; Garaj- Vrhovac et al., 2008; Thomas & Fenech, 2011).

4. GENOTOXICITY OF P,P'-DDT IN HUMAN PERIPHERAL BLOOD LYMPHOCYTES

4.1 Methodology

4.1.1 Chemicals

Chromosome kit P was from Euroclone, Milan, Italy; RPMI 1640, was from Invitrogen, Carlsbad, CA, USA; cytochalasin B, histopaque-1119, ethidium bromide, low melting point (LMP) and normal melting point (NMP) agaroses were from Sigma, St Louis, MO, USA; heparinised vacutainer tubes from Becton Dickinson, Franklin Lakes, NJ, USA; acridine orange form Heidelberg, Germany; Giemsa from Merk, Darmastad, Germany; All other reagents used were laboratory-grade chemicals from Kemika, Zagreb, Croatia. DDT was administered as *p,p'*-DDT (Sulpeco, Bellefonte, PA, USA) in final concentration of 0.025 mg/L at different time points.

4.1.2 Blood sampling and treatment

Whole blood samples were taken from a healthy female donor who had not been exposed to ionizing radiation, vaccinated or treated with drugs for a year before blood sampling. Whole venous blood was collected under sterile conditions in heparinised vacutainer tubes containing lithium heparin as anticoagulant. The comet assay and the micronucleus assay were conducted on whole blood cultivated at 37 °C in an atmosphere with 5% CO_2 (Heraeus Heracell 240 incubator, Langenselbold, Germany). The whole blood was treated with 0.025 mg/L *p,p'*-DDT for 1, 2, 4, 8, 24 and 48 h for the cytotoxicity assay and alkaline comet assay and 24 and 48 h for the CBMN assay. In each experiment, a non treated control was included.

4.1.3 Cell viability (cytotoxicity) assay

The indices of cell viability and necrosis were established by differential staining of human peripheral blood lymphocytes (HPBLs) with acridine orange and ethidium bromide using fluorescence microscopy (Duke & Cohen, 1992). Lymphocytes were isolated using a modified Ficoll-Histopaque centrifugation method (Singh, 2000). The slides were prepared using 200 µL of HPBLs and 2 µL of stain (acridine orange and ethidium bromide). The suspension mixed with dye was covered with a cover slip and analyzed under the epifluorescence microscope (Olympus AX 70, Tokyo, Japan) using a 60× objective and fluorescence filters of 515–560 nm. A total of 100 cells per repetition were examined. The nuclei of vital cells emitted a green fluorescence and necrotic cells emitted red fluorescence.

4.1.4 The alkaline comet (SCGE) assay

The alkaline comet assay was carried out as described by Singh et al. (1988). Briefly, after the exposure to *p,p'*-DDT, 5 µL of whole blood was mixed with 100 µL of 0.5% LMP agarose and added to fully frosted slides pre-coated with 0.6% NMP agarose. After solidifiying, the slides were covered with 0.5% LMP agarose, and the cells were lysed (2.5 M NaCl, 100 mM $EDTANa_2$, 10mM Tris, 1% sodium sarcosinate, 1% Triton X-100, 10% dimethyl sulfoxide, pH 10) overnight at 4°C. After the lysis the slides were placed into alkaline solution (300 mM NaOH, 1 mM $EDTANa_2$, pH 13) for 20 min at 4°C to allow DNA unwinding and subsequently electrophoresed for 20 min at 1 V/cm. Finally, the slides were neutralized in 0.4 M Tris buffer (pH 7.5) for 5 minutes 3 times, stained with EtBr (20 µg/mL) and analyzed at 250× magnification using an epifluorescence microscope (Zeiss, Göttingen, Germany) connected through camera to an image analysis system (Comet Assay II; Perceptive Instruments Ltd., Haverhill, Suffolk, UK). The tail length parameter was used to measure the level of DNA damage and a total of 100 randomly captured nuclei were examined from each slide. Tail length (i.e. the length of DNA migration) is related directly to the DNA fragment size and is presented in micrometers (µm). It was calculated from the centre of the nucleus.

4.1.5 Cytokinesis-blocked micronucleus (CBMN) assay

The micronucleus assay was performed in agreement with guidelines by Fenech & Morley, (1985). After the exposure to *p,p'*-DDT the whole blood (500 µL) was incubated in a Euroclone medium at 37 °C in an atmosphere of 5% CO_2 in air. Cytochalasin-B was added at a final concentration of 3 µg/mL 44 h after the culture was started. The cultures were harvested at 72 h. The lymphocytes were fixed in methanol-acetic acid solution (3:1), air- dried and stained with 5% Giemsa solution. All slides were randomised and coded prior to analysis. Binuclear lymphocytes were analyzed under a light microscope (Olympus CX41, Tokyo, Japan) at 400×

magnification. Micronuclei, nucleoplasmic bridges and nuclear buds were counted in 1000 binucleated cells and were scored according to the HUMN project criteria published by Fenech et al. (2003).

4.1.6 Statistics

For the results of the comet assay measured after treatment with *p,p'*-DDT statistical evaluation was performed using Statistica 5.0 package (StaSoft, Tulsa, OK, USA). Multiple comparisons between groups were done by means of ANOVA on log-transformed data. Post hoc analyses of differences were done by using the Scheffé test. Differences in the frequency of micronuclei, nucleoplasmic bridges, and nuclear buds were assessed using the chi-square test. $P<0.05$ was considered significant.

4.2 Results

4.2.1 Vital staining using ethidium bromide and acridine orange

The viability of HPBLs exposed to aqueous *p,p'*-DDT (0.025 mg/L) for different lengths of time as determined by acridine orange and ethidium bromide staining, using fluorescence microscopy was not significantly affected (data not shown). Changes were determined according to the different staining of the nucleus (Figure 2).

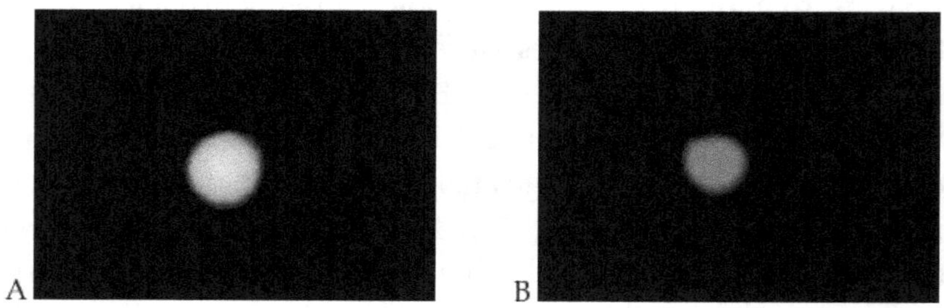

Fig. 2. Cell viability microphotographs represent viable lymphocyte from the unexposed sample (A; green), and dead lymphocyte (B; red) from sample treated with aqueous solution of *p,p'*-DDT.

4.2.2 Induction of DNA strand breaks

The whole blood was exposed to aqueous solution of *p,p'*-DDT (0.025 mg/L) and the DNA damage in HPBLs was determined with the alkaline comet assay. Figure 3 represents different levels of DNA fragmentation between non-exposed sample and sample exposed to aqueous solution of *p,p'*-DDT. Statistically significant ($P<0.05$) increase in the amount of DNA strand breaks was observed after all exposure times to *p,p'*-DDT (Figure 4).

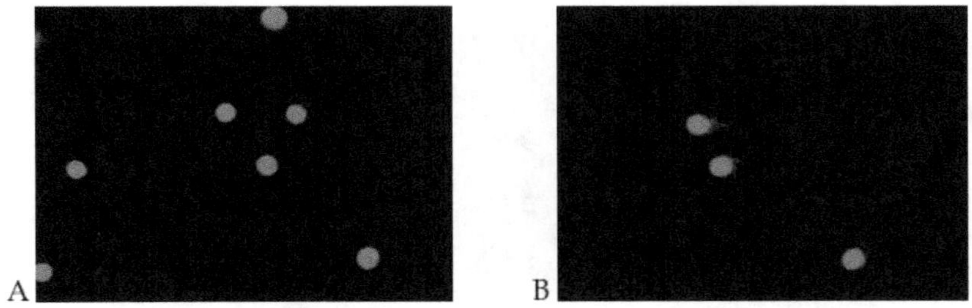

Fig. 3. Comet assay microphotographs represent undamaged lymphocytes from the un- exposed sample (A). Image (B) represents damaged lymphocytes that have comet appearance after the treatment with aqueous solution of *p,p'*-DDT.

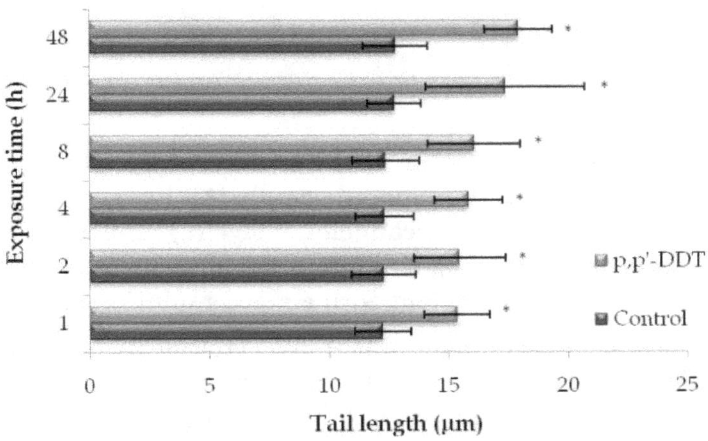

Fig. 4. Tail length (length of DNA migration) as comet assay parameter in human peripheral blood lymphocytes after exposure to low concentration of aqueous *p,p'*-DDT. * Statistically significant compared to corresponding control (P<0.05).

4.2.3 Induction of micronuclei, nucleoplasmic bridges and nuclear buds

The genotoxic activity of *p,p'*-DDT (0.025 mg/L) was further evaluated using the CBMN assay. Figure 5 represents binucleated lymphocytes from the non-exposed sample and samples exposed to aqueous solution of *p,p'*-DDT. Following *p,p'*-DDT treatment for 24 and 48 h, increase in the frequency of MNi was detected for both exposure times (Table 8). Additionally, statistically significant induction (P<0.05) of NBPs and NBUDs was also observed following *p,p'*-DDT treatment for 24 and 48 h (Table 9).

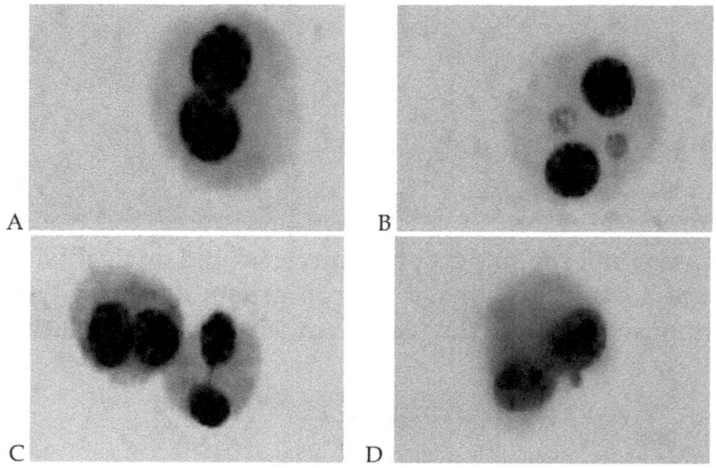

Fig. 5. Cytokinesis-block miconucleus assay microphotographs represent binucleated lymphocyte from the un-exposed sample (A). Image (B) represents binucleated lymphocyte with two micronuclei, image (C) binucleated lymphocyte with nucleoplasmic bridge (right), and image (D) binucleated lymphocyte with nucler bud after the treatment with aqueous solution of *p,p'*-DDT.

Table 8. Incidence of micronuclei (MNi) as cytokinesis-block micrunucleus assay parameter in human peripheral blood lymphocytes after exposure to low concentration of aqueous *p,p'*-DDT, * Statistically significant compared to corresponding control (*P*<0.05).

Exposure time (h)	Sample	1 MN	2 MNi	3 MNi	Total no. of MNi
24	Control	0.00±0.00	0.00±0.00	0.00±0.00	0.00±0.00
	p,p'-DDT	28.50±13.44*	1.50±2.12	0.50±0.71	33.00±19.80*
48	Control	6.00±1.41	0.00±0.00	0.00±0.00	6.00±1.41
	p,p'-DDT	30.50±9.19*	1.50±2.12	0.50±0.71	35.00±15.56*

Table 9. Incidence of nucleoplasmic bridges (NPBs) and nuclear buds (NBUDs) as cytokinesis-block micrunucleus assay parameters in human peripheral blood lymphocytes after exposure to low concentration of aqueous *p,p'*-DDT, * Statistically significant compared to corresponding control (*P*<0.05).

Exposure time (h)	Sample	Total no. of NPBs	Total no. of NBUDs
24	Control	0.00±0.00	0.00±0.00
	p,p'-DDT	10.50±0.71*	10.50±4.95*
48	Control	0.00±0.00	2.50±0.71
	p,p'-DDT	14.00±11.31*	20.50±3.54*

5. DISCUSSION

Due to uncontrolled use for several decades, many pesticides, among them DDT, probably the best known and the most useful insecticide in the world has damaged wild life and might have adverse effects on human health. Because of its current use in countries of the Third World, DDT still enters environment and in that way it can still represent health risk for human population, even in countries that have banned its use almost 40 years ago (ATSDR, 2002; Crinnion, 2009; Ecobichon, 2000, 2001; Eskenazi et al., 2009; Gajski et al., 2007; Turusov et al., 2002).

Large number of epidemiological studies regarding health risk of DDT confirmed that it represents major threat for wild life and human health (Aronson et al., 2010; Demers et al., 2000; Donato & Zani, 2010; Ecobichon, 1995; Martin et al., 2002; van Wendel de Joode et al., 2001; Wojtowicz et al., 2004, 2007; Woolcoot et al., 2001). Although tested in several genotoxicity studies on bacterial and animal models (Amer et al., 1996; Binneli et al., 2007, 2008a, 2008b; Canales-Aquirre, et al., 2011; Donnato et al., 1997a, 1997b; Galindo Reyes et al., 2002; Gauthier et al., 1999; Uppala et al., 2005) there is still necessity for conducting cytogenetic research regarding its genotoxicity employing sensitive methods to reveal the exact mechanisms of action of this chemical.

Combining different cytogenetic methods may play an important role in assessing genotoxic damage from different environmental chemical or physical agents. With these methods it is possible to evaluate the level of primary DNA damage or the dynamics of its repair even after short-term exposure to these agents (Garcia-Sagredo, 2008; Gajski & Garaj-Vrhovac, 2008). The comet assay is a sensitive method for measuring and analyzing DNA damage at the single cell level, and can be used both in *in vivo* and *in vitro* (Collins, 2004; Collins et al., 2008; Dušinská & Collins, 2008). The comet assay detects single and double stranded breaks at the level of DNA molecule, sites of incomplete repair, alkali labile sites, and DNA-DNA and DNA-protein cross-links (Piperakis, 2009). Furthermore, micronucleus assay can indicate cellular and nuclear dysfunction caused by *in vitro* or *in vivo* exposure to toxic substances. It is a reliable method for measuring chromosomal instability and altered cellular viability (Fenech, 2009; Fenech et al., 2003). It includes micronuclei, which are biomarkers of chromosome breakage and whole chromosome loss, nucleoplasmic bridges, which are biomarkers of DNA misrepair and telomere end-fusions, and nuclear buds, which are biomarkers of elimination of amplified DNA and DNA repair complexes (Fenech, 2007; Garaj-Vrhovac et al., 2008).

Considering the lack of data on the effect of DDT on the cellular genome, and taking into account its usage in some countries of the Third World and its environmental persistence, the aim of this study was to evaluate the genotoxic potential of a low concentration of aqueous *p,p'*-DDT upon *in vitro* exposure of HPBLs of different duration, by using alkaline comet assay and CBMN assay. Our results showed that exposure of HPBLs to aqueous *p,p'*- DDT increased DNA damage in time dependent fashion as measured by the comet assay. In addition, CBMN assay parameters revealed a wider scale of chromosomal alterations after *p,p'*-DDT treatment.

Cytogenetic studies of DDT are mainly based on *in vitro* research on the animal models and its genotoxicity was evaluated in a variety of test systems. Results obtained by studying cytogenetic effects of DDT on DNA of shrimp larvae (*Litopenaeus stylirostris*) indicated that DDT causes DNA adducts and/or breaks (Galindo Reyes et al., 2002). DDT and its metabolites DDE and DDD showed a clear genotoxic effect on haemocytes of zebra mussel (*Dreissena polymorpha*) specimens in different concentrations that have been found in several aquatic ecosystems worldwide, with a greater genotoxic potential of the DDE in respect to the other two chemicals (Binneli et al., 2008a, 2008b). DDT has also the ability to induce chromosomal aberrations in mouse spleen indicating its genotoxicity (Amer at al., 1996). In addition, DDT was genotoxic towards lymphocytes and mammary epithelial cells of female rats showing an increase in lipid peroxidation, the outcome of the growth level of free oxygen radicals, which lead to an oxidative stress (Canales-Aquirre et al., 2011). DDT also induces cellular and chromosomal alterations in the rat mammary gland, which is consistent with the hypothesis that it can induce early events in mammary carcinogenesis (Uppala et al., 2005). Additionally, beluga whales (*Delphinapterus leucas*) inhabiting the St. Lawrence estuary are highly contaminated with environmental pollutants including DDT which can induced significant increases of micronucleated cells in skin fibroblasts of an Arctic beluga whale (Gauthier et al., 1999).

Regarding human test system, the cytogenetic effect of DDT was investigated both *in vitro* and *in vivo*. *In vitro*, certain DDT concentrations have the effects on human leukocyte functions (Lee et al., 1979), are causing chromosomal aberrations (Lessa et al., 1976), DNA strand breaks (Yáñez et al., 2004), and apoptosis induction which is preceded by an increase in the levels of reactive oxygen species (Pérez-Maldonado et al., 2004, 2005). *In vivo*, DDT is able to induce chromatid lesions (Rabello et al., 1975), increase in chromosomal aberrations and sister chromatid exchanges (Rupa et al., 1989, 1991), DNA strand breaks (Yáñez et al., 2004; Pérez-Maldonado et al., 2006), apoptosis (Pérez-Maldonado et al., 2004) as well as cell cycle delay and decrease in mitotic index (Rupa et al., 1991).

Before 1973 when it was banned, DDT entered the air, water and soil during its production and use as an insecticide. DDT is present at many waste sites and from these sites it might continue to contaminate the environment. DDT still enters the environment because of its current use in other areas of the world. DDT may be released into atmosphere in countries where it is still manufactured and used; it can also enter the air by evaporation from contaminated water and soil and than it can be deposited on land or surface water. This cycle of evaporation and deposition may be repeated many times and as a result, DDT can be carried long distances in the atmosphere (ATSDR, 2002; Crinnion, 2009; Donato & Zani, 2010; Gajski et al., 2007; Torres-Sánchez & López-Carrillo 2007). These chemicals have been found in bogs, snow and animals even in the Artic and Antarctic regions, far from where they were ever used. DDT can last in the soil for a very long time, potentially for hundreds of years. Most DDT breaks down slowly into DDE and DDD, generally by the action of microorganisms and can be deposited in other places like in the surface layers of soil; it may get into rivers and lakes in runoff or get into groundwater. In surface waters, DDT will bind to particles in the water, settle and be deposited in the sediment. DDT is than taken up by small organisms and fish in the water. It accumulates to high levels in adipose tissue of fish and marine mammals, reaching levels many thousands of times higher than in water. DDT can also be absorbed by some plants and by the animals that can directly impact human population and like that represent a major health threat (ATSDR, 2002; Beard, 2006; Gajski et al., 2007; Gauthier et al., 1999).

All of these findings suggests that DDT is still present not only in poorly developed countries of the Third World but it can still be found in other countries that have banned its use almost 40 years ago due to its stability and long persistence in the environment.

6. CONCISE SUMMARY

Substantial concentrations of DDT and its metabolites persist in biological samples such as serum, adipose tissue, and maternal milk in non-occupationally exposed populations. Individuals are typically exposed to DDT by consumption, inhalation, or skin contact. Furthermore, there is evidence of health detriments, particularly with reproductive health, and more recently, cellular damage, along with alterations in the psychomotor development of infants exposed in utero. Despite existing studies on the detrimental impacts of pesticide exposure, there remains a significant necessity to elucidate the precise mechanisms and health ramifications associated with DDT exposure and its metabolites. Our studies, together with other existing data on pesticide genotoxicity, have demonstrated that DDT promotes DNA strand breaks in human peripheral blood cells both in vitro and in vivo. In our investigation, this impact was observed even following treatment with a very low quantity of aqueous DDT. These results also confirm prior findings that DDT generates abnormalities in cellular ultrastructure and causes DNA damage through single strand breaking and adduct formation in the DNA molecule. The current work demonstrates that the combination of sensitive approaches, such as the alkaline comet test and cytokinesis-block micronucleus assay, is effective for evaluating cellular and DNA modifications following exposure to environmental mutagens and carcinogens. The findings of this research underscore the necessity for enhanced environmental and dietary monitoring, as well as cytogenetic studies employing sensitive techniques to detect primary genomic damage following DDT exposure, in order to ascertain the effects of such chemicals on the human genome and health.

7. REFERENCES

Agency for Toxic Substances and Disease Service. (2002). *Toxicological profile for DDT, DDE and DDD*, U.S. Department of Health and Human Services, Public Health Service, ATSDR, Atlanta, USA

Albertini, R.J.; Anderson, D.; Douglas, G.R.; Hagmar, L.; Hemminki, K.; Merlo, F.; Natarajan, A.T.; Norppa, H.; Shuker, D.E.G.; Tice, R.; Waters, M.D. & Aitio, A. (2000). IPCS guidelines for the monitoring of genotoxic effects of carcinogens in humans. *Mutation Research*, Vol.463, No.2, pp. 111–172, ISSN 1383-5742

Amer, S.M.; Fahmy, M.A. & Donya, S.M. (1996). Cytogenetic effect of some insecticides in mouse spleen. *Journal of Applied Toxicology*, Vol.16, No.1, pp. 1-3, 1099-1263, ISSN 0260-437X

Angelis, K.J.; Dušinská, M. & Collins, A.R. (1999). Single cell gel electrophoresis: detection of DNA damage at different levels of sensitivity. *Electrophoresis,* Vol.20, No.10, pp. 2133-2138, ISSN 0173–0835

Aronson, K.J.; Wilson, J.W.; Hamel, M.; Diarsvitri, W.; Fan, W.; Woolcott, C.; Heaton, J.P.; Nickel, J.C.; Macneily, A. & Morales, A. (2010). Plasma organochlorine levels and prostate cancer risk. *Journal of Exposure Science and Environmental Epidemiology*, Vol.20, No.5, pp. 434-445, ISSN 1559-0631

Arutyunyan, R.; Rapp, A.; Greulich, K.O.; Hovhannisyan, G.; Haroutiunian, S. & Gebhart, E. (2005). Fragility of telomeres after bleomycin and cisplatin combined treatment measured in human leukocytes with the Comet–FISH technique. *Experimental Oncology,* Vol.27, No.1, pp. 38-42, ISSN 1812–9269

Ashman, C.R. & Davidson, R.L. (1981). Bromodeoxyuridine mutagenesis in mammalian cells is related to deoxyribonucleotide pool imbalance, *Molecular and cellular biology*, Vol.1, No.3, pp.254–260, ISSN 0270-7306

Aubé, M., Larochelle, C. & Ayotte, P. (2008). 1,1-dichloro-2,2-bis(p-chlorophenyl) ethylene (*p,p'*-DDE) disrupts the estrogen-androgen balance regulating the growth of hormone-dependent breast cancer cells. *Breast Cancer Research*, Vol.10, No.1, R16, ISSN 1465-5411

Bakić, J. (2011) Seven centuries of the prevention of spreading of the infectious diseases into Croatia - Overview of the 60th anniversary of establishment of modern pest control in Croatia. *Hrvatski Časopis za Javno Zdravstvo*, Vol.7, No.25, ISSN 1845-3082

Bartsch, H. (1996). DNA adducts in human carcinogenesis: Etiological relevance and structure-activity relationship. *Mutation Research,* Vol.340, No.2, pp. 67-79, ISSN 1383-5742

Beard, J, & Australian Rural Health Research Collaboration. (2006). DDT and human health. *Science of the Total Environment*, Vol.355, No.1-3, pp. 78-89, ISSN 0048-9697

Benameur, L.; Orsière, T.; Rose, J. & Botta, A. (2011). Detection of environmental clastogens and aneugens in human fibroblasts by cytokinesis-blocked micronucleus assay associated with imunofluorescent staining of CENP-A in micronuclei. *Chemosphere,* Vol.84, No.5, pp, 676-680, ISSN 0045-6535

Binelli, A.; Riva, C.; Cogni, D. & Provini, A. (2008a). Genotoxic effects of p,p'-DDT (1,1,1- trichloro-2,2-bis-(chlorophenyl)ethane) and its metabolites in Zebra mussel (D. polymorpha) by SCGE assay and micronucleus test. *Environmental and Molecular Mutagenesis*, Vol.49, No.5, pp. 406-415, ISSN 1098-2280

Binelli, A.; Riva, C.; Cogni, D. & Provini A. (2008b). Assessment of the genotoxic potential of benzo(a)pyrene and pp'-dichlorodiphenyldichloroethylene in Zebra mussel (Dreissena polymorpha). *Mutation Research*, Vol.649, No.1-2, pp. 135-145, ISSN 1383-5718

Binelli, A.; Riva, C. & Provini, A. (2007). Biomarkers in Zebra mussel for monitoring and quality assessment of Lake Maggiore (Italy). *Biomarkers*, Vol.12, No.4, pp. 349-368, ISSN 1354-750X

Boiteux, S. & Radicella, P. (2000). The human OGG1 gene: structure, functions, and its implication in the process of carcinogenesis. *Archives of Biochemistry and Biophysics,* Vol. 377, No. 1, pp. 1– 8, ISSN 0003–9861

Boller, K. & Schmid, W. (1970). Chemical mutagenesis in mammals. The Chinese hamster bone marrow as an in vivo test system. Hematological findings after treatment with trenimon. *Humangenetik,* Vol.11, No.1, pp. 35–54, ISSN 0018–7348

Bolognesi, C. (2003). Genotoxicity of pesticides: a review of human biomonitoring studies. *Mutation Research*, Vol.543, No.3, pp. 251–272, ISSN 1383-5742

Bošnir, J.; Puntarić, D.; Smit, Z.; Klarić, M.; Grgić, M. & Kosanović, L.M. (2007). Organochlorine pesticides in freshwater fish from the Zagreb area. *Arhiv za Higijenu Rada i Toksikologiju,* Vol.58, No.2, pp. 187-193, ISSN 0004-1254

Bošnir, J.; Novoselac, V.; Puntarić, D.; Klarić, I.; Miškulin, M. (2010). Organochlorine pesticide residues in cow's milk from Karlovac County, Croatia. *Acta Alimentaria.* Vol.39, No.3, pp. 317-326, ISSN 0139-3006

Brooks, G.T. (1986). Insecticide metabolism and selective toxicity. *Xenobiotica*, Vol.16, No.10- 11, pp. 989-1002, ISSN 0049-8254

Buckton, K.E.; Court Brown, W.M. & Smith, P.G. (1967). Lymphocyte survival in men treated with X-rays for ankylosing spondylitis. *Nature*, Vol.241, No.5087, pp. 470- 473, ISSN 0028-0836

Bulut, S.; Akkaya, L.; Gok, V.; & Konuk, M. (2010). Organochlorine pesticide residues in butter and kaymak in afyonkarahisar, Turkey. *Journal of Animal and Veterinary Advances*, Vol.9, No.22, pp. 2797-2801, ISSN 1680-5593

Canales-Aguirre, A.; Padilla-Camberos, E.; Gómez-Pinedo, U.; Salado-Ponce, H.; Feria- Velasco, A. & De Celis, R. (2011). Genotoxic effect of chronic exposure to DDT on lymphocytes, oral mucosa and breast cells of female rats. *International Journal of Environmental Research and Public Health*, Vol.8, No.2, pp. 540-553, ISSN 1661-7827

Carson, R. (2002). *Silent Spring*, Mariner Books, ISBN 0618249060, New York, USA

Charlwood, D. (2003). Did Herodotus describe the first airborne use of mosquito repellents. *Trends in Parasitology*, Vol. 19, No. 12, pp. 555-556, 1471-4922

Collins, A.R. (2004). The comet assay for DNA damage and repair: principles, applications, and limitations. *Molecular Biotechnology*, Vol.26, No.3, pp. 249-261, ISSN 1073–6085

Collins, A.R. (2009). Investigating oxidative DNA damage and its repair using the comet assay. *Mutation Research*, Vol.681, No.1, pp. 24-32, ISSN 1383-5742

Collins, A.R.; Dobson, V.L.; Dušinská, M.; Kennedy, G. & Stetina, R. (1997). The comet assay: what can it really tell us? *Mutation Research*, Vol.375, No.2, pp. 183-193, ISSN 0027- 5107

Collins, A.R.; Oscoz, A.A.; Brunborg, G.; Gaivão, I.; Giovannelli, L.; Kruszewski, M.; Smith, C.C. & Stetina, R. (2008). The comet assay: topical issues. *Mutagenesis*, Vol.23, No.3, pp. 143-151, ISSN 1464-3804

Countryman, P.I. & Heddle, J.A. (1976). The production of micronuclei from chromosome aberrations in irradiated cultures of human lymphocytes. *Mutation Research*, Vol.41, No.2-3, pp. 321–332, ISSN 0027–5107

Covaci, A.; Gheorghe, A. & Schepens, P. (2004). Distribution of organochlorine pesticides, polychlorinated biphenyls and a-HCH enantiomers in pork tissues. *Chemosphere*, Vol.56, No.8, pp. 757–766, ISSN 0045-6535

Crinnion, W.J. (2009). Chlorinated pesticides: threats to health and importance of detection. *Alternative Medicine Review*, Vol. 14, No. 4, pp. 347-359, ISSN 1089-5159.

Crosby, D.G. (1982). Pesticides as environmental mutagens, In: *Genetic Toxicology: An agricultural Perspective,* R.A. Fleck & A. Hollander, (Eds.), 201-218, Plenum Press, ISBN 9780306411359, New York, USA

Dawood, A.W.A.; Abd El-Maaboud, R. M.; Helal, M.A.; Mohamed, S.A. & Waleed, H.A. (2004). Detection of organochlorine pesticide residues in samples of cow milk collected from Sohag and Qena overnorates. *Assiut University Bulletin for Environmental Researches*, Vol.7, No.2, pp. 105-116, ISSN 1110-6107

Degrassi, F. & Tanzarella, C. (1988). Immunofluorescent staining of kinetochores in micronuclei: a new assay for the detection of aneuploidy. *Mutation Research*, Vol.203, No.5, pp. 339–345, ISSN 0165-7992

Demers, A.; Ayotte, P.; Brisson, J.; Dodin, S.; Robert, J. & Dewailly, E. (2000). Risk and aggressiveness of breast cancer in relation to plasma organochlorine concentrations. *Cancer Epidemiology, Biomarkers and Prevention*, Vol.9, No.2, pp. 161- 166, ISSN 1055-9965

Doetsch, P.W.; Henner, W.D.; Cunningham, R.P.; Toney, J.H. & Helland, D.E. (1987). A highly conserved endonuclease activity present in *Escherichia coli*, bovine, and human cells recognizes oxidative DNA damage at sites of pyrimidines. *Molecular and Cellular Biology*, Vol.7, No.1, pp. 26–32, ISSN 0270-7306

Donato, M.M.; Jurado, A.S.; Antunes-Madeira, M.C. & Madeira, V.M. (1997a). Bacillus stearothermophilus as a model to evaluate membrane toxicity of a lipophilic environmental pollutant (DDT). *Archives of Environmental Contamination and Toxicology*, Vol.33, No.2, pp. 109-116, ISSN 0090-4341

Donato, M.M.; Jurado, A.S.; Antunes-Madeira, M.C. & Madeira, V.M. (1997b) Effects of a lipophilic environmental pollutant (DDT) on the phospholipid and fatty acid contents of Bacillus stearothermophilus. *Archives of Environmental Contamination and Toxicology*, Vol.33, No.4, pp. 341-349, ISSN 0090-4341

Donato, F. & Zani, C. (2010). Chronic exposure to organochlorine compounds and health effects in adults: diabetes and thyroid diseases. *Annali di Igiene: Medicina Preventiva e di Comunita*, Vol.22, No.3, pp. 185-198, ISSN 1120-9135

Drevenkar, V. & Fingler, S. (2000). Persistent organochlorine compounds in Croatia. *Arhiv za Higijenu Rada i Toksikologiju*, Vol.51, No.Suppl, pp. 59–73, ISSN 0004-1254

Drevenkar, V.; Fingler, S. & Fröbe, Z. (1994). Some organochlorine pollutants in the water environment and their influence on drinking water quality, In: *Chemical Safety, International Reference Manual,* Richardson, M.L. (Ed.), 297-310, Wiley-VCH, ISBN 9783527286300, Weinheim, Germany

Dugački, V. (2005). Dr. Rudolf Battara operation in Nin in 1902, the first systematic battle attempt against malaria in Croatia. *Medica Jadertina*, Vol.35, No.Suppl, pp. 33-40, ISSN 0351-0093

Duke, R.C. & Cohen, J.J. (1992). Morphological and biochemical assays of apoptosis, In: *Current Protocols in Immunology*, J.E. Coligan & A.M. Kruisbeal, (Eds.), 1-3, John Willey & Sons, ISBN 9780471522768, New York, USA

Dušinská, M. & Collins, A. (1996). Detection of oxidised purines and UV-induced photoproducts in DNA of single cells, by inclusion of lesion-specific enzymes in the comet assay. *Alternatives to Laboratory Animals*, Vol.24, No.3, pp. 405–411, ISSN 0261-1929

Dušinská, M. & Collins, A.R. (2008). The comet assay in human biomonitoring: gene- environment interactions. *Mutagenesis*, Vol.23, No.3, pp. 191-205, ISSN 1464-3804

Eastmond, D. A. & Tucker, J. D. (1989). Identification of aneuploidyinducing agents using cytokinesis-blocked human lymphocytes and an anti-kinetochore antibody. *Environmental and Molecular Mutagenesis*, Vol.13, No.1, pp. 34-43, ISSN 0893-6692

Ecobichon, D.J. (1995). Toxic effects of pesticides, In: Casarett and Doull's toxicology: the Basic Science of Poisons, C.D. Klaassen, M.O. Amdur & J. Doull, (Eds.), 643-689, Macmillan Co., ISBN 0071054766, New York, USA

Ecobichon, D.J. (2000). Our changing perspectives on benefits and risks of pesticides: a historical overview. *Neurotoxicology*, Vol.21, No.1-2, pp. 211-218, ISSN 0161-813X

Ecobichon, D.J. (2001). Pesticide use in developing countries. *Toxicology*, Vol.160, No.1-3, pp. 27-33, ISSN 0300-483X

El-Zein, R.A.; Monroy, C.M.; Cortes, A.; Spitz, M.R.; Greisinger, A. & Etzel, C.J. (2010). Rapid method for determination of DNA repair capacity in human peripheral blood lymphocytes amongst smokers. *Bio Med Central Cancer*, Vol.10, No.1, pp. 439, ISSN 1471-2407

Eskenazi, B.; Chevrier, J.; Rosas, L.G.; Anderson, H.A.; Bornman, M.S.; Bouwman, H.; Chen, A.; Cohn, B.A.; de Jager, C.; Henshel, D.S.; Leipzig, F.; Leipzig, J.S.; Lorenz, E.C.; Snedeker, S.M. & Stapleton, D. (2009). The Pine River statement: human health consequences of DDT use. *Environmental Health Perspectives*, Vol.117, No.9, pp. 1359-1367, ISSN 0091-6765

Fenech, M. (1991). Optimisation of micronucleus assays for biological dosimetry. *Progress in Clinical and Biological Research*, Vol.372, pp. 373-386, ISSN 0361-7742

Fenech, M. (2000) The in vitro micronucleus technique. *Mutation Research*, Vol.455, No.1-2, pp. 81-95, ISSN 0027-5107

Fenech, M. (2006). Cytokinesis-block micronucleus assay evolves into a "cytome" assay of chromosomal instability, mitotic dysfunction and cell death. *Mutation Research*, Vol.600, No.1-2, pp. 58-66, ISSN 0027-5107

Fenech, M. (2007). Cytokinesis-block micronucleus cytome assay. *Nature Protocols*, Vol.2, No.5, pp. 1084-1104, ISSN 1750-2779

Fenech, M. (2009). A lifetime passion for micronucleus cytome assays - reflections from Down Under. *Mutation Research*, Vol.681, No.2-3, pp. 111-117, ISSN 1383-5742

Fenech, M.; Perepetskaya, G. & Mikhalevic, L. (1997). A more comprehensive application of the micronucleus technique for biomonitoring of genetic damage rates in human populations. Experiences from the Chernobyl catastrophe, *Environmental and Molecular Mutagenesis,* Vol.30, No.2, pp.112–118, ISSN 0893-6692

Fenech. M.; Chang, W.P.; Kirsch-Volders, M.; Holland, N.; Bonassi, S.; Zeiger, E. & HUman MicronNucleus project. (2003). HUMN project: detailed description of the scoring criteria for the cytokinesis-block micronucleus assay using isolated human lymphocyte cultures. *Mutation Research*, Vol.534, No.1-2, pp. 65-75, ISSN 1383-5718

Fenech, M. & Crott, J.W. (2002). Micronuclei, nucleoplasmic bridges and nuclear buds induced in folic acid deficient human lymphocytes-evidence for breakage-fusion- bridge cycles in the cytokinesis-block micronucleus assay. *Mutation Research,* Vol.504, No.1-2, pp. 131-136, ISSN 0027-5107

Fenech, M.; Kirsch-Volders, M.; Natarajan, A.T.; Surralles, J.; Crott, J.W.; Parry, J.; Norppa, H.; Eastmond, D.A.; Tucker, J.D. & Thomas, P. (2011). Molecular mechanisms of micronucleus, nucleoplasmic bridge and nuclear bud formation in mammalian and human cells. *Mutagenesis*, Vol.26, No.1, pp. 125-132, ISSN 0267-8357

Fenech, M. & Morley, A. A. (1985). Measurement of micronuclei in lymphocytes. *Mutation Research*, Vol.147, No.1-2, pp. 29-36, ISSN 0027-5107

Fenech, M. & Morley, A.A. (1986). Cytokinesis-block micronucleus method in human lymphocytes: effect of in vivo ageing and low dose X- irradiation. *Mutation Research*, Vol.161, No.2, pp. 193–198, ISSN 0027–5107

Fenech, M. & Morley, A.A. (1989). Kinetochore detection in micronuclei: an alternative method for measuring chromosome loss. *Mutagenesis*, Vol.4, No.2, pp. 98-104, ISSN 0267-8357

Fernández, J.L.; Vázquez-Gundín, F.; Rivero, M.T.; Genescá, A.; Gosálvez, J. & Goyanes, V. (2001). DBD-fish on neutral comets: simultaneous analysis of DNA single- and double-strand breaks in individual cells. *Experimental Cell Research*, Vol.270, No.1, pp. 102-109, ISSN 0014-4827

Fingler, S.; Drevenkar, V.; Tkalčević, B. & Šmit, Z. (1992). Levels of polychlorinated biphenyls, organochlorine pesticides, and chlorophenols in the Kupa river water and in drinking waters from different areas in Croatia, *Bulletin of Environmental Contamination and Toxicology*, Vol. 49, No. 6, pp. 805-812 ISSN 1432-0800

Food and Agriculture Organization of United Nations. (2005). *International Code of Conduct on the Distribution and Use of Pesticides*, FAO, ISBN 9251054118, Rome, Italy

Fučić, A.; Markučič, D.; Mijić, A. & Jazbec, A.M. (2000). Estimation of genome damage after exposure to ionising radiation and ultrasound used in industry. *Environmental and molecular mutagenesis*, Vol.36, No.1, pp. 47–51, ISSN 0893-6692

Gajski, G. & Garaj-Vrhovac, V. Application of cytogenetic endpoints and comet assay on human lymphocytes treated with atorvastatin in vitro. *Journal of Environmental Science and Health Part A Toxic/Hazardous Substances & Environmental Engineering*, Vo.43, No.1, pp. 78-85, ISSN 1093-4529

Gajski, G.; Garaj-Vrhovac, V. & Oreščanin V. (2008). Cytogenetic status and oxidative DNA- damage induced by atorvastatin in human peripheral blood lymphocytes: standard and Fpg-modified comet assay. *Toxicology and Applied Pharmacology*, Vol.231, No.1, pp. 85-93, ISSN 0041-008X

Gajski, G.; Ravlić, S.; Capuder, Ž. & Garaj-Vrhovac, V. (2007). Use of sensitive methods for detection of DNA damage on human lymphocytes exposed to p,p'-DDT: Comet assay and new criteria for scoring micronucleus test. *Journal of Environmental Science and Health Part B*, Vol.42, No.6, pp. 607-613, ISSN 0360-1234

Galassi, S.; Bettinetti, R.; Neri, M.C.; Falandysz, J.; Kotecka, W.; King, I.; Lo, S.; Klingmueller, D. & Schulte-Oehlmann, U. (2008). *p,p'*-DDE contamination of the blood and diet in central European populations. *Science of the Total Environment*, Vol. 390, No.1, pp. 45-52, ISSN 0048-9697

Galindo Reyes, J.G.; Leyva, N.R.; Millan, O.A. & Lazcano, G.A. (2002). Effects of pesticides on DNA and protein of shrimp larvae Litopenaeus stylirostris of the California Gulf. *Ecotoxicology and Environmental Safety*, Vol.53, No.2, pp. 191-195, ISSN 0147- 6513

Garaj-Vrhovac, V.; Gajski, G. & Ravlić, S. (2008). Efficacy of HUMN criteria for scoring the micronucleus assay in human lymphocytes exposed to a low concentration of p,p'- DDT. *Brazilian Journal of Medical and Biological Research*, Vol.41, No.6, pp. 473-476, ISSN 1414-431X

Garcia-Sagredo, J.M. (2008). Fifty years of cytogenetics: a parallel view of the evolution of cytogenetics and genotoxicology. *Biochimica et Biophysica Acta*, Vol.1779, No.6-7, pp. 363-375, ISSN 0006-3002

Gauthier, J.M.; Dubeau, H. & Rassart, E. (1999). Induction of micronuclei in vitro by organochlorine compounds in beluga whale skin fibroblasts. *Mutation Research*, Vol.439, No.1, pp. 87-95, ISSN 1383-5718

Glynn, A.; Lignell, S.; Darnerud, P.O.; Aune, M.; Ankarberg, E.H.; Bergdahl, I.A.; Barregård, L. & Bensryd, I. (2011). Regional differences in levels of chlorinated an brominated pollutants in mother's milk from primiparous women in Sweden. *Environment International, Vol.* 37, No. 1, pp. 71-79, ISSN 0160-4120

Gold, B. & Brunk, G. (1982). Metabolism of 1,1,1-trichloro-2,2-bis(p-chlorophenyl)-ethane and 1,1-dichloro-2,2-bis(p-chlorophenyl)ethane in the mouse. *Chemico-Biological Interactions*, Vol.41, No.3, pp. 327-339, ISSN 0009-2797

Gold, B. & Brunk, G. (1983). Metabolism of 1,1,1-trichloro-2,2-bis(p-chlorophenyl) ethane (DDT), 1,1-dichloro-2,2-bis(p-chlorophenyl)ethane, and 1-chloro-2,2-bis(p- chlorophenyl)ethene in the hamster. *Cancer Research*, Vol.43, No.6, pp. 2644-2647, ISSN 0008-5472

Hagmar, L.; Brogger, A.; Hansteen, I.L.; Heim, S.; Hogstedt, B.; Knudsen, L.; Lambert, B.; Linnainmaa, K.; Mitelman, F.; Nordenson, I.; Reuterwall, C.; Salomaa, S.I.; Skerfving, S. & Sorsa, M. (1994). Cancer risk in human predicted by increased levels of chromosomal aberrations in lymphocytes: Nordic Study Group on the Health Risk of Chromosome Damage. *Cancer research*, Vol.54, No.11, pp. 2919–2922, ISSN 0008-5472

Hagmar, L.; Bonassi, S.; Stromberg, U.; Brogger, A.; Knudsen, L.; Norppa, H. & Reuterwall, C. (1998). The European Study Group on Cytogenetic Biomarkers and Health, Chromosomal aberrations in lymphocytes predict human cancer. A report from the European Study Group on Cytogenetic Biomarkers and Health (ESCH), *Cancer Research*, Vol.58, No.18, pp. 4117–4121, ISSN 0008-5472

Hamel, D. (2003). Inventarizacija perzistentnih organskih onečišćavala – pesticidi, Available from http://www.cro-cpc.hr/projekti/pops/Pest_Izvjestaj.pdf

Hansen, W.K. & Kelley, M.R. (2000). Review of mammalian DNA repair and translational implications. *The Journal of Pharmacology and Experimental Therapeutics*, Vol.295, No.1, pp. 1-9, ISSN 0022-3565

Harreus, U.A.; Kleinsasser, N.H.; Zieger, S.; Wallner, B.; Reiter, M.; Schuller, P. & Berghaus, A. (2004). Sensitivity to DNA-damage induction and chromosomal alterations in mucosa cells from patients with and without cancer of the oropharynx detected by a combination of Comet assay and fluorescence in situ hybridization. *Mutation Research*, Vol.563, No.2, pp. 131-138, ISSN 1383-5718

Hartmann, A.; Fender, H. & Speit, G. (1998). Comparative biomonitoring study of workers at a waste disposal site using cytogenetic tests and the Comet (Single-Cell Gel) assay. *Environmental and Molecular Mutagenesis,* Vol.32, No.1, pp. 17-24, ISSN 0893- 6692

Heck, M.C.; Sifuentes dos Santos, J.; Bogusz Junior, S.; Costabeber, I & Emanuelli, T. (2006). Estimation of children exposure to organochlorine compounds through milk in Rio Grande do Sul, Brazil. *Food Chemistry,* Vol.102, No.1, pp. 288-294, ISSN 0308-8146

Heddle, J.A. (1973). A rapid in vivo test for chromosome damage. *Mutation Research,* Vol.18, No.2, pp. 187–190, ISSN 0027-5107

Hennig, U.G.G.; Rudd, N.L. & Hoar, D.I. (1988). Kinetochore immunofluorescence in micronuclei: a rapid method for the in situ detection of aneuploidy and chromosome breakage in human fibroblasts. *Mutation Research,* Vol.203, No.6, pp. 405-414, ISSN 0165-7992

Horváthová, E.; Dusinská, M.; Shaposhnikov, S. & Collins, A.R. (2004). DNA damage and repair measured in different genomic regions using the comet assay with fluorescent in situ hybridization. *Mutagenesis,* Vol.19, No.4, pp. 269-276, ISSN 0267– 8357

International Agency for Research on Cancer. (2011). *Agents reviewed by the IARC monographs,* Volumes 1-101, IARC, Lyon, France

International standard HRN EN ISO 6468. (2002). *Water quality-Determination of certain organochlorine insecticides, policlorinated biphenyls and chlorbenzenes-Gas chromatographic method after liqid-liquid extraction*

International standard ISO 10382. (2002). *Soil quality: Determination of organochlorine pesticides and polychlorinated biphenyls; Gaschromatographic method with electron capture detection*

Ji, B.T.; Silverman, D.T. & Stewart, P.A.; Blair, A.; Swanson, G.M.; Greenberg, R.S.; Hayes, R.B.; Brown, L.M.; Lillemoe, K.D.; Schoenberg, J.B.; Pottern, L.M.; Schwartz, A.G. & Hoover, R.N. (2001). Occupational exposure to pesticides and pancreatic cancer. *American Journal of Industrial Medicine,* Vol.39, No.1, pp. 92–99, ISSN 0271-3586

Johansen, P.; Muir, D.; Asmund, G. & Riget, F. (2004). Human exposure to contaminants in the traditional Greenland diet. *Science of the Total Environment,* Vol.331, No. 1-3, pp. 189–206, ISSN 0048-9697

Kamrin, M.A. (2000). *Pesticide Profiles, Toxicity, Environmental Impact, and Fate,* Lewis Publisher, ISBN 0849321794, New York, USA

Kirsch-Volders, M. & Fenech, M. (2001). Inclusion of micronuclei in non-divided mononuclear lymphocytes and necrosis/apoptosis may provide a more comprehensive cytokinesis block micronucleus assay for biomonitoring purposes. *Mutagenesis,* Vol.16, No.1, pp. 51-58, ISSN 0267-8357

Kirsch-Volders, M.; Sofuni, T.; Aardema, M.; Albertini, S.; Eastmond, D.; Fenech, M.; Ishidate, M.Jr.; Lorge, E.; Norppa, H.; Surrallés, J.; von der Hude, W. & Wakata, A. (2000). Report from the In Vitro Micronucleus Assay Working Group.

Environmental and Molecular Mutagenesis, Vol.35, No.3, pp. 167-172, ISSN 0893-6692

Kirsch-Volders, M.; Vanhauwaert, A.; De Boeck, M. & Decordier, I. (2002). Importance of detecting numerical versus structural chromosome aberrations. *Mutation Research*, Vol.504, No.1-2, pp. 137–148, ISSN 0027-5107

Kodba, Z.C. & Voncina, D.B. (2007). A rapid method for the determination of organochlorine, pyrethroid pesticides and polychlorobiphenyls in fatty foods using GC with electron capture detection. *Chromatographia*, Vol.66, No.7-8, pp. 619–624, ISSN 0009-5893

Krauthacker, B. & Reiner, E. (2001). Organochlorine compounds in human milk and food of animal origin in samples from Croatia. *Arhiv za Higijenu Rada i Toksikologiju*, Vol.52, Vol.2, pp. 217–227, ISSN 0004-154

Lander, B.F.; Knudsen, L.E.; Gamborg, M.O.; Jarventaus, H. & Norppa, H. (2000). Chromosome aberrations in pesticide-exposed greenhouse workers. *Scandinavian Journal of Work, Environment & Health*, Vol. 26, No.5, pp. 436–442, ISSN 0355-3140

Lee, T.P.; Moscati, R. & Park, B.H. (1979). Effects of pesticides on human leukocyte functions. *Research Communication in Chemical and Pathological Pharmacology*, Vol.23, No.3, pp. 597-609, ISSN 0034-5164.

Lessa, J.M.; Beçak, W.; Nazareth Rabello, M.; Pereira, C.A. & Ungaro, M.T. (1976). Cytogenetic study of DDT on human lymphocytes in vitro. *Mutation Research*, Vol.40, No.2, pp. 131-138, ISSN 0027-5107

Lindberg, H.K.; Wang, X.; Järventaus, H.; Falck, G.C.; Norppa, H. & Fenech, M. (2006). Origin of nuclear buds and micronuclei in normal and folate-deprived human lymphocytes. *Mutation Research*, Vol.617, No.1-2, pp. 33-45, ISSN 0027-5107

Martin, S.A.Jr.; Harlow, S.D.; Sowers, M.F.; Longnecker, M.P.; Garabrant, D.; Shore, D.L. & Sandler, D.P. (2002). DDT metabolite and androgens in African-American farmers. *Epidemiology*, Vol.13, No.4, pp. 454-458 ISSN 1044-3983

Mateuca, R.; Lombaert, N.; Aka, P.V.; Decordier, I. & Kirsch-Volders, M. (2006). Chromosomal changes: induction, detection methods and applicability in human biomonitoring. *Biochimie*, Vol.88, No.11, pp. 1515–1531, ISSN 0300-9084

McGlynn, K.A.; Quraishi, S.M.; Graubard, B.I.; Weber, J.P.; Rubertone, M.V. & Erickson, R.L. (2008). Persistent organochlorine pesticides and risk of testicular germ cell tumors. *Journal of the National Cancer Institute*, Vol. 100, No. 9, pp. 663-671, ISSN 0027-8874.

McKenna, D.J.; Gallus, M.; McKeown, S.R.; Downes, C.S. & McKelvey- Martin, V.J. (2003). Modification of the alkaline Comet assay to allow simultaneous evaluation of mitomycin C-induced DNA cross-link damage and repair of specific DNA sequences in RT4 cells. *DNA Repair*, Vol. 2, No.8, pp. 879–890, ISSN 1568–7864

Meinert, R.; Schuz, J.; Kaatsch, P. & Michaelis, J. (2000). Leukemia and non-Hodgkin's lymphoma in childhood and exposure to pesticides: results of a register-based cite- control study in Germany. *American Journal of Epidemiology*, Vol.151, No.7, pp. 639– 646, ISSN 0002-9262

Mellon, I.; Bohr, V.A.; Smith, C.A. & Hanawalt, P.C. (1986). Preferential DNA repair of an active gene in human cells, *Proceedings of the National Academy of Sciences of the United States of America,* Vol. 83, No.23, pp. 8878–8882, ISSN 0027–8424

Miller, GT. (2004). *Living in the environment*, Brooks/Cole Publishing, ISBN 978-0534997298, Pacific Grove, USA

Morgan, J.C.; Irving, H.; Okedi, L.M.; Steven, A. & Wondji, C.S. (2010). Pyrethroid resistance in an *Anopheles funestus* population from Uganda. *PLoS ONE,* Vol. 5, No. 7, e11872, ISSN 1932-6203

Morgan, D. & Roan C. (1974). The metabolism of DDT in man, In: *Essays in Toxicology,* W.Jr. Hayes, (Ed.), 39-97, Academic Press, ISBN 978-0121076054, New York, USA

Moriya, M. (1993). Single-stranded shuttle phagemid for mutagenesis studies in mammalian cells: 8-Oxoguanine in DNA induces targeted G.C->T.A transversions in simian kidney cells. *Proceedings of the National Academy of Sciences of the United States of America*, Vol.90, No. 3, pp. 1122–1126, ISSN 0027-8424

Moroy, Y.; Peebles, C.; Fritzler,M.J.; Steigerwald, J. & Tan,E.M. (1980). Autoantibody to centromere (kinetochore) in scleroderma sera. *Proceedings of the National Academy of Sciences of the United States of America,* Vol.77, No.3, pp. 1627–1631, ISSN 0027-8424

Natarajan, A.T.; Boei, J.J.; Darroudi, F.; Van Diemen, P.C.; Dulout, F.; Hande, M.P. & Ramalho, A.T. (1996). Current cytogenetic methods for detecting and effects of mutagens and carcinogens. *Environmental Health Perspectives*, Vol.104, No.3, pp. 445–448, ISSN 0091–6765

Natarajan, A.T. & Obe, G. (1982). Mutagenicity testing with cultured mammalian cells: cytogenetic assays, In: *Mutagenicity: New Horizons in Genetic Toxicology*, J.A. Heddle, (Ed.), 171-213, Academic Press, ISBN 978-0123361806, New York, USA

National Implementation Plan for the Stocholm Convention, Republic of Croatia. (2004). Available from http://www.cro-cpc.hr/projekti/pops/NIP_eng.pdf

Nevein, S.A. & Eman, M.S.Z. (2009). Detection of some organochlorine pesticides in raw milk in Giza Governorate. *Journal of Applied Sciences Research*, Vol.5, No.12, pp. 2520-2523, ISSN 1816-157X

Okonkwo, J.O.; Mutshatshi, T.N.; Botha, B. & Agyei, N. (2008). DDT, DDE and DDD in human milk from South Africa. *Bulletin of Environmental Contamination and Toxicology*, Vol. 81, No. 4, pp. 348-354, ISSN 0007-4861.

Olive, P.L. (1989). Cell proliferation as a requirement for development of the contact effect in Chinese hamster V79 spheroids. *Radiation Research*, Vol.117, No.1, pp. 79–92, ISSN 0033–7587

Olive, P.L.; Banáth, J.P. & Durand, R.E. (1990). Heterogeneity in radiation-induced DNA damage and repair in tumor and normal cells measured using the "comet" assay. *Radiation Research*, Vol.122, No.1, pp. 69-72, ISSN 0033–7587

Östling, O.; Johanson, K.J. (1984). Microelectrophoretic study of radiationinduced DNA damages in individual mammalian cells. *Biochemical and Biophysical Research Communications*, Vol.123, No.1, pp. 291–298, ISSN 0006–291X

Pérez-Maldonado, I.N.; Athanasiadou, M.; Yáñez, L.; González-Amaro, R.; Bergman, A. & Díaz-Barriga, F. (2006). DDE-induced apoptosis in children exposed to the DDT metabolite. *Science of the Total Environment*, Vol.370, No.2-3, pp. 343-351, ISSN 0048- 9697

Pérez-Maldonado, I.N.; Díaz-Barriga, F.; de la Fuente, H.; González-Amaro, R.; Calderón, J. & Yáñez, L. (2004). DDT induces apoptosis in human mononuclear cells in vitro and is associated with increased apoptosis in exposed children. *Environmental Research*, Vol.94, No.1, pp. 38-46, ISSN 0013-9351

Pérez-Maldonado, I.N.; Herrera, C.; Batres, L.E.; González-Amaro, R.; Díaz-Barriga, F. & Yáñez, L. (2005). DDT-induced oxidative damage in human blood mononuclear cells. *Environmental Research*, Vol.98, No.2, pp. 177-184, ISSN 0013-9351

Peterson, J. & Robison, W. (1964). Metabolic products of p,p'-DDT in the rat. *Toxicology and Applied Pharmacology*, Vol.6, pp. 321-327, ISSN 0041-008X

Picer, M.; Picer, N.; Holoubek, I.; Klanova, J. & Hodak Kobasić, V. (2004). Chlorinate hydrocarbons in the atmosphere and surface soil in the areas of the city Zadar and Mt. Velebit Croatia. *Fresenius Environmental Bulletin*, Vol.13, No.8, pp. 712–718, ISSN 1018-4619

Pimentel, D. (2005). Environmental and economic costs of the application of pesticides primarily in the United States. *Environment, Development and Sustainability*, Vol.7, No.2, pp. 229-252, ISSN 1387-585X

Piperakis, S.M. (2009). Comet assay: a brief history. *Cell Biology and Toxicology*, Vol.25, No.1, pp. 1-3, ISSN 1573-6822

Popescu, N.C. (1999). Sister chromatid exchange formation in mammalian cells is modulated by deoxyribonucleotide pool imbalance. *Somatic Cell and Molecular Genetics*, Vol.25, No.2, pp. 101–108, ISSN 0740-7750

Rabello, M.N.; Dealmeida, W.F.; Pigati, P.; Ungaro, M.T.; Murata, T.; Perira, C.A. & Beçak, W. (1975). Cytogenetic study on individuals occupationally exposed to DDT. *Mutation Research*, Vol.28, No.3, pp. 449-454, ISSN 0027-5107

Ribas, G.; Surrales, J.; Carbonell, E.; Xamena, N.; Creus, A. & Marcos, R. (1996). Genotoxicity of the herbicides alachlor and maleic hydrazide in cultured human lymphocytes. *Mutagenesis*, Vol.11, No.3, pp. 221–227, ISSN 0267-8357

Rignell-Hydbom, A.; Lidfeldt, J.; Kiviranta, H.; Rantakokko, P.; Samsioe, G.; Agardh, C.D. & Rylander, L. (2009). Exposure to *p,p'*-DDE: a risk factor for type 2 diabetes. *PLoS ONE,* Vol.4, No.10, e7503, ISSN 1932-6203

Rupa, D.S.; Reddy, P.P. & Reddi, O.S. (1989). Frequencies of chromosomal aberrations in smokers exposed to pesticides in cotton fields. *Mutation Research*, Vol.222, No.1, pp. 37-41, ISSN 0027-5107

Rupa, D.S.; Reddy, P.P.; Sreemannarayana, K. & Reddi, O.S. (1991). Frequency of sister chromatid exchange in peripheral lymphocytes of male pesticide applicators. *Environmental and Molecular Mutagenesis*, Vol.18, No.2, pp. 136-138, ISSN 1098-2280

Rydberg, B.; Johanson K.J. (1978). Estimation of DNA strand breaks in single mammalian cells, In: DNA Repair Mechanisms, P.C. Hanawalt, E.C. Friedberg & C.F. Fox, (Eds.), 465-468, Academic Press, ISBN 9781555813192, New York, USA

Salama, S.A.; Serrana, M. & Au, W.W. (1999). Biomonitoring using accessible human cells for exposure and health risk assessment. *Mutation Research*, Vol.436, No.1, pp. 99- 112, ISSN 1383-5742

Sannino, A.; Mambriani, P.; Bandini, M. & Bolzoni, L. (1996). Multiresidue method for determination of organochlorine insecticide residues in fatty processed foods by gel permeation chromatography. *Journal of AOAC International*, Vo.79, No.6, pp. 1434- 1446, ISSN 1060-3271

Santos, S.J.; Singh, N.P. & Natarajan, A.T. (1997). Fluorescence in situ hybridization with comets. *Experimental Cell Research*, Vol.232, No.2, pp. 407–411, ISSN 0014-4827

Schaeferhenrich, A.; Sendt, W.; Scheele, J.; Kuechler, A.; Liehr, T.; Claussen, U.; Rapp, A.; Greulich, K.O. & Pool-Zobel, B.L. (2003). Putative colon cancer risk factors damage global DNA and TP53 in primary human colon cells isolated from surgical samples, *Food and Chemical Toxicology,* Vol.41, No.5, pp. 655–664, ISSN 0278–6915

Shaposhnikov, S.A.; Salenko, V.B.; Brunborg, G.; Nygren, J. & Collins, A.R. (2008) Single-cell gel electrophoresis (the comet assay): loops or fragments? *Electrophoresis*, Vol.29, No.14, pp. 3005-3012, ISSN 0173-0835

Shinmura, K. & Yokota, J. (2001). The OGG1 gene encodes a repair enzyme for oxidatively damaged DNA and is involved in human carcinogenesis. *Antioxidants & Redox Signaling,* Vol.3, No.4, pp. 597–609, ISSN 1523-0864

Singh N.P. (2000). A simple method for accurate estimation of apoptotic cells. *Experimental Cell Research*, Vol.256, No.1, pp. 328-337, ISSN 0014-4827

Singh, N.P.; McCoy, M.T.; Tice, R.R. & Schneider, E.L. (1988). A simple technique for quantitation of low levels of DNA damage in individual cells. *Experimental Cell Research*, Vol.175, No.1, pp. 184-191, ISSN 0014-4827

Smith, C.C.; O'Donovan, M.R. & Martin, E.A. (2006). hOGG1 recognizes oxidative damage using the comet assay with greater specificity than FPG or ENDOIII. *Mutagenesis,* Vol.21, No.3, pp. 185–190, ISSN 0267-8357

Surrallés, J.; Falck, G. & Norppa, H. (1996). In vivo cytogenetic damage revealed by FISH analysis of micronuclei in uncultured human T lymphocytes. *Cytogenetics and Cell Genetics*, Vol.75, No.2-3, pp.151–154, ISSN 0301-0171

Sviežená, B.; Gálová, E.; Kusenda, B.; Slaninová, M.; Vlček, D.; Dušinská, M. (2004). The Comet assay and the troubles with its application in the green alga *Chlamydomonas reinhardtii. Czech Phycology,* Vol.4, No.1, pp. 163-174, ISSN 1802-5439

The European Standard EN 1258/1-4. (1996). *Fatty food: Determination of pesticides and polychlorinated biphenyls (PCBs), CEN/TC 275 N0360 Food analysis - Horizontal methods*

Thoday, J.M. (1951). The effect of ionizing radiation on the broad bean root. Part IX. Chromosome breakage and the lethality of ionizing radiations to the root meristem, *The British Journal of Radiology,* Vol.24, No.286, pp.276–572, ISSN 0007–1285

Thomas, P. & Fenech, M. (2011). Cytokinesis-block micronucleus cytome assay in lymphocytes. *Methods in Molecular Biology*, Vol.682, No.3, pp. 217-234, ISSN 1940-6029

Thomas, P.; Umegaki, K. & Fenech, M. (2003). Nucleoplasmic bridges are a sensitive measure of chromosome rearrangement in the cytokinesis-block micronucleus assay. *Mutagenesis*, Vol.18, No.2, pp. 187-194, ISSN 0267-8357

Thompson, E. J. & Perry, P. E. (1988). The identification of micronucleated chromosomes: a possible assay for aneuploidy. *Mutagenesis,* Vol.3, No.5, pp. 415-418, ISSN 0267- 8357

Tice, R.R. (1995). Applications of the single cell gel assay to environmental biomonitoring for genotoxic pollutants, In: *Biomonitors and Biomarkers as Indicators of Environmental Change,* F.M. Butterworth, (Ed.), 314-327, Plenum Press, ISBN 9780306451904, New York, USA

Tice, R.R.; Agurell, E.; Anderson, D.; Burlinson, B.; Hartmann, A.; Kobayashi, H.; Miyamae, Y.; Rojas, E.; Ryu, J.C. & Sasaki, Y.F. (2000). Single cell gel/comet assay: guidelines for in vitro and in vivo genetic toxicology testing. *Environmental and Molecular Mutagenesis,* Vol.35, No.3, pp. 206-221, ISSN 0893–6692

Tirukalikundram, S.; Kumaravel, T.S. & Bristow, R.G. (2005). Detection of genetic instability at HER-2/neu and p53 loci in breast cancer cells using Comet- FISH. *Breast Cancer Research and Treatment,* Vol.91, No.1, pp. 89–93, ISSN 0167–6806

Tompić, T.; Šimunić-Mežnarić, V.; Posedi, M. & Božidar, B. (2011). Pesticides in food, *Hrvatski Časopis za Javno Zdravstvo*, Vol. 7, No. 25, ISSN 1845-3082

Torres-Sánchez, L. & López-Carrillo, L. (2007). Human health effects and p,p'-DDE and p,p'- DDT exposure: the case of Mexico. *Ciência & Saúde Coletiva*, Vol.12, No.1, pp. 51-60, ISSN 1413-8123

Turusov, V.; Rakitsky, V. & Tomatis, L. (2002). Dichlorodiphenyltrichloroethane (DDT): ubiquity, persistence, and risks. *Environmental Health Perspectives*, Vol.110, No.2, pp. 125-128, ISSN 0091-6765

United Nations Environment Programme/International Atomic Energy Agency. (1982). *Determination of DDTs, PCBs in selected marine organisms by gas-liquid chromatography*, Reference Methods for Marine Pollution Studies, No.17, UNEP

Uppala, P.T.; Roy, S.K.; Tousson, A.; Barnes, S.; Uppala, G.R. & Eastmond, D.A. (2005). Induction of cell proliferation, micronuclei and hyperdiploidy/polyploidy in the mammary cells of DDT- and DMBA-treated pubertal rats. *Environmental and Molecular Mutagenesis*, Vol.46, No.1, pp. 43-52, ISSN 1098-2280

Vaglenov, A.; Nosko, M.; Georgieva, R.; Carbonell, E.; Creus, A. & Marcos, R. (1999). Genotoxicity and radioresistance in electroplating workers exposed to chromium. *Mutation Research*, Vol.446, No.1, pp. 23–34, ISSN 1383-5718

Valko, M.; Izakovic, M.; Mazur, M.; Rhodes, C.I. & Telser, J. (2004). Role of oxygen radicals in DNA damage and cancer incidence. *Molecular and Cellular Biochemistry*, Vol.266, No.1-2, pp. 37-56, ISSN 0300-8177

van Wendel de Joode, B.; Wesseling, C.; Kromhout, H.; Monge, P.; García, M. & Mergler, D. (2001). Chronic nervous-system effects of long-term occupational exposure to DDT. *Lancet*, Vol.357, No.9261, pp. 1014-1016 ISSN 0140-6736

Wójtowicz, A.K.; Gregoraszczuk, E.L.; Ptak, A. & Falandysz, J. (2004). Effect of single and repeated in vitro exposure of ovarian follicles to o,p'-DDT and p,p'-DDT and their metabolites. *Polish Journa of Pharmacology*, Vol.56, No.4, pp. 465-472 ISSN 1230-6002

Wójtowicz, A.K.; Milewicz, T. & Gregoraszczuk, E.L. (2007). DDT and its metabolite DDE alter steroid hormone secretion in human term placental explants by regulation of aromatase activity. *Toxicology Letters*, Vol.173, No.1, pp. 24-30, ISSN 0378-4274

Wong, W.C.; Szeto, Y.T.; Collins, A.R. & Benzie, I.F.F. (2005). The comet assay: a biomonitoring tool for nutraceutical research. *Current Topics in Nutraceutical Research,* Vol. 3, No.1, pp. 1–14, ISSN 1540-7535

Woolcott, C.G.; Aronson, K.J.; Hanna, W.M.; SenGupta, S.K.; McCready, D.R.; Sterns, E.E. & Miller, A.B. (2001). Organochlorines and breast cancer risk by receptor status, tumor size, and grade (Canada). *Cancer Causes and Control*, Vol.12, No.5, pp. 395- 404, ISSN 0957-5243

World Health Organization. (1990). *Public health impact of pesticides used in agriculture,* WHO, ISBN 9241561394, Geneva, Switzerland

World Health Organization. (2007). *Insecticide-treated mosquito nets: a WHO Position Statement*, WHO, Available from http://www.who.int/malaria/publications/atoz/itnspospaperfinal/en/index.html

World Health Organization. (2009a). *The WHO recommended classification of pesticides by hazard and guidlines to classification*, WHO, ISBN 9789241547963, Geneva Switzerland

World Health Organization. (2009b). *Global insecticide use for vector - borne disease control* (4th ed), WHO, ISBN 9789241598781, Lyon, France

Yáñez, L.; Borja-Aburto, V.H.; Rojas, E.; de la Fuente, H.; González-Amaro, R.; Gómez, H.; Jongitud, A.A. & Díaz-Barriga, F. (2004). DDT induces DNA damage in blood cells. Studies in vitro and in women chronically exposed to this insecticide. *Environmental Research*, Vol.94, No.1, pp. 18-24, ISSN 0013-9351

CHAPTER 7

Insecticide-Based Vector Control

1. INTRODUCTION: AN OVERVIEW

By the conclusion of the 19th century, it was ascertained that specific bug species, various arthropods, and freshwater snails were implicated in the spread of certain diseases significant to public health. In the absence of viable vaccinations or medications for the prevention or treatment of many diseases, transmission control primarily depended on vector management. The control programs encompassed, among other measures, the utilization of mosquito nets, drainage of gutters, and the filling of potholes and other water bodies that serve as breeding sites for insects. The 1940s witnessed the development of DDT insecticide (dichlorodiphenyltrichloroethane), a significant advancement in the management of vector-borne diseases. DDT demonstrated efficacy and cost-effectiveness in the management of various biting flies (including tsetse flies, simulium, and sand flies), midges, and infestations of fleas, lice, bedbugs, and triatomine bugs.

The initial extensive success of the control campaign was ephemeral, as the vectors gained resistance to the employed pesticides, necessitating the application of other, more costly chemicals. Interest in alternatives to insecticides, such as environmental management (source reduction) and biological control, has been rekindled due to rising resistance among significant vector species (e.g., malaria) to commonly utilized insecticides, as well as concerns regarding the environmental impact of DDT and other insecticides. For numerous vector species, environmental cleanliness through source reduction and health education is the primary strategy of management; alternative techniques should act as a complement, not a replacement. In recent years, vector control strategies have advanced, with environmental management and modification becoming prominent for disease control as well as agricultural and commercial objectives; this field is complicated and multidisciplinary. The successful implementation of any control method must rely on a comprehensive understanding of the ecology, bionomics, and behavior of the target vector species, as well as its relationship with its host and the environment. Efficient vector control necessitates meticulous training and oversight of pest management activities, along with regular assessment of the effects of the control strategies.

In recent years, there has been a diminished reliance on a singular approach of chemical control; a transition towards integrated vector control has emerged, incorporating various forms of environmental management, several chemical control methods, and the application of pharmaceuticals. Increased emphasis has been placed on community involvement, a fundamental aspect of primary health care (PHC), in eradicating vector breeding places, particularly the removal of vegetation near residential areas, and in diminishing vector populations.

Ultimately, there is a necessity to consistently furnish knowledge regarding singular, effective, and accepted ways for vector source reduction and personal protection to individuals and families inside the community at a fair cost.

This chapter will address just a limited number of significant vector species, and even this discussion will be succinct, focusing solely on the critical characteristics of the arthropods that demonstrate their impact on public health..

2. MALARIA AND ITS VECTORS

Human malaria is a number one public enemy and an illness caused by the bite of an infective female anopheles mosquito which transfers parasites called plasmodium from person to person. Four plasmodium parasites exists (P. falciparum, P. ovale, P. malariae and P. vivax) but only one (P. falciparum) is of vital importance in disease transmission in Nigeria. The important vectors in Nigeria are anopheles gambiae and A. funestus. These mosquitoes breed readily in ditches and collections of water in empty receptacles around the houses. The disease is endemic in Nigeria and over 90% of the population is at risk. Fifty percent of the population will have at least one attack per year, 300, 000 children and 11% of pregnant women die of malaria each year respectively and millions of dollars is lost each year in treatment of malaria.

The table below shows some of these vectors and diseases transmitted by them.

Table 1. Pathogens and vectors of some human diseases (sg) indicates definite involvement of salivary gland.

Type	Designation	Disease	Vectors	Mode of transmission
Protozoa	Plasmodium spp Trypanosome spp Wucheria	Malaria African sleeping sickness Filariasis	Mosquitoes Tsetse fly Mosquitoes	Bite (sg) Bite (sg) Bite
Filarial nematodes	Bancrofti Onchocerca Volvula	Onchocerciasis	Black flies	Bite

Type	Designation	Disease	Vectors	Mode of transmission
Viruses	YF Cholera/ diarrhea	Yellow Fever Cholera	Aedes	Bite (sg)
Bacteria	Typhoid fever Bacillary dysentery	Typhoid fever Dysentery	Cockroaches	Mechanical carriers

The results of several surveys showed that the most prevalent species of malaria parasites is p. falciparum with a prevalent rate of between 80% and 100% of all positive blood films. P. vivax is conspicuously absent in West Africans who because of their Duffy negative erythrocyte membrane are immune to P. vivax.

It is estimated that there are more than 400 species of anopheles in the world, but only 40 of these are important vectors (as transmitters or carriers) of malaria. Only the female anopheles mosquito transmits malaria to humans. Some anopheles mosquitoes can also carry other human diseases like Filariasis and some viruses. There are two other genera of mosquitoes which are important carriers of other mosquito-borne diseases: Aedes is a vector of viral diseases- Yellow fever and Dengue while culex mosquito is the vector of Filariasis and Japanese encephalitis. For practical purposes and entomologically, it is useful to be able to differentiate between these three types of mosquitoes; so the main distinguishing features in each stage of the life cycle are shown below in the diagram.

Basically, the mosquito has four stages in its life cycle: eggs → larva→ pupa → adult. It is useful to understand the life cycle and natural history of the anopheles mosquito not only for epidemiological reasons but most importantly for the fact that all the four stages are targets to control mosquito vector. There are several features of the behaviour of anopheles mosquitoes which are important in understanding malaria epidemiology and also for planning mosquito control.

Malaria is holo-endemic in most regions of Nigeria; ideal climatic conditions for the propagation and transmission of the infection are a temperature between 26°C – 30°C and relative humidity of over 60 percent.

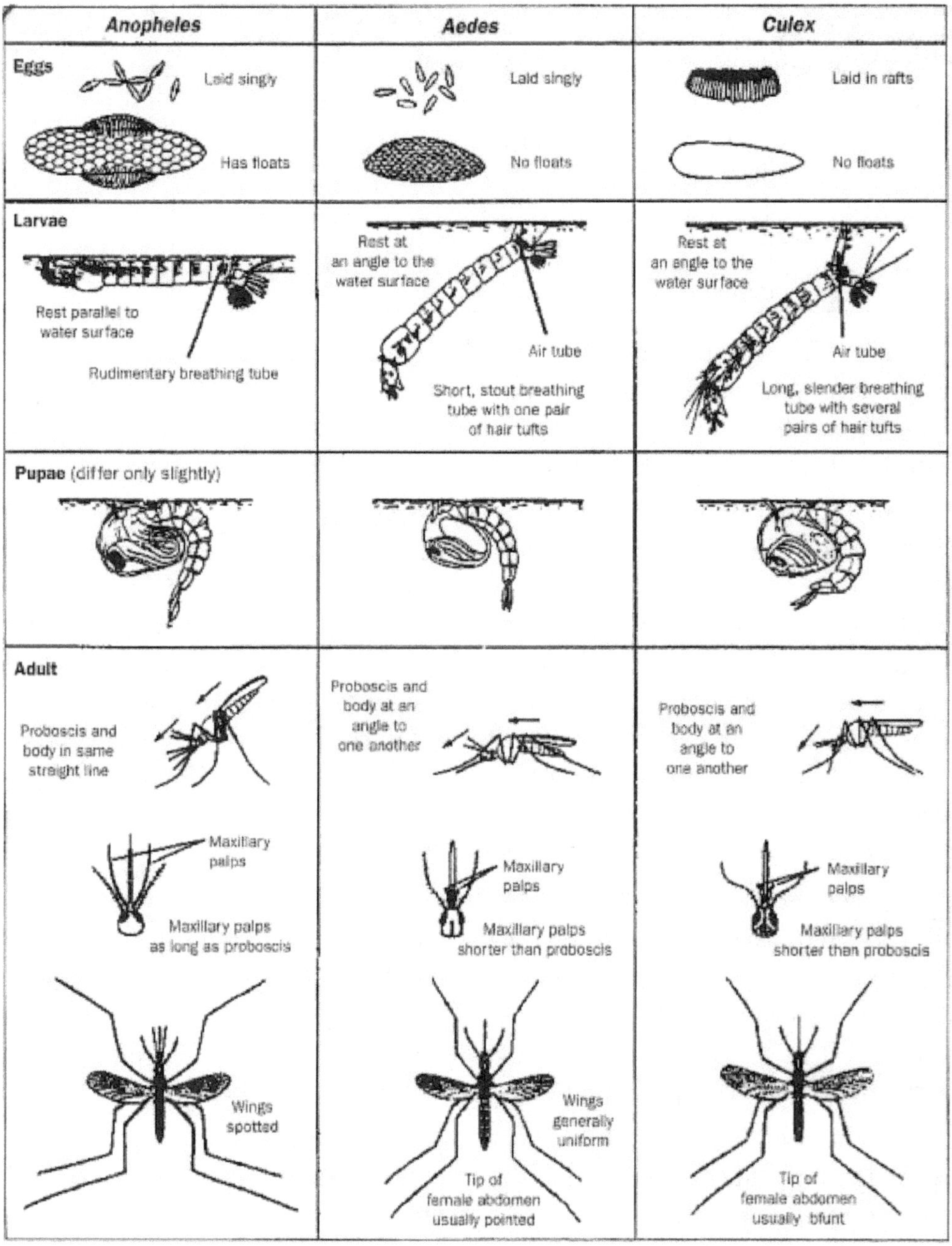

Fig. 1. Life cycles of different mosquito genera (online, 2011).

3. DISEASE CONTROL

The general methods of malaria control can be grouped into three measures directed against the parasite in man, measures directed against the vector and measures designed to prevent mosquito-man contact, these are summarized in table 2.

Table 2. Interventions to control malaria5.

Principal goal	Interventions
Treatment	Outpatient treatment of uncomplicated malaria Inpatient treatment of severe and complicated malaria
Prevention	Home treatment
(a) Inhibit mosque ito breeding	i) Source reduction (drainage, filing in ditches) ii) Chemical larviciding iii) Management of agricultural, industrial and urban development to avoid breeding sites
(b) Kill adult mosquitoes	i) IRS (indoor residual spraying) ii) ITMs (insecticide treated materials – bed nets, curtains, etc)
(c) Prevent mosquito contact	i) ITMs ii) Repellants, sprays, coils, etc
(d) Reduce malaria infection and morbidity in humans	i) IPT (intermittent preventive treatment of pregnant women) ii) Chemoprophylaxis

The first involves the use of appropriate anti-malarial agents to treat clinical malaria and the use of chemoprophylaxis among the vulnerable groups. The problem facing most tropical countries is serious as majority of patients with clinical malaria are either untreated or treated inadequately by self-medication. Also, their governments cannot afford to buy sufficient anti-malarial drugs for their needs and most people cannot afford to purchase effective treatments. The annual per capita expenditure on anti-malarial drugs in most of sub-Saharan Africa is still <US$10. An adult (60kg) course of Chloroquine costs US$0.08 but the new artemisinin based combinations (ACTs) cost more than five times as shown in table 3.

Table 3. Average cost of a full course of adult outpatient treatment.

Drug	US$
Chloroquine	0.13
SP	0.14
Amodiaquine	0.20
Artemisinin-based combinations (ACTs)	1-3

This is currently unaffordable to most patients surviving barely on less than a US$1 per day. The treatment of malaria has become quite challenging, and the emergence of resistant strains of the parasite. In Nigeria where at least 80% of the people live in rural areas and are supperstitious illiterates, early recognition and the right treatment are not adhered to. This encourages drug resistance.

The Abuja declaration on Roll Back Malaria on 25 April 2000 and agreed to by African Heads of State sets an ambitious goals to reduce the burden of malaria (insecticide-treated nets, prompt access to treatment and prevention of malaria in pregnancy) by the year 2010. Achieving high coverage in both IPT and use of ITNs among the vulnerable groups and the general population has remained elusive for many countries in sub-Saharan Africa. A major barrier to net ownership is poverty as the price of a net represents a large proportion of the income of a poor household, this has been reported in various studies.

Environmental control (source reduction) offers the best practical and easy measure to control disease vector as it eliminates the breeding places. The filling of mosquito breeding sites with soil, ash or rubbish and is most suitable for reducing breeding in small depressions, water holes or pools, which does not require much filling material. On the other hand, drainage of water can be achieved by constructing open ditches; however, the drainage systems used in agriculture or for the transportation of sewage and rainwater in cities often promote breeding because of poor design and maintenance.

4. THE USE OF INSECTICIDES

The first house-spraying campaigns after the Second World War, showed the capacity of this interventions to produce profound reductions in malaria transmission in a wide variety of circumstances. In Africa, the intervention was used in 1960s and 1970s but later abandoned except in some countries in southern and eastern Africa where residual insecticide spraying (IRS) remained the cornerstone of malaria control strategy. Since 2005, however, there has been renewed interest in large scale IRS programmes. To date, 25 out of the 42 malaria endemic countries in the WHO Africa region have included IRS in their control strategy for malaria control. Of these, 17 countries routinely implement IRS as a major malaria control intervention, six including Nigeria are piloting the IRS in a few districts while 2 are planning to pilot. At the end of 2006, the National Malaria Control Programme (NMCP) and its partners initiated a pilot IRS project in 3 local government areas/ districts in 3 states (Lagos, Plateau, Borno states) using the WHOPES approved insecticides:

a. Lambdacyhalothrin CS 10% (0.030g/m^2)
b. Alphacypermethrin WP10% (0.030g/m^2)
c. Bifenthrin WP 10% (0.050g/m^2)

The evaluation of the local context of the 3 chemicals confirmed that residual effectiveness of the insecticides lasted for at least 4 months.

Residual spraying of houses involves the treating of interior walls and ceilings using a handheld compression spraying and is effective against mosquitoes that favour indoor resting before or after feeding. For a more detailed discourse of notable insecticide formulations, spray pumps, spraying techniques and maintenance of equipment, see Jan A. Rozendaal.

The figure 2, below summarizes and describes the IRS management principles;

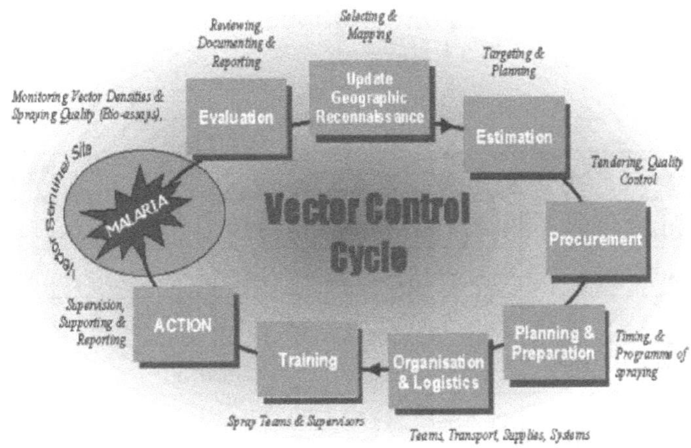

Fig. 2. IRS Management cycle.

Mosquito nets are an old technology and most people in sub-Saharan Africa are aware of the existence of nets. In some countries such as Nigeria and the Gambia nets have been in regular use for over hundred years. Similarly, in some parts of Africa, net ownership is a well established social norm and nets are widely available. For instance, recently, it was found that 70% of households (HHs) in Dar es Salaam (Tanzania) owned a mosquito net and around 35% of HHs were found to own a net in urban Burkina Faso.

Insecticide treated mosquito nets have had significant impact in reducing morbidity and mortality among children under-five years old and pregnant women where ITNs have been appropriately and extensively used in malaria endemic areas. The potential epidemiological advantages and public health benefits of treating nets with insecticide for protection against malaria were recognized in the mid-1980s. Specifically, the efficacy of insecticide treated nets (ITNs) for the control of malaria in children under-5 years of age has recently been demonstrated by several large scale studies which find reductions in all cause – mortality, ranging from 16% to 63%. These insecticides which have been approved by WHOPES (World Health Organization Pesticides Evaluation Scheme) are safe. They have the following properties; provide personal protection from mosquito bites, effective against other insects: bedbugs, flies and cockroaches, community and household mass effective which may be more important in some contexts and mosquito nuisance effect. There have been reports also of dead insects on the nets and on the floor, less mosquito noise and that ITNs providing a better night's sleep than a net alone.

Most nets (and other materials) need to be treated and retreated with insecticides to increase their effectiveness. There are several insecticides that can be used which have proved to be safe. The insecticides recommended by the National Malaria and Vector Control Programme for treatment/retreatment of nets are those approved by WHOPES and registered by NAFDAC (National Foods and Drugs Administration Commission) are shown in table 4 below:

Table 4.

Generic name	Trade name		Dose per 1 net	Manufacturer
Alphacypermethrin 10%.SC[a]	Fendona		6ml	BASF
Deltamelthrin 1%.SC	K-Othrine,	Ko-	40ml	Aventis
	Tab			
Etofenprox 10%.EW[b]	Vectron		30ml	Mitsui
Permethrin 10%.EC[c]	Peripel,		75ml	Aventis
	Imperator			
Cyfluthrin 5% EW	Solfac EW		15ml	Bayer
Lambdacyhalothrin 2.5% CS[d]	Icon		10ml	Syngenta

SC[a] = Suspension concentrate

EW[b] = Emulsion oil in water

EC[c] = Emulsifiable concentrate

CS[d] = Capsule suspension

For a detailed description of preparation of insecticide mixtures and treatment methods, the reader should consult guidelines on the use of insecticide-treated mosquito nets.

A single impregnation of a cotton or nylon mosquito net will provide protection for 1 year. Nylon tends to retain permethrin and deltamethrin better than cotton. The impregnated nets can be washed and can tolerate small tears/holes without markedly reducing the protective effect. Recently, long lasting nets (LLNs) have been developed and have the advantage of retaining insecticidal activity for years (so the nets will not lose its potency with repeated washings). Despite various government policies, cost of the nets remains a significant barrier and a long obstacle to the Roll Back Malaria goal of universal coverage - defined as one long lasting insecticide-treated nets for every two people in the household with 80% usage. ITN development is a public good. The development of insecticide resistance in the 1950s and recently by vectors has been a cause for concern. There have been reports of resistance to DDT in a wide range of sub-Saharan African countries26, but this has not reached an operationally significant level. With the exception of the Gezira region of Sudan, widespread loss of vector susceptibility is not yet a big problem in Nigeria. A worrying new development is the emergence of knockdown resistance to pyrethroid insecticides in natural populations of anopheles mosquitoes in Cote d'Ivoire and Burkina Faso where insecticides are widely used in cotton production.

At present, pyrethroids are the sole insecticides employed for net treatment and are increasingly utilized for spraying, posing a risk that if widespread resistance emerges, these interventions may progressively diminish in cost-effectiveness over time.

It is essential to elucidate the WHO integrated vector management (IVM) concept. Vector control is ideally suited for integrated strategies due to vectors' role in multiple diseases and the efficacy of interventions, such as insecticides, against various vectors. Consequently, the concept of Integrated Vector Management (IVM) emerged from insights gained through integrated pest management practices in agriculture.

Integrated vector management (IVM) is a crucial element of the worldwide initiative to combat malaria. The updated strategic plan for RBM proposed that from 2006 to 2010, 80% of the at-risk population must be safeguarded by effective vector control methods. IVM fosters synergy across several vector-borne illness management initiatives. The application of a singular treatment could be enhanced to manage many vector-borne diseases; for instance, insecticide-treated nets (ITNs) can mitigate malaria, lymphatic filariasis, and to a certain degree, leishmaniasis. IVM functions within the framework of inter-sectoral collaboration. The implementation of Integrated Vector Management concepts in vector control will promote the prudent use of pesticides and prolong their efficacy.

5. REFERENCES

World Health Organization. Chemical methods for the control of arthropod vectors and pests of Public Health importance. WHO, Geneva; 1984.

Najera J A. Malaria control: achievements, problems and strategies. Parasitologica 2001, 43:1-89

Molineaux I, Grammiccia G: The Garki project; WHO 1980

Garnham P.c.C. (1971).Progress in parasitology. The Atlone Press, London

Kara Hanson, Catherine Goodman, Jolines, Sylvia Meek, David Bradley and Anne Mills. Global Forum for health Research: The Economics of Malaria Control Interventions, 2004.

Stages of life cycle of vector parasites, available at www.who.int/docstore/water_control_htlm (cited 11/05/2011)

Miller, L.H., Mason, S J, Clyde, DF and McGinnis M H. Resistance factors to plasmodium vivax: Duffy genotype Fyfy. New England Journal of Medicine: 1976; 255:302-304

Denis MB, Tsuyuoka R, Poravuth Y et al. Surveillance of the efficacy artersunate and mefloquine combination for the treatment of uncomplicated malaria in Cambodia. Tropical Medicine International health 2006; 11:1160-1366

WHO. The Africa Malaria Report 2003, WHO/CDS/MAL/2003/1903:31-37

Salako A A. Environmental health education in schools and communities in Nigeria: The essence. The Nigerian Clinical Review Journal (2008): 14; 12:17-25

The Challenges of Malaria Infection in Nigeria: The Nigerian clinical Review. March/April 2011; 89:3-6.

Anna Maria van Eijk, Jenny Hill, Victor A. Alegana et al. Coverage of Malaria protection in pregnant women in sub-Saharan Africa: A synthesis and analysis of National survey data. Medi-Link Journal. April 2011: 12:4:20-42

Aliyu A A, Alti-Mu'azu M, Insecticide-treated net usage and malaria episodes among boarding students in Zaria, Northern Nigeria. Annals of African Medicine, 2009; 8:2:85-89.

Aniefiok M. Acceptance and use of insecticide treated bed nets (ITN in the Roll Back Malaria Programme in Kuje Area Council, Abuja. Journal of Environmental health 2005; 2:26-33

Insecticide-treated nets. New findings from Kenya, published by Centers for Disease Control and Prevention (CDC), Atlanta, USA, Confirm effectiveness in areas of intense transmission. Africa Health (Mera) July 2003; Issue 6

Jan A. Rozendaal: Vector control (Methods used by individuals and communities) 1997.

Jones C. Bed nets and malaria. Postgraduate Doctor. May 2002; vol. 24 No. 2:23-25.

Guiguemde T R et al. Household expenditure on Malaria prevention and treatment of families in the town of Bobo-Diolasso, Burkina Faso. Trans. of the Royal Society of Tropical Medicine and Hygiene 1994: 88: (3)285- 287

Alonso P L, Lindsay SW, Schellenberg J. et al. A malaria control net trial using insecticide-treated bed nets and targeted chemoprophylaxis in a rural area of the Gambia, West Africa. The impact of the interventions on mortality and morbidity from malaria. Transactions of the Royal Society of Tropical Medicine and Hygiene, 1993; 87(S):37-44.

D'Alessandro U, Olaleye BO, McGuire W et al. Mortality and Morbidity from malaria in Gambian children after introduction of an impregnated bed net programme. Lancet 1995:345 (8948).479-83

Binka F N, Kubaje A, Adjuik M et al. Impact of permethrin impregnated bed nets on child mortality in Kassena-Nankana district, Ghana: A randomized controlled trial. Tropical Medicine and International health 1996; 1(2):147-54.

Nevill CG, Same ES, Mung'ala VO et al. Insecticide-treated bed nets reduce mortality and severe morbidity from malaria among children on the Kenyan Coast. Tropical Medicine and International Health 1996: 1(2) 139-46

Nicholas J. White. Malaria Gordon C. Cook and Alimuddin I Zumla (Eds). Manson's Tropical Diseases! 22nd edition, China, Saunders Elsevier: 2009; 1201-1300

WHO Guidelines on the use of insecticide-treated nets for the prevention and control of malaria in Africa.

Lindsay SW, Gibson MF. Bed nets revisited – old idea, new angle. Parasitology Today 1988:4:270-272

Alonso PI, Lindsay SW, Armstrong IRM, et al. The effects of insecticide-treated bed nets on mortality of Gambian children. Lancet 1991;337:1499-1502

WHO Vector resistance to pesticides, Fifteenth Report of the WHO Expert Committee of Vector Biology and Control, Geneva; 1992

El Gaddal A A, Haridi A A M, Hassan F T, Hussein H. Malaria Control in Gezira-Managil Irrigated Scheme of the Sudan. Journal of Tropical Medicine and Hygiene 1985; 88:153-157

Martinez Torres D, Chandre F, Willamson M S et al. Molecular characterization of pyrethroid knockdown resistance (kdr) in the major malaria vector anopheles Gambiae SS. Insect Molecular Biology 1998; 7(2):174-184

Elissa N, Morchet J, Riviere F, Mevnier J-Y, Yaok. Resistance of Anopheles gambiae SS to pyrethroids in Cote d'Ivoire. Annales de la Societe Belge de Medecine Tropicale 1993; 73:291-294.

CHAPTER

8

Susceptibility of Aedes aegypti to Insecticides in Colombia

1. INTRODUCTION: AN OVERVIEW

Dengue is a public health concern in Colombia because to its effects on morbidity and mortality. This illness has endemic characteristics in the regions of the country situated below 2200 meters above sea level. From 1978 to 2010, around 1,011,852 cases of dengue were registered in Colombia. For over five decades, vector populations of Aedes aegypti in the country have been persistently targeted with insecticides, which are the predominant method employed to disrupt the transmission of the dengue virus during outbreaks and epidemics in the absence of a vaccine (Figure 1). The selection pressure imposed by these chemicals has resulted in the creation of resistant populations of the vector to organochlorines, organophosphates, pyrethroids, and carbamates.

Fig. 1. Application of insecticides in the home by fogging during outbreaks or epidemics.

Since 2007, the documentation of vector resistance in the country has significantly escalated, coinciding with the establishment of the national network for monitoring susceptibility to pesticides employed in public health against A. aegypti and the primary malaria vector in Colombia. This information has established a baseline for several Colombian departments in the region and has generated fundamental knowledge for decision-making regarding the operational aspects of vector control initiatives. Further investigation into the enzymatic mechanisms and molecular origins of resistance in vector populations is necessary, as the majority of findings thus far have stemmed from diagnostic doses utilizing standardized methodologies established by the Centers for Disease Control and Prevention (CDC) and the World Health Organization (WHO). This chapter presents the results and discusses associated concepts.

2. STUDY AREA

The continental territory of the Republic of Colombia is located in the northwest corner of South America on the equator. Colombia is located in the north to 12°26’46” north latitude at the place called Punta Gallinas in the peninsula of Guajira. To the south, is located at 4° 12’ 30” de latitude south, the East is located at 60°50’54” west longitude from Greenwich, by the West is located at 79°02’ 33” west longitude from Greenwich.

Colombia territory also includes the archipelago of San Andres and Providencia in the Caribbean sea between 12° and 16°30′north latitude and 78° and 82° west longitude of Greenwich. Colombia has a land area of 1.141.748 km2 ranking fourth among the countries of South America.

Colombia is a country with a wide variety of climates, from the coldest to 0°C to temperatures exceeding 28°C especially in the Caribbean regions, Pacific, Amazon and the Orinoco, accounting for nearly 80% of Colombia territory. Colombia is divided administratively into 32 states, these in turn are divided into 1102 municipalities.

3. THE SITUATION OF DENGUE IN COLOMBIA

In Colombia, dengue is a disease of public health interest for its impact on morbidity and mortality. Its endemic behavior is due to endemic multiple factors including the re-emergence and intense viral transmission with a rising trend, the behavior of epidemic cycles getting shorter, the increased frequency of outbreaks and serious of the disease, the simultaneous circulation of the four serotypes, infestation by *A. aegypti* in over 90% of the national territory located below the 2,200 meters, and other social factors such as uncontrolled increase and lack of basic sanitation in urban centers (Escobar, 2009). The epidemiological pattern of disease in the last decades has been upward, characterized by exponential increase in endemic areas during the different decades. Cyclical behavior has been characterized by epidemic peaks every three or four years, dealing with the reentry of new serotypes to the country.

In Colombia from 1978 to 2010, has officially been recorded a total of 1.011.852 cases of dengue. For this time period there is record history of the following epidemics in the country:

1977: First recorded epidemic of dengue in Colombia.

2002: Epidemic in 76,996 cases, of which 5269 corresponded to Dengue hemorrhagic fever. 2007: Epidemic in 43,227 cases, of which 4665 corresponded to Dengue hemorrhagic fever. 2010: It is recorded in the country the worst epidemic of this disease with 157,152 confirmed cases and 217 deaths.

The departments that have historically had more transmission of dengue in the country are: Atlántico, Santander, Norte de Santander, Valle del Cauca, Antioquia, Tolima, Huila, Cundinamarca, and Casanare.

4. RESISTANCE TO INSECTICIDES

4.1 Definition

Resistance is defined as the development of the ability to tolerate doses of toxics, which can be lethal to most individuals in a normal population of the same species and is the result of positive selection pressure exerted by the insecticidal over genes initially in low frequency (WHO, 1957).

4.2 Mechanisms of resistance

The two main resistance mechanisms are alterations in the target site and metabolic resistance, also called increase in the rate of detoxification of insecticides. However, there are other less frequent as there are the resistance behavior and decreased penetration. (Miller, 1988; Bisset, 2002; Fonseca and Quiñones, 2005)

4.2.1 Altered target site

This is generated when no silent mutations occur in structural genes that generate a disturbance of amino acids responsible of anchor of the insecticide at a specific site.

4.2.1.1 GABA Receptor

In bugs this is an hetero multi numeric channel acting as a site of action for cyclodiene and avermectins. The resistance is given by *Rdl* gene (Resistance to dieldrin), which encodes for RDL, a GABA receptor subunit. This type of resistance was first identified in *Drosophila* first identified this type of resistance (Fonseca and Quiñones, 2005).

4.2.1.2 Voltage-gated sodium channel

They are the target of action of pyrethroids and DDT. The sodium para-channel protein is a complex of over 2000 amino acids, composed of 4 homologous domains separated by hydrophilic links, each domain contains six segments. The first mutation in the sodium channel that conferred resistance was detected in *Musca domestica*. subsequently mutations in other insects have been identified including Culicidae resistant to pyrethroids (Martínez et al, 1998; Brengues et al, 2003; Anstead, 2005; Saavedra et al, 2007; Brooke, 2008; Chang et al 2009; Yanola et al, 2010).

4.2.1.3 Insensitive acethylcholinesterase

Insecticide resistance, attributed to Ache insensitivity is found in a number of *Anopheles* and *Culex* species. In general, this mechanism produces a broad spectrum resistance to most organophosphates and carbamates, although more pronounced to carbamatos. (Bisset, 2002).

4.2.2 Metabolic resistance

It is conferred by an increase in detoxification insecticide or an inability to metabolize the toxic compound. The most important form of metabolic resistance is given by detoxificating enzymes of the type glutathione S-transferase, mixed function oxidases and esterases (Bisset 2002).

4.2.2.1 Carboxylesterases

They Catalyze the hydrolysis of carboxylic esters and changes in their expression levels, this is the resistance mechanism that occurs most often in insects. In mosquitoes, high levels of these enzymes have been associated with resistance to organophosphate and pyrethroid insecticides (Cui et al, 2007; Rodriguez et al, 2007). Esterases is a family of six proteins Esterases are a family of six proteins grouped in α and β hydrolases superfamily. In the Diptera these enzymes are encoded by a gene cluster on the same chromosome, each gene may suffer modification that confers resistance to the insecticide (Fonseca and Quiñones, 2005; Santacoloma, 2008).

4.2.2.2 P450 Enzymes or mixed function oxidases

P450 enzymes are encoded by cytochrome P450 genes. P450 enzymes also have other names such as: cytochrome P450 monooxygenases, mixed function oxidases, monooxygenases with polisustrato, microsomal oxidases and heme-thiolate proteins. In insects these enzymes are involved in growth and development through the processing of fatty acids, hormones and pheromones; in the metabolism of secondary plants products and synthetic chemicals such as insecticides. The cytochrome P-450 is implicated as the major factor in many cases of metabolic resistance to carbamates and detoxify organophosphates, pyrethroids and DDT as well. It expresses in tissues of the gut, fat, reproductive tract and Malpighian tubes (Feyereisen, 1999; Poupardin et al, 2008).

4.2.2.3 Glutatione s-transferase

The glutathione s-transferases (GSTs, EC 2.5.1.18) are phase II enzymes involved in xenobiotics detoxification in many organisms. These enzymes metabolize a wide range of toxic hydrophobic compounds such as drugs, insecticides and endogenous toxic substrates, catalyzing the conjugation of glutathione to the hydrophilic center of toxic substances, allowing the increase in compounds solubility. The GSTs are divided into three main groups: the cytosolic, microsomal and mitochondrial. The mitochondrial GSTs have not been found in insects species including mosquitoes. The cytosolic GSTs in insects are grouped into six different kinds: Delta, Epsilon, Omega, Sigma, Theta and Zeta. Most GSTs involved in the metabolism of xenobiotics in insects belong to the class Delta Epsilon. GSTs in insects have been implicated in the insecticide resistance through its direct metabolism, sequestration or protection against secondary toxic effects such as increased lipid peroxidation induced by exposure to insecticides. Glutathione S- transferase mediated detoxification has been reported for insecticides of the organophosphate type and DDT. The main role of GSTs in the resistance to organophosphates is the detoxification of the insecticide by a conjugation reaction. The specific GSTs also catalyze the metabolism of DDT to a non-toxic substance: 1,1-dichloro-2 ,2-bis-(p-chlorophenyl) ethane (DDE), through the process of dehydrochlorination. Epsilon class of GST has been involved in resistance to DDT in *Anopheles gambiae* and *A. aegypti* (Lumjuan et al, 2007).

4.3 Type of resistance

4.3.1 Cross resistance

It occurs when a single gene confers resistance to a number of chemicals in the same group, as it is the case of phosphotriesterases that provides resistance to several organophosphates or *kdr* gene that confers resistance DDT and the pyrethroids (WHO, 1957).

4.3.2 Multiple resistance

It occurs when two or more resistance mechanisms are operating in the same insect. The multiple resistance term does not necessarily involve the term cross-resistance, because a bug may be resistant to 2 or more insecticides and each resistance can be attributed to different mechanisms (Bisset, 2002).

4.4 Techniques used to detect resistance insecticides in mosquitoes

4.4.1 Biological test of the World Health Organization (WHO)

Technique standardized by the World Health Organization (WHO). In adult mosquitoes the technique consists of exposing individuals of the populations we intent to evaluate to papers impregnated with a single and specific dose of an insecticide during a determined time (1 hour with the exception of the organophosphate fenitrothion, to which will be exposed for 2 hours). The mosquitoes are then transfered to a paper that hasn't been impregnated, where they rest under controlled conditions of temperature, relative humidity and an energy source consisting of a sugar solution (Figure 2A). The mortality reading is done after 24 hours of exposure to the insecticide (WHO 1981a).

The susceptibility of larvae is measured by exposing individuals to a diagnostic concentration or several temephos concentrations to determine the lethal concentration (LC_{50} or CL_{95}) to calculate resistance factor using water as solvent (Figure 2B). Individuals remain exposed for 24 hours. (WHO 1981b) (Santacoloma, 2008).

Fig. 2A. Biological test of the World Health Organization (WHO) (mosquitoes).

4.4.2 Test for biological control center and prevention (CDC)

The impregnated bottles technique was standardized by the Center for Diseases Control and Prevention (CDC) (Brogdon and McAllister, 1998). The objective of this technique is to measure the time it takes dose of an insecticide to reach the target site of action in the mosquito. Involves subjecting a sample of mosquitoes between 15 and 25 individuals of the population to be evaluated on a glass surface pre-impregnated with the dose of insecticide. (Santacoloma, 2008) (Figure 3).

Fig. 2B. Biological test of the World Health Organization (WHO) (Larvae).

Fig. 3. Test for biological control center and prevention (CDC).

4.4.3 Synergists

These are chemicals that specifically inhibit insecticide-metabolizing enzymes, enhancing their action. Among the synergists more used to detect resistance mechanisms in Insects are the S, S, S - tributilfosforotioato (DJF) and esterases inhibitor and of the enzyme glutathione transferase (GST), the triphenyl phosphate (TFF) specific esterase inhibitor, piperonyl butoxide (PB), an inhibitor of monooxygenases, and ethacrynic acid (EA), a specific inhibitor of the enzyme glutathione transferase (GST) (Rodríguez, 2008).

4.4.4 Biochemical test

The biochemical assays are used to define metabolic mechanisms that may be responsible for the physiological resistance in an insect population (WHO, 1992). The metabolic mechanisms include tests to determine the target enzyme decreased sensitivity or the increased enzyme activity. For the first mechanism in particular, measures the change in acetylcholinesterase associated with resistance to carbamates and organophosphates. For the second, evaluates the increased activity of esterases, mixed function oxidases and glutathione s-transferase for kidnapping or increased detoxification of insecticides (Santacoloma, 2008).

4.4.5 Molecular tests

These tests consist in the amplification of specific gene sequences through polymerase chain reaction technique (PCR) to detect mutations.

4.5 Current state of susceptibility to insecticides Aedes aegypti in Colombia

Since the late forties, when first reported resistance to DDT in *Aedes tritaeniorhynchus* (Weidemann) and *Aedes solicitans* (Walker) resistance has been recorded in over a hundred species for one or more insecticides of public health use worldwide (Brown, 1986; Fonseca & Quiñones, 2005). For *A. aegypti* in America resistance has been reported to organochlorines, organophosphates, pyrethroids and carbamates in Argentina, Brazil, Mexico, El Salvador, Peru, Panamá, Venezuela, Cuba, Puerto Rico, among other Caribbean countries, whose resistance mechanism in some of these stocks has been associated with altered levels of alpha esters, beta esterases, mixed function oxidases, glutathione s- transferase, as well as mutations in the voltage-gated sodium channel. (Rawlins, 1998; Bisset et al, 2001; Brengues et al, 2003; Macoris et al, 2003; Aparecida et al, 2004; Rodríguez et al, 2004; Chavez et al; 2005, Flores et al, 2005; Pereira da-Cunha et al, 2005, Alvarez et al, 2006, Pereira-Lima et al, 2006; Beserra et al, 2007; Saavedra et al, 2007; Bisset et al, 2009; Martins et al, 2009; Albrieu- Llinas et al, 2010; Polson et al, 2010).

Colombia has applied insecticides for control of vector insects for over five decades. The DDT was the first applied to control malaria and during the campaign for the eradication of *A. aegypti* conducted in the early 1950. This insecticide was banned in the late 60's, due to the findings of resistance worldwide (Brown, 1986). Since 1970, organophosphates including temephos were applied and from the early 90's the use of pyrethroids was started. From that time on, the country has been rotating the application of molecules for mosquito control such as: deltamethrin, lambda-cyhalothrin, malathion, fenitrothion and in the last three years cyfluthrin and pirimiphos-methyl. However, the resistance to these insecticides has been documented gradually, making it difficult to take control actions within programs of Vector Borne Diseases in different regions of the country.

In Colombia until the 1990's there were few works assessing the state of susceptibility in Culicidae populations of interest in public health. Between 1959 and 1987 the first cases of DDT resistance were registered in populations of *Anopheles albimanus* (Wiedemann) in the municipalities of El Carmen (Bolivar); Codazzi, Robles and Valledupar (Cesar); Acandí (Choco) and *An. darlingi* (Root) in some locations of Quibdó municipality (Chocó) (Quiñones et al, 1987). Later Suarez et al, (1996) recorded in the 90's decade the first case of resistance to temephos in the *A. aegypti* species in Cali, Valle del Cauca, and Bisset et al, (1998) evaluated the susceptibility in a strain of *Culex quinquefasciatus* (Say) from Medellin, Antioquia, encountering resistance to organophosphates malathion, primifos-methyl, chlorpyrifos, temephos, fenthion and pyrethroids deltamethrin and permethrin.

In the absence of enough studies in Colombia on susceptibility status of populations of *A. aegypti* to several insecticides of use in public health and in compliance with public policies enshrined in the American continent resolutions CD39.R11 1996, CD43R4 2001 of the Pan American Health Organization, during the years of 2005 and 2007 a national project was conducted, this was funded by Colciencias (Colombian Science and Research Organization) and implemented by the Learning and Control of Tropical Diseases Program (PECET) of the University of Antioquia, the International Centre for Training and Medical Research (CIDEIM), the National University in Colombia, the National Institute of Health (NIH) and 12 departments of health seeking to generate baseline susceptibility of vector populations in Colombia. This multicentered project gave rise to the national surveillance network susceptibility to insecticides for *A. aegypti* and main vectors of malaria led by the National Institute of Health (INS). Since then the record of resistance to *A. aegypti* in Colombia expanded through biological tests by the CDC and WHO as well as the determination of impairment of enzymes involved in resistance.

With these results it has been observed for Colombia widespread resistance to DDT (Figure 4A-4B) and variability in susceptibility to the following insecticides: temephos, lambda- cyhalothrin, deltamethrin, permethrin, cyfluthrin, etofenprox, malathion, fenitrothion, pirimiphos methyl, bendiocarb and propoxur in different regions, with deterioration in some populations in the levels of nonspecific esterases, mixed function oxidases and in smaller proportion to glutathione s-transferesas (Figure 5A, 5B, 5C, 5D) (Rojas et al, 2003; Cadavid et al, 2008; Fonseca et al, 2006; Fonseca et al, 2007; Orjuela et al; 2007, Salazar et al, 2007; Santacoloma et al, 2008; Fonseca et al, 2009; Maestre et al, 2009; Maestre et al, 2010; Ardila and Brochero, 2010, Gomez et al, 2010; Maestre et al, 2010, Fonseca et al, 2011).

For temephos resistance has been observed in Cundinamarca, Guaviare, Meta, Santander, Cauca, Valle del Cauca, Nariño, Huila, Caldas, Sucre, Atlantico, La Guajira. (Anaya et al, 2007; Santacoloma, 2008; Maestre et al, 2009; Ocampo et al, 2011) (Figure 6).

Pyrethroids that despite being used in Colombia more recently compared to organophosphates, have shown higher levels of resistance despite the increased use time of organophosphate (Figures 7A, 7B, 7C, 7D, 7E), (8A, 8B, 8C). Among the pyrethroids lambda is the insecticide wich displays higher frequency of resistance in vector populations in the country. However, there are pyrethroids such as permethrin and etofenprox that despite having no use in public health have resistance generated in populations of *A. aegypti* from Casanare, Antioquia, Chocó and Putumayo (Ardila and Brochero, 2010; Fonseca et al, 2011). For the carbamate propoxur discordance in susceptibility results was observed between the WHO technique in which resistance and CDC susceptibility is recorded. Further studies are required to determine the state of populations' susceptibility to the insecticide (Fig. 9 AB). Furthermore, few studies in the country have evaluated the susceptibility of the vector to the insecticide Bendiocarb to which there has been resistance registered in the populations in the department of Cauca, Valle del Cauca, Huila and Nariño (Figure 10 AB) (Ocampo et al, 2011). Currently it has not been registered for the country mutations in voltage-gated sodium channel gene. It is therefore recommended studies to perform studies to explain resistance observed to most of the pyrethroids evaluated in different country populations and may be related to a crossed resistance to DDT. It is also recommended to perform studies to determine cross resistance to other molecules such as organophosphate and carbamate, as well as multi resistance studies.

For Colombia it is recommended to maintain a system of permanent time and space surveillance that allows health authorities to use insecticides with technical criteria to maintain effective control interventions in vector populations.

Fig. 4A. Susceptility status of *Aedes aegypti* populations to DDT in Colombia (CDC test).

Fig. 4B. Susceptility status of *Aedes aegypti* populations to DDT in Colombia (OMS test).

Fig. 5A. Non-specific esterases (NSE).

Fig. 5B. Mixed-function oxidases (MFO).

Fig. 5C. Glutathione-S-transferases (GST).

5D

Fig. 5. Biochemical mechanism of resistance in population of *Aedes aegypti* in Colombia: Non-specific esterases (NSE) (5A); Mixed-function oxidases (MFO) (5B); Glutathione-*S*- transferases (GST) (5C); acethylcholinesterase (AChE) (5D).

Fig. 6. Susceptility status of *Aedes aegypti* populations to Temephos in Colombia (OMS test).

Fig. 7A. Susceptility status of *Aedes aegypti* populations to lambda-cyhalothrin in Colombia (CDC test).

Fig. 7B. Susceptility status of *Aedes aegypti* populations to lambda-cyhalothrin in Colombia (OMS test).

Fig. 7C. Susceptility status of *Aedes aegypti* populations to Deltamethrin in Colombia (CDC test).

Fig. 7D. Susceptility status of *Aedes aegypti* populations to Deltamethrin in Colombia (OMS test).

Fig. 7E. Susceptility status of *Aedes aegypti* populations to cyfluthrin in Colombia (OMS test).

Fig. 7F. Susceptility status of *Aedes aegypti* populations to Permethrin in Colombia (CDC test).

Fig. 7G. Susceptility status of *Aedes aegypti* populations to Permethrin in Colombia (OMS test).

Fig. 7H. Susceptility status of *Aedes aegypti* populations to etofenprox in Colombia (OMS test).

Fig. 8A. Susceptility status of *Aedes aegypti* populations to Malathion in Colombia (CDC test).

Fig. 8B. Susceptility status of *Aedes aegypti* populations to Malathion in Colombia (OMS test).

Fig. 8C. Susceptility status of *Aedes aegypti* populations to Fenitrothion in Colombia (CDC test).

Fig. 8D. Susceptility status of *Aedes aegypti* populations to Fenitrothion in Colombia (OMS test).

Fig. 8E. Susceptility status of *Aedes aegypti* populations to pirimiphos methyl in Colombia (CDC test).

Fig. 8F. Susceptility status of *Aedes aegypti* populations to pirimiphos methyl in Colombia (OMS test).

Fig. 9A. Susceptility status of *Aedes aegypti* populations to propoxur in Colombia (CDC test).

Fig. 9B. Susceptility status of *Aedes aegypti* populations to propoxur in Colombia (OMS test).

Fig. 10A. Susceptility status of *Aedes aegypti* populations to bendiocarb in Colombia (CDC test).

Fig. 10B. Susceptility status of *Aedes aegypti* populations to bendiocarb in Colombia (OMS test).

5. CONCISE SUMMARY

The selection pressure imposed by insecticides over more than fifty years on A. aegypti populations in Colombia has resulted in extensive resistance to DDT and variability in susceptibility to organophosphates, carbamates, and pyrethroids. Alterations in nonspecific esterase levels, mixed function oxidases, and s-glutathione transferases have been shown as potential contributors to metabolic resistance to these pesticides. It remains uncertain whether modifications in the target action sites, particularly in voltage-dependent sodium channel genes, account for the widespread resistance to DDT and most pyrethroids in the country.

6. REFERENCES

Albrieu-Llinás, G.; Seccacini, E.; Gardenal, CN. & Licastro, S. (2010). Current resistance status to temephos in *Aedes aegypti* from different regions of Argentina. *Mem Inst Oswaldo Cruz*, Vol.105, No.1, pp.113-116.

Álvarez, L.; Briceño, A. & Oviedo, M. (2006). Resistencia al temefos en poblaciones de *Aedes aegypti* (Diptera: Culicidae) del occidente de Venezuela. *Revista Colombiana de Entomología*, Vol.32, No.2, pp.172-5.

Anaya, Y.; Cochero, S.; Rey, G. & Santacoloma, L. (2007). Evaluación de la Susceptibilidad a insecticidas en *Aedes aegypti* (Diptera: Culicidae) capturados en Sincelejo. *Biomedica*, Vol.27, No.2, pp.257-8.

Anstead, JA.; Williamson, MS. & Denholm, I. (2005). Evidence for multiple origins of identical insecticide resistance mutations in the aphid Myzus persicae. *Insect Biochemistry and Molecular Biology*, Vol.35, pp.249–256.

Aparecida-Braga, I.; Pereira-Lima, J.; Da Silva-Soares, S. & Valle, D. (2004). *Aedes aegypti* resistance to temephos during 2001 in several municipalities in the states of Rio de Janeiro, Sergipe and Alagoas, Brazil. *Memórias do InstitutoOswaldo Cruz*, Vol.99, No.2, pp.199-203.

Ardila-Roldan, S. & Brochero, HL. (2010).Status of insecticide resistance in natural population from *Aedes aegypti* and CAP surveys of dengue vector in Casanare, Colombia. *Journal of American Mosquito Control Association,* Vol.26, No.3, pp.306-20.

Beserra, E.; Fernández, C.; De Quiroga, M. & De Castro, F. (2007). Resistance of *Aedes aegypti* (L.) (Diptera: Culicidae)populations to organophosphates temephos in the Paraíba State, Brazil. *Neotropical Entomology*, Vol.36, No.2, pp.303-7.

Bisset, JA.; Rodríguez, MM.; Molina, D.; Díaz, C. & Soca, LA. (2001). Esterasas elevadas como mecanismo de resistencia a insecticidas organofosforados en cepas de *Aedes aegypti*. *Revista Cubana de Medicina Tropical*, Vol.53, No.1, pp.37-43.

Bisset J. (2002).Uso correcto de insecticidas: control de la resistencia. *Revista Cubana de Medicina Tropical*, Vol.54, No.3, pp.202-219.

Bisset-Lazcano JA.; Rodríguez, MM.; San Martín, JL.; Romero, JE. & Montoya, R. (2009).Evaluación de la resistencia a insecticidas de una cepa de *Aedes aegypti* de El Salvador. *Revista Panamericana de Salud Publica*, Vol.26, No.3, pp.229-34.

Brogdon, W. & Mcallister, J. (1998).Simplification of adult mosquito bioassays through use of time mortality determinations in glass bottles. *Journal of the American Mosquito Control Association*, Vol.14, No2, pp.159-64.

Brengues, C.; Hawkes, NJ.; Chandre, F.; McCarroll, L.; Duchon, S. & Guillet, P. *et al.* (2003). Pyrethroid and DDT cross-resistance in *Aedes aegypti* is correlated with novel mutations in the voltagegated sodium channel gene. *Med Vet Entomol.*, Vol.17, pp.87–94.

Brooke B. (2008). *kdr*: can a single mutation produce an entire insecticide resistance phenotype?. *Transactions of the Royal Society of Tropical Medicine and Hygiene*, Vol.102, pp.524—525.

Brown, A. (1986). Insecticide resistance in mosquitoes: a pragmatic review. *Journal of American Mosquito Control Association*, Vol.2, pp.123-40.

Cadavid, JM.; Valderrama-Hernández, R.; Saenz-Osorio, O.; Quintero-Ruiz, B.; Rodríguez Ríos, C. & Contreras-Samper, A. (2008). Susceptibility of *Aedes aegypti* to insecticides in three high dengue transmission localities of Antioquia, Colombia *Journal of American Mosquito Control Association,* Vol.22, No.4, pp.748.

Chang, Ch.; Shen, WK.; Wang, TT.; Lin, YH.; Hsu, EL. & Dai, SM. (2009). A novel amino acid substitution in a voltage-gated sodium channel is associated with knockdown resistance to permethrin in *Aedes aegypti. Insect Biochemistry and Molecular Biology*, Vol.39, pp. 272–278

Chávez, J.; Vargas, J. & Vargas, F. (2005). Resistencia a deltametrina en dos poblaciones de *Aedes aegypti* (Diptera: Culicidae) del Perú. *Revista Peruana de Biología*, Vol.12, No1, pp.161-4.

Cui, F.; Weill, M.; Berthomieu, A.; Raymond, M. & Qiao, Ch. (2007). Characterization of novel esterases in insecticide-resistant mosquitoes. *Insect Biochemistry and Molecular Biology*,Vol.37, pp.1131–37.

Escobar JP. (2009). Políticas y orientaciones técnicas OPS/OMS para prevención y control del dengue y dengue hemorrágico. *Biomédica*, Vol.29, No1, pp.123-5.

Feyereisen R. (1999). Insect P450 enzymes. *Annu. Rev. Entomol.*, Vol.44, pp.507-33.

Flores, A.; Albeldaño-Vásquez, W.; Fernández-Salas, I.; Badii, M.; Loaiza-Becerra, H.; Ponce-Garcia, G.; Lozano-Fuentes, S.; Brogdon, W.; Black, W.; Beaty, B. (2005). Elevated á- esterase levels associated withpermethrin tolerance in *Aedes aegypti*

(L.) from Baja California, México. *Pesticide Biochemistry and Physiology*, Vol.82, pp.66-78.

Fonseca, I. & Quiñones, M. (2005). Resistencia a insecticidas en mosquitos (Diptera: Culicidae): Mecanismos, detección y vigilancia en salud pública. *Revista Colombiana de Entomología*, Vol.31, No2, pp.107-115.

Fonseca I., Brant T.,Quiñones ML., Brogdonm, WG. (2006). Increased glutathione S transferases associated to DDT resistance in *Aedes aegypti* from Colombia. *Journal of American Mosquito Control Association,* Vol.22, No4, pp.738.

Fonseca, I.; Bolaños, D.; Gomez, W. & Quiñones, ML. (2007). Evaluación de la susceptibilidad de larvas de *Aedes aegypti* a insecticidas en el deprtamento del Antioquia. *Biomedica*, Vol.27, No.2, pp.176-7.

Fonseca-González, I. (2008). Estatus de la resistencia a insecticidas de los vectores primarios de malaria y dengue en Antioquia, Chocó, Norte de Santander y Putumayo, Colombia. [Tesis de doctorado]. Medellín: Universidad de Antioquia.

Fonseca, I. & Bolaños, D. (2009).Variación temporal en la susceptibilidad a malatión y Lambdacialotrina en *Aedes aegypti* (L) de Quibdo, Colombia. *Biomédica*, Vol.29, No.1, pp.218-9.

Fonseca-Gonzalez, I.; Quiñones, ML.; Lenhartc, A. & Brogdon, WG. (2011). Insecticide resistance status of *Aedes aegypti* (L.) from Colombia. *Pest Management Science*, Vol.67, No.4, pp.430-7.

Gómez-Camargo DE.; Maestre, RY.; Pisciotti, I.; Malambo, DI. & Gómez-Alegría, CJ. (2010). Insecticide susceptibility of *Aedes aegypti* in Cartagena (Colombia). *The American Journal of Tropical Medicine and Hygiene*, Vol.83, No.5, pp.298-9.

Hemingway, J. & Black IV, WC. (2007). A mutation in the voltage-gated sodium channel gene associated with pyrethroid resistance in Latin American *Aedes aegypti. Insect Molecular Biology*, Vol.16, No.6, pp.785-98.

Lumjuan, N.; Stevensona, BJ.; Prapanthadarab, L.; Somboonc, P.; Brophyd, PM.; Loftuse, BJ.; Seversonf, DW. & Ransona, H. (2007). The *Aedes aegypti* glutathione transferase family. *Insect Biochemistry and Molecular Biology*, Vol.37, pp.1026-35.

Macoris, M.; Andrighetti, M.; Takaku, L.; Glasse, C.; Garbeloto, V. & Bracco, J. (2003). Resistance of *Aedes aegypti* from the state of Sao Paulo, Brazil to Organophosphates insecticidas. *Memórias do Instituto Oswaldo Cruz*, Vol.98, No.5, pp.703-8.

Maestre, R.; Rey, G.; De las Salas, J.; Vergara, C.; Santacoloma, L.; Goenaga, S. & Carrasquilla, MC. (2009). Susceptibilidad de *Aedes aegypti* (Diptera: Culicidae) a temefos en Atlántico-Colombia. *Revista Colombiana de Entomología,* Vol.35, No.2, pp.202-205.

Maestre-Serrano, R.; Goenaga, S.; Carrasquilla-Ferro MC.; Rey-Vega G.; Santacoloma-Baron L. & Vergara-Sánchez C. (2009). Estado de la susceptibilidad/resistencia a insecticidas en cuatro poblaciones de *Aedes aegypti* del departamento del Atlántico. *Biomedica*, Vol.29, No.1, pp.231-2.

Maestre, RY.; Florez, Z.; Cabrera, C.; Goenaga, S.; Gómez, D. & Gómez, C. (2010). Susceptibility status of *Aedes aegypti* to insecticides in La Guajira (Colombia). *The American Journal of Tropical Medicine and Hygiene,* Vol.83, No.5, pp.53.

Maestre, R.; Rey, G.; De las Salas, J.; Vergara, C.; Santacoloma, L.; Goenaga, S. & Carrasquilla, MC. (2010). Estado de la susceptibilidad de *Aedes aegypti* a insecticidas en Atlántico (Colombia). *Revista Colombiana de Entomología*,Vol.36, No.2, pp.242-248.

Maestre-Serrano, R.; Goenaga, S.; Flores, Z.; Cabrera C.; Gómez, D. & Gómez, C. (2010). *The Amercican Journal of Tropical Medicine and Hygiene* Vol.83, No.5, pp.53.

Martins, AJ.; Pereira-Lima, JB.; Peixoto, AA. & Valle, D. (2009). Frequency of Val1016Ile mutation in the voltage-gated sodium channel gene of *Aedes aegypti* Brazilian populations. *Tropical Medicine and International Health*, Vol.14, No.11, pp.1351-55.

Martinez-Torres, D.; Chandre, F.; Williamson, MS.; Darriet, F.; Berge, JB.; Devonshire, AL.; Guillet, P.; Pasteur, N. & Pauron, D. (1998). Molecular characterization of pyrethroid knockdown resistance (kdr) in the major malaria vector Anopheles gambiae s.s. *Insect Molecular Biology,* Vol.7, pp.179–184.

Miller TA. 1988. Mechanisms of resistance to pyrethroid insecticides. *Parasitology Today*, Vol.4, pp.8-12.

Ocampo, C.; Salazar-Terreros, M.; Mina, N.; McAllister, J. & Brogdon, W. (2011). Insecticide Resistance status of *Aedes aegypti* in 10 localities of Colombia. *Acta Trop.*, Vol.118 No.1, pp.37-44.

Orjuela, L.; Herrera, M.; Quintana, N. & Quiñones, M. (2007). Estado de la susceptibilidad a insecticidas del mosquito *Aedes aegypti* en lo municipios de Chinchiná y La Dorada, Caldas. *Biomedica*, Vol.27, No.2, pp.176.

Pereira-da-Cunha, M.; Pereira-Lima JB.; Brogdon, WG.; Efrain-Moya G. & Valle D. (2005). Monitoring of resistance to the pyrethroid cypermethrin in Brazilian *Aedes aegypti* (Diptera: Culicidae) populations collected between 2001 and 2003. *Memorias do Instituto Oswaldo Cruz*, Vol.100, No.4, pp.441-44.

Pereira-Lima, E.; Da Oliveira-Filho, A.; Da Oliveira-Lima, J.; Ramos-Junior, A.; Da Goes- Cavalcanti, L. & Soares-Pontes, R. (2006). *Aedes aegypti* resistance to temephos in counties of Ceará State. *Revista da Sociedade Brasileira de Medicina Tropical,* Vol.39, No.3, pp.259-63.

Polson, KA.; Rawlins, SC.; Brogdon, WG. & Chadee, DD. (2010). Biochemical mechanisms involved in DDT and pyrethroid resistance in Trinidad and Tobago strains of *Aedes aegypti. The American Journal of tropical medicine and hygiene*, Vol.83, No.5, pp.51.

Poupardin, R.; Reynaud, S.; Strode, C.; Ranson, H.; Vontas, J. & David JP. (2008). Cross- induction of detoxification genes by environmental xenobiotics and insecticides in the mosquito Aedes aegypti: Impact on larval tolerance to chemical insecticides. *Insect Biochemistry and Molecular Biology*, Vol.38, pp.540-51.

Quiñones, ML.; Suarez, MF. & Fleming, GA. (1987). Estado de la susceptibilidad al DDT de los principales vectores de malaria en Colombia y su implicación epidemiológica. *Biomédica*, Vol.7, pp.81-6.

Rawlins S. (1998). Distribución espacial e importancia de la resistencia a insecticidas de poblaciones de *Aedes aegypti* en el Caribe. *Revista Panamericana de Salud Pública,* Vol.4, No.4, pp.243-51.

Rodríguez, M.; Bisset, J.; Fernández, D. & Pérez, O. (2004). Resistencia a insecticidas en larvas y adultos de *Aedes aegypti*: prevalencia de esterasa A4 asociada con la resistencia a temefos. *Revista Cubana de Medicina Tropical*, Vol.56, No.1, pp.54-60.

Rodríguez, MM.; Bisset, JA. & Fernández, D. (2007). Determinación *in vivo* del papel de las enzimas esterasas y glutation transferasa en la resistencia a piretroides en *Aedes aegypti* (Diptera: Culicidae). *Revista Cubana de Medicina Tropical,* Vol.59, No.3, pp.209-12.

Rodríguez, M. (2008). Estudio de la resistencia a insecticidas en *Aedes aegypti* (Diptera: *Culicidae*). Tesis presentada al grado científico de Doctor en Ciencias de la Salud. Laboratorio de Toxicología y genética. Instituto de Medicina Tropical "Pedro Kourí". 128p.

Rojas, W.; Gonzalez, J.; Amud, MI.; Quiñones, ML. & Velez, ID. (2003). Evaluación de la susceptibilidad de Aedes aegypti del municipio de Barrancabermeja Santander a los insecticidas Malatión Fenitrotion, temefos, Lambdacihalotrina, deltametrina, permetrina, propoxur y DDT. *Biomédica,* Vol.23, No.1, pp.56

Rojas-Alvarez, DP. (2010). Comportamineto epidemiologico del dengue en Colombia, año 2009. *Informe quincenal Epidemiologico Nacional*, Vol.15, No.9, pp.129-37.

Saavedra-Rodriguez, K.; Urdaneta-Marquez, L.; Rajatileka, S.; Moulton, M.; Flores, AE.; Fernandez-Salas, I.; Bisset, J.; Rodriguez, M.; Mccall, PJ.; Donnelly, MJ.; Ranson, H.;

Salazar, M.; Carvajal, A.; Cuellar, ME.; Olaya, A.; Quiñones, J.; Velásquez, OL.; Viveros, A. & Ocampo, C. (2007). Resitencia a insecticidas en poblaciones de *Aedes aegypti* y *Anopheles* spp en los departamentos de Huila, valle Cauca y Nariño. *Biomédica*, Vol.27, No.2, pp.177.

Santacoloma, L. (2008). Estado de la susceptibilidad a insecticidas de poblaciones naturales de *Aedes aegypti* Linnaeus 1762 vector del dengue y *Anopheles darlingi* root 1926 vector primario de malaria (Diptera: Culicidae) en cinco departamentos de Colombia. [Tesis de maestría]. Bogotá: Universidad Nacional de Colombia.

Santacoloma-Varon, L.; Chavez-Cordoba, B. & Brochero, HL. (2010). Susceptibilidad de *Aedes aegypti* a DDT, Deltametrina y Lambdacialotrina en Colombia. *Revista Panamericana de Salud Publica*, Vol.27, No.1, pp.66-73. http://www.todacolombia.com/geografia/ubicacion.html

Valderrama-Eslava, EI.; Gomzalez, R. & Jaramillo, GI. (2008). Evaluación de la susceptibilidad de *Aedes aegypti* (L) (Diptera: Culicidae) a un insecticida Organosfosforado y un piretroide en cuatro poblaciones del Valle del Cauca, mediante dos tipos de bioensayos. *Boletin del Museo de Entomología de la Universidad del Valle*, Vol.9, No.2, pp.1-11.

World Health Organization (WHO). (1957). Seventh report Expert Committee on insecticides *WHO Tech Report Ser*, Vol.125, pp.37.

World Health Organization (WHO). (1981a). Instructions for determining the susceptibility or resistance of adult mosquitoes to organoclorine, organophosphorus and carbamate insecticides. Establishment of the base-line. WHO/ VBC/81.805.

World Health Organization (WHO). (1981b). Instructions for determining the susceptibility or resistance of mosquitoes larvae to insecticide. Unpublished document.WHO/ VBC/81.807.

Yanola, J.; Somboon, P.; Walton, C.; Nachaiwieng, W. & Prapanthadara L. (2010). A novel F1552/C1552 point mutation in the *Aedes aegypti* voltage-gated sodium channel gene associated with permethrin resistance. *Pesticide Biochemistry and hysiology*, Vol.96, pp.127-13.

CHAPTER

9

Behavioral Reactions of Mosquitoes to Insecticides

1. INTRODUCTION: AN OVERVIEW

A significant number of individuals residing in tropical and subtropical regions face a substantial risk of contracting various vector-borne diseases, particularly malaria and dengue. Globally, an estimated 50-100 million individuals are at risk of dengue virus infection (the causative agents of dengue fever and dengue hemorrhagic fever), while between 100-300 million reside in malaria-endemic regions (World Health Organization [WHO], 2009). Dengue viruses are primarily transmitted by Aedes aegypti, an urban, diurnal mosquito that typically inhabits areas near human residences and preferentially feeds on humans (Gubler, 1997); in contrast, the four human malaria parasites (Plasmodium) are transmitted by various Anopheles species (Service & Townson, 2002). The dengue vector has demonstrated remarkable resilience to control methods due to its intimate interaction with and exploitation of domestic and peridomestic human environments (Reiter & Gubler, 1997). Conversely, malaria vectors exhibit a broader spectrum of host-seeking behaviors, preferences, biting patterns, and larval breeding habitats (Pates & Curtis, 2005; Sinka et al., 2011). Despite decades of comprehensive research, effective and commercially feasible vaccines for these two significant vector-borne diseases remain unavailable. Consequently, the prevention and control of dengue and malaria rely on diverse vector control strategies to mitigate transmission risk; in certain cases, this necessitates the application of various chemical insecticides, including larvicides, space sprays, and indoor residual spraying (IRS), as well as the utilization of insecticide-impregnated bed nets to manage adult mosquito blood feeding. (Roberts & Andre, 1994; WHO, 1999; Reiter & Gubler, 1997; Grieco et al., 2007).

Chemical insecticides, including organochlorines, organophosphates, carbamates, and synthetic pyrethroids, have long been used with great effect in public health vector control programs worldwide (Reiter & Gubler, 1997; Roberts & Andre, 1994; WHO, 1992). Although DDT used ceased in many countries several decades ago, the chemical has returned for use in malaria control IRS programs in Africa because of some its superior attributes (Roberts & Tren, 2010). The dramatic impact of

DDT on mosquito populations in terms of both toxicity and behavior suppressing disease transmission is well known but in some instances the actual mechanisms at work remain unclear and poorly understood. Most studies on insecticides have placed attention exclusively on the direct toxicological (knockdown and killing) effects on mosquito populations; whereas far less research has focused on the behavioral responses and outcomes as a result of chemical exposure (Roberts, 1993; Roberts et al., 1997; Chareonviriyaphap et al., 1997). Research indicates that numerous chemical compounds affect insect locomotor behavior, frequently leading to significant excitation and premature movement away from treated surfaces or areas, with one or more kinetic mechanisms causing what is termed "avoidance behavior" (Kongmee et al., 2004; Miller et al., 2009). The term "excito-repellency" refers to behavior induced by direct contact with a chemical, leading to abnormal excitation (often called 'irritancy') and spatial repellency occurring without physical contact with the chemical (Roberts et al., 1997). This chapter delineates the diverse behavioral responses to insecticides in significant malaria vectors (Anopheles species) (Table 1), dengue vectors (Aedes aegypti) (Table 2), and the tropical house mosquito, Culex quinquefasciatus (Table 3), employing two laboratory systems and field assays utilizing experimental huts.

2. HISTORY OF TEST SYSTEMS TO STUDY MOSQUITO BEHAVIORS

Behavioral responses of mosquitoes to chemical compounds can be directly demonstrated by using various laboratory devices and field assay systems. For laboratory assays, many of the variations have been reviewed (Roberts et al., 1997). The WHO developed the first test box using plywoood in its construction attempting to access the excitation ("irritability") of exposed mosquitoes following physical contact with insecticides (WHO, 1970). This system was subsequently referred to as an "excito-repellency" test box (Rachou et al., 1963). Subsequently, the test system was further modified by other investigators interested in behavioral avoidance responses exposed to DDT and some of the early synthetic pyrethroids (Charlwood & Paraluppi, 1978; Roberts et al., 1984; Bondareva et al., 1986; Rozendaal et al., 1989; Quinones & Suarez, 1989; Ree & Loong, 1989). Years later, a lightproof test chamber was designed to study the irritant response of *Anopheles gambiae*, an importent malaria vector in Africa, to several chemical compounds (Evans, 1993). One key concern with all these test systems was associated with the technical difficulties of the test boxes for introducing and removing test specimens. Other concerns were controling for the various physiological conditions of wild-caught mosquitoes, and selecting the ideal range of concentrations for chemical compounds. In addition, no single or set of statistical methods for analysis of data has been fully accepted nor has any test system been specifically designed to truely discriminate between contact excitation and noncontact repellency responses (Roberts et al., 1997).

An improved excito-repellency test device that was able to clearly differentiate between excitation and spatial repellency was developed and initially tested against several field populations of *Anopheles albimanus* from Central America (Roberts et al., 1997; Chareonviriyaphap et al., 1997). Unfortunately, this fixed prototype was cumbersome to handle and required considerable time for attaching the chemcial-treated test papers. Several years on, a more field-friendly test system was designed that was both collapsible and easily transportable (Chareonviriyaphap & Aum-Aong, 2000; Chareonviriyaphap et al., 2002) and was used to investigate the behavioral responses of various mosquito species and geographical populations from Thailand and a few populations from elsewhere in Asia (Chareonviriyaphap et al., 2001; Sungvornyothin et al., 2001; Kongmee et al., 2004; Pothikasikorn et al., 2005). Recently, a novel modular assay system was developed for mass screening of chemical actions; including contact irritancy, spatial repellency, or toxicity responses, on adult mosquitoes (Grieco et al., 2005). This modular system is substantially reduced in size compared to the previous excito-repellency box and minimizes the treated surface area and therefore the amount of chemical required for testing. In field observations, numerous attempts have been made to determine behavioal responses of mosquitoes using specially constructed experimental huts (Smith, 1965; Roberts et al., 1984, 1987; Roberts & Alecrim, 1991; Rozendaal et al., 1989; Bangs, 1999; Grieco et al., 2000; Grieco et al., 2007; Polsomboon et al., 2008; Malaithong et al., 2010). Most experimental hut studies have been conducted to observe the behavior of *Anopheles* mosquitoes with far fewer studies on other genera. Grieco et al., (2007) successfully demonstrated all 3 chemical actions using *Ae. aegypti* as a model system. The results obtained from both laboratory and field studies can help faciliate the choice of the most effective chemical measures to control house-frequenting adult mosquitoes.

Two standard systems used in the laboratory, the excito-repellency box (ERB) and the high throughput screening system (HITSS), and the experimental hut in the field to help characterize the behavioral responses of mosquitoes to chemical compounds are discussed herein.

2.1 Excito-Repellency Box (ERB) test system

Given the complexities of insect behavioral research, the excito-repellency testing design and testing methodology along with methods of analyzing and interpreting test data, have yet to be universally accepted. However, tremendous improvement has occurred in recent years that effectively addressed a number of the previous drawbacks, opening up significance progress in behavioral studies involving innate responses to insecticides. The ERB test system (Fig.1) was developed to distinguish irritancy (Fig. 2) and repellency (Fig. 3) (Roberts, et al., 1997) and was first used to study the avoidance behavior of *Anopheles albimanus* to DDT and synthetic pyrethroids in Belize, Central America in which consistent and reliable results were generated measuring the escape behavior of female mosquitoes (Chareonviriyaphap et al., 1997). Over time, the initial system was modified into a collapsible test chamber with identical dimensions and operational attributes to help alleviate some of the previous handicaps (Chareonviriyaphap et al., 2002). This improved version has been used extensively in the evaluation of behavioral responses by laboratory and field mosquito populations in Thailand.

In 2006, a more field compatible device was designed with a substantial reduction in chamber size to minimize the amount of chemical and treated paper required, provide greater ease of test preparation and allow use of a small number of test specimens (15 vs. 25 females) compared to the larger version (Tanasinchayakul et al., 2006). Using this system, consistent and reliable results have been produced (Muenworn et al., 2006; Polsomboon et al., 2008; Thanispong et al., 2009; Mongkalagoon et al., 2009).

The system comprises of 4 outer stainless steel sheet metal walls (Fig. 1). Each wall is constructed with an aluminum sliding rib on each end and socket providing a surface for the test paper holder in the middle. The test paper holder consists of 2 sides: a sheet of fine mesh iron screen attached on one side and a panel to hold the test papers in place on top on the opposite side. There is a 0.8 cm gap between the test paper and screen barrier to prevent mosquitoes from making physical contact ('noncontact') on the treated paper surface during the repellency assay. The paper holder simply has to be inverted to provide the proper conditions to expose the paper in the contact test. On one end of the chamber a portal is made of overlapping sheets of dental dam and is used for placing mosquitoes inside the chamber and later for removing them after the exposure test period. A Plexiglas door serves to seal the chamber and allow the investigator to view the exposure chamber before and after the actual test period. A stainless steel outer rear door cover is used to shut off all external light inside the chamber when the experiment is being performed.

Each tests consists of enclosing 15 female mosquitoes in each of 4 chambers lined with either insecticide-treated or untreated (control) papers. On the opposite side of the Plexiglas portal is a single exit portal for mosquitoes to escape to a receiving cage. At the beginning of a test, a 3-min rest period allows mosquitoes to adjust to test chamber conditions before the exit portal is opened to initiate the observation period (30 or 60 min depending on experiment). The numbers of mosquitoes escaping from the chamber into receiving cage are recorded at 1 min intervals until test completion.

Fig. 1. Excito-repellency test chamber showing side of exit portal (Roberts et al., 1997).

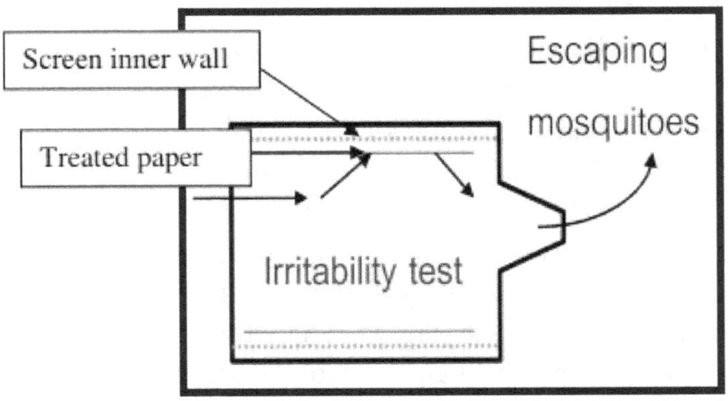

Fig. 2. Excito-repellency test chamber: Irritability test design (Roberts et al., 1997).

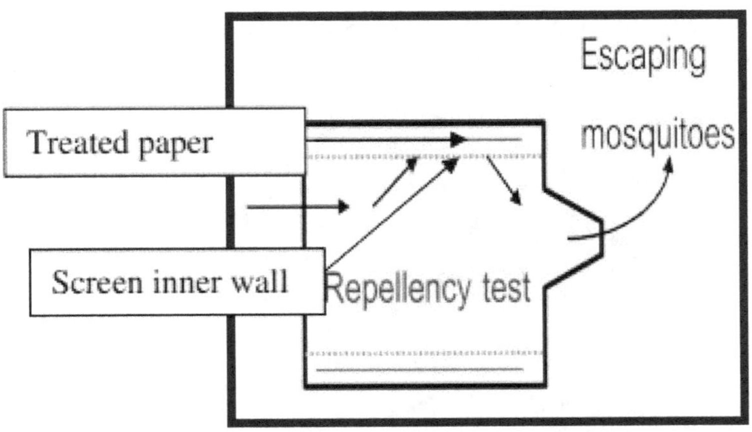

Fig. 3. Excito-repellency test chamber: Repellency test design (Roberts et al., 1997)

2.2 High Throughput Screening System (HITSS)

Even though an ERB test system has been found valuable to evaluate the innate behavioral responses of mosquito to chemical compounds, this system is relatively resource intensive requiring a comparatively large amount of chemical to be used for treating papers and having to use a large number of test mosquitoes. Moreover, a current ERB design is not conducive for mass screening the candidate chemical compounds. With the development of the HITSS assay (Grieco et al., 2005), a smaller amount of chemical and a lower number of mosquitoes is required per test. In addition, the HITSS can also be configured to test each of the three actions of insecticide compounds, namely contact irritancy assay (CIA) (Fig. 4), spatial repellency assay (SRA) (Fig. 5) and the toxicity assay (TOX) (Fig. 6). This modular system is made from a variety of durable materials including a thick, clear plastic cylinder, metal chambers, hard plastic end caps and a butterfly valve to control the opening and closing of the door. For the CIA, a single clear plastic cylinder is connected to a metal chamber lined with either treated or untreated netting material using a butterfly value placed in the open position (Fig. 4). For the SRA, an assay comprises a single clear plastic cylinder and 2 metal chambers on either end containing either treated or untreated (control) netting. The plastic chamber is positioned between the 2 metal chambers using a butterfly value as a linking system (Fig. 5). For the TOX assay, a single metal chamber is equipped with plastic end caps (solid cap on one side and tunnel cap on the other end). Netting material treated with either chemical active ingredient (treatment) or acetone carrier only (control) lines the inner chamber (Fig. 6). The HITSS has been standardized and used to evaluate the 3 behavioral responses of *Ae. aegypti* against DEET, Bayrepel®, and SS220 (Grieco et al., 2005, 2007). Recently, HITSS has been used to define the behavioral responses among six field populations of *Ae. aegypti* from Thailand against three synthetic pyrethroids (Thanispong et al., 2010). From this study, it was clearly shown that the HITSS assay is an effective and easy to use tool for distinguishing the three actions of chemicals and screening new compounds.

Fig. 4 High throughput screening system (HITSS): (CIA) Contact Irritant Assay (Grieco et al., 2005)

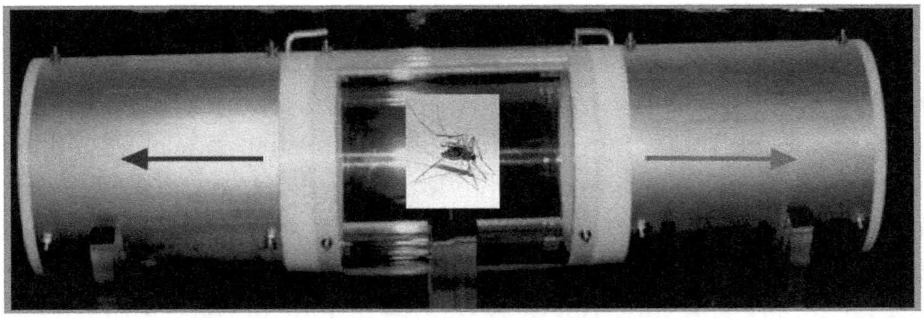

Fig. 5. High throughput screening system (HITSS): (SRA) Spatial Repellency Asay (Grieco et al., 2005)

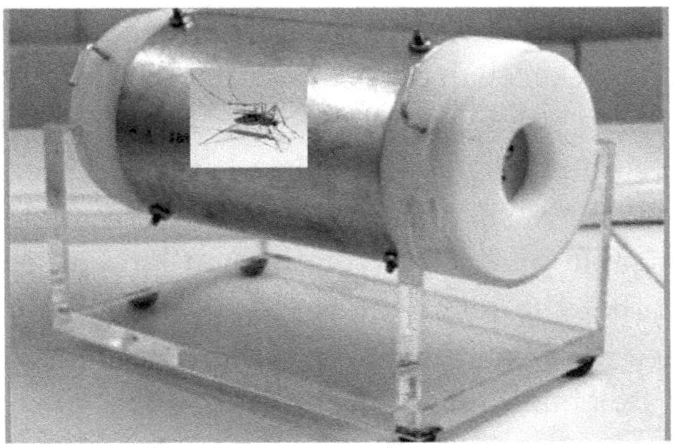

Fig. 6 High throughput screening system (HITSS): (TOX) Toxicity Assay (Grieco et al., 2005)

2.3 Field-based assay: experimental huts

To better understand the behavior of a mosquito exposed to a residual avoidance of under more natural, realistic conditions, field studies should be performed using experimental huts. So-called hut studies provide valuable information that can facilitate vector control operations by helping select the most appropraite and effective tool in combating disease vectors. Since the 1940s, most attention has been directed to the study of *Anopheles* mosquitoes, yet few investigations have been carried out to describe the innate behaviors of mosquitoes entering, resting, biting (blood feeding) and exiting human dwellings (Gahan & Lindquist, 1945; Giglioli, 1948; Smith, 1965; Roberts et al., 1984; Bangs, 1999; Grieco et al., 2000; Pates & Curtis, 2005; Roberts et al., 1997). Another important vector species, *Ae. aegypti*, has received even less attention using of experimental hut assays (Suwannachote et al., 2009). The discovery of using natural pyrethrum extract to prevent human-vector contact inside homes was well known prior to 1945 (Muirhead-Thomson, 1951) and the first report of behavioral responses of malaria vectors to DDT was documented in 1947 (Kennedy, 1947, Brown, 1983). Strong behavioral responses to DDT were progressively reported up until 1975 (Muirhead-Thomson, 1960; Elliott & de Zulueta, 1975; Brown, 1983). For example, de Zulueta and Cullen (1963) observed significant reductions in the numbers of biting mosqutioes on human indoors and resting on walls in houses sprayed with DDT. However, there was no clear explanation on how (mechanism) DDT functioned by either repelling or preventing mosquitoes from locating host stimuli inside sprayed houses. A significant scientific observation on behavioral avoidance of *Anopheles ganbiae* was carried out in the mid 1960s (Smith & Webley, 1968) wherein 2 huts, 1 control and one treated with DDT were equipped with verandah traps and gas chromatography was used to evaluate the subsequent response of mosquitoes. A hut treated with DDT was observed to prevent between 60 and 70% of *An. gambiae* from normally entering indoors as an outflow of DDT was continuously decreasing. Additionally, those females that did enter the sprayed hut became stimulated (irritated) and escaped the hut much quicker than those mosqutioes present in the unsprayed hut. Another study comparing biting patterns of *Anopheles minimus* in Thailand showed a 71.5% decline in attempted blood feeding inside the DDT treated hut and a 42.8% reduction in blood feeding success in a deltamethrin treated hut (Polsomboon et al., 2008). Using the same huts, observing human landing patterns of *Anopheles dirus* complex found that the relative risk (odds) of female mosquitoes entering and attempting to feed were half the number when exposed to DDT compared with the deltamethrin treated hut (Malaithong et al., 2010).

Of the possible responses a mosquto exposed to a chemcial can preform, spatial repellency is prehaps the most interesting and important. Observing true repellency presents problems of accurate messurement in the natural field situation. One of classic methods is to use the experimental huts wherein 2 experimental huts, 1 untreated control and 1 treated with an active ingredient is fitted with entrance (measure of repellency) and exit (irritancy) traps as illustrated in (Figs. 7-9). A landmark study was conducted by Grieco et al., (2007) on the movement patterns of *Ae. aegpyti* into and out of the experimental huts alternatively equipped with entrance (Fig. 8) and exit traps (Fig. 9) to describe the 3 chemical actions and mosquito responses as previously proposed in the mathematical framework for understanding the impact and relevance of repellency, irritability and toxicity on mosquito populations and disease transmission (Roberts et al. 2000). This study clearly showed that the 3 actions described and help to validate the model of actions. It was concluded that contact irritancy is the predominant action of synthetic pyrethroids, whereas spatial repellency is the primary action of DDT. Dieldrin, a once commonly used cyclodiene compound, exhibited primarily a toxic action only.

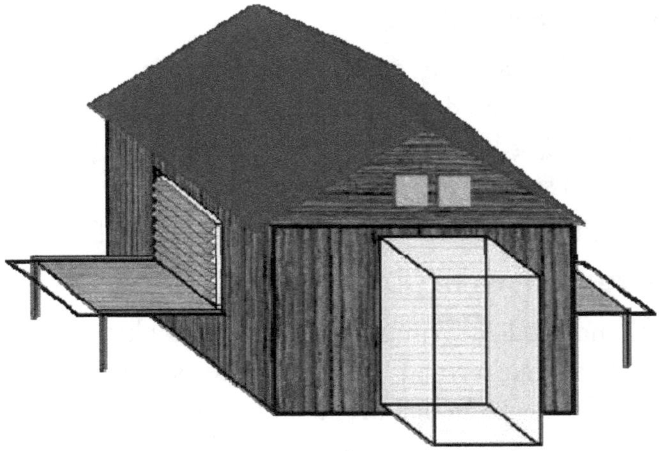

Fig. 7 Experimental hut with traps attached to exterior walls.

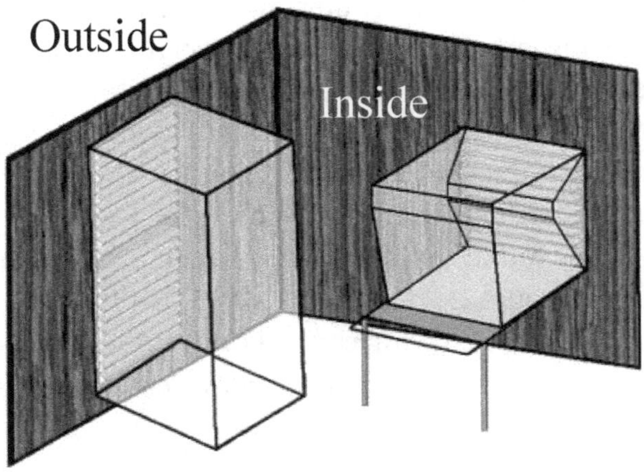

Fig. 8 Experimental hut with entrance traps attached to interior walls.

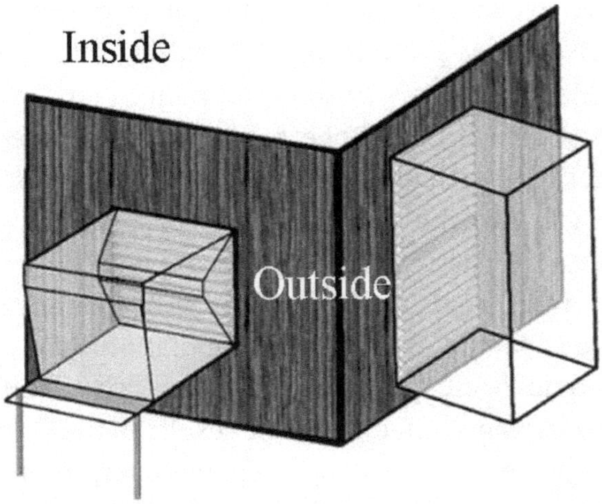

Fig. 9. Experimental hut with exit traps attached to exterior walls.

3. BEHAVIORAL RESPONSES

To facilitate the study of behavioral responses of mosquitoes to chemical compounds, several test systems have been developed (Roberts et al., 1997; Chareonviriyaphap et al., 2001). The following is a brief review of important historical findings and more recent work from studies in Thialand on behavioral responses of mosquitoes to DDT and various synthetic pyrethroids commonly used in vector control.

3.1 Behaviroral responses to DDT

Some agricultural and medically important insects, including vectors of malaria, demonstrate what has been termed "behavioral resistance" to DDT (Lockwood et al., 1984). However, the term "behavioral avoidance" as an innate response rather than a permanent genetic shift in behavior is preferred since the development of behavioral changes due to insecticide selective pressure in nature has not been adequately documented (Muirhead- Thomson, 1960). The behavioral responses of mosquitoes have been investigated using either specially constructed experimental huts and/or ERB test system. The first study on the irritant effect of DDT residual deposits was conducted on *Anopheles quadrimaculatus* where females were found to be irritated after short contact with treated surfaces with many quickly escaping the DDT treated house without taking a blood meal (Gahan & Lindquist, 1945). Subsequent observations found that *An. quadrimaculatus* had received a lethal dose and perished within 24 hours (Metcalf et al., 1945). Unfortunately, these studies made the observations without having control (untreated houses) for comparison. Moreover, the high mortality seen with *An. quadrimaculatus* may have been caused by further contacts with toxic active ingredients while attempting to leave a treated house through a small outlet (Muirhead-Thomson, 1960). In the studies with *Anopheles albimanus* in Panama, Trapido (1954) concluded that wild-caught mosquitoes lacking re-exposure to DDT for a long period of time, showed the same susceptibility levels to DDT as those from a laboratory colony with no a history of previous exposure. Malaria vectors in some countries (e.g., Brazil, Thailand) have never developed resistance to DDT (Roberts et al., 1984; Chareonviriyaphap et al., 2001), suggesting that the particular mosquito population possibly avoids making direct physical contact with the chemical, thereby precluding any selection for resistance. Table 1 lists *Anopheles* species tested and levels of behavioral responses to DDT and synthetic pyrethroids. DDT was found to be a strong contact irritant and more moderately as a repellent among 3 test populations of *An. albimanus* and that most specimens that quickly escaped DDT exposure survived (Chareonviriyaphap et al., 1997). This finding is in agreement with the results of the field studies by Roberts and Alecrim (1991) who reported a strong repellent action of DDT with *Anopheles darlingi* in a sprayed house. In a similar study, both irritancy and repellency escape responses were observed in 2 populations of *Anopheles minimus* with a major action of DDT being contact excitation (Chareonviriyaphap et al., 2001). Another study was made on the 2 complex species within the Minimus Subgroup; *An. minimus* (species A) and *Anopheles harrisoni* (species C). DDT produced a rapid and striking irritancy response in both species. Additionally, repellency was more pronounced with DDT on *An. minimus* but was seen to be much weaker with *An. harrisoni* (Pothikasikorn et al. 2005). With *Ae. aegypti*, results demonstrated that the higher the degree of physiological resistance to DDT, the greater the apparent suppression of both contact irritant and noncontact repellency responses, yet avoidance behavior was still a significant event (Thanispong et al., 2009).

Table 1. Degree of behavioral responses of female *Anopheles* species to synthetic pyrethroids and DDT at an operational field dose (mg/cm2).

Species	Strain	Permethrin		Deltamethrin		Cypermethrin		λ-cyhalothrin		Bifenthrin		DDT		References
		I	R	I	R	I	R			I	R	I	R	
An. albimanus	Lab	+	-	−	-							−	-	Chareonviriyaphap et al., 1997
An. albimanus	Field	−+++	-	−+++	-							+−++	++	Chareonviriyaphap et al., 1997
An. albimanus	Field	−+++	-	−+++	-							+−++	++	Chareonviriyaphap et al., 1997
An. albimanus	Field	I I I	-	I I I	-							I I I	I I	Chareonviriyaphap et al., 1997
An. minimus	Lab			I I I	I			I I	-			I I	I	Chareonviriyaphap et al., 2001
An. minimus	Field			++	+			+−++	-			++−	+	Chareonviriyaphap et al., 2001
An. minimus	Lab			−+++	++−									Chareonviriyaphap et al., 2004
An. minimus	Lab			I I I	I I									Chareonviriyaphap et al., 2004
An. dirus	Lab			I I	I I									Chareonviriyaphap et al., 2004
An. maculatus	Field			−+++	+									Chareonviriyaphap et al., 2004
An. sawadwongporni	Field			−+++	++									Chareonviriyaphap et al. 2004
An. dirus	Field			−++	++									Chareonviriyaphap et al. 2004
An. minimus	Field			I I I	I I			I I I	-			I I I	I I	Potikasikorn et al., 2005
An. harrisoni	Field			−+++	-			+−++	-			+−++	-	Potikasikorn et al., 2005
An. maculatus	Field	+++	-									−	-	Muenworn et al., 2006
An. sawadwongporni	Field	+++	-									++−	-	Muenworn et al., 2006
An. minimus	Field					−+	+							Pothikasikorn et al., 2007
An. minimus	Field									+	-			Tisgratog et al., 2011
harrisoni	Field									+−+	−			Tisgratog et al., 2011

I: Irritancy (Excitation and movement away from the source following direct contact with chemical stimulant)
R: Repellency (noncontact, spatial detection of chemical that results in movement away from the source)
++++: 81-100% escaped from treated chamber
+++: 61-79% escaped from treated chamber
++: 41-59% escaped from treated chamber
+: 21-40% escaped from treated chamber
-: ≤ 20% escaped from treated chamber but statistically significant ($P < 0.05$) from the matched control

3.2 Behavioral response to synthetic pyrethroids

A number of synthetic pyrethroids, i.e. allethrin, deltamethrin, permethrin, cypermethrin, alpha-cypermethrin, cyfluthrin among others, are commonly used by home owners, private business and government sectors to control both household mosquitoes and those as important vectors. These pyrethroids have found to exhibit a moderate to strong repellent effect for many agricultural and medically important insects (Lockwood et al., 1984) and were observed to cause mosquitoes to move away ('avoidance') from sprayed areas (Miller, 1990; Lindsay et al., 1991). The extensive and continuing use of pyrethroids should be a major stimulus to intensify observations on the significance of pyrethroid-induced avoidance behavior in mosquito vectors and other arthropods. Given the role of indoor residual spraying of homes as a means of controlling malaria transmission, the precise role of excitation and repellent actions of pyrethriods should be well defined for specific malaria vectors prior to beginning any large scale control program. Following the refinement of the ERB test system allowing separation of the 2 types of primary behavioral actions (Roberts et al., 1997), a series of important findings on excito-repellency behavior in *Anopheles* mosquitoes have been subsequently reported (Chareonviriyaphap et al., 1997, 2001, 2004; Pothikasikorn et al., 2005; Muenworn et al., 2006; Pothikasikorn et al., 2007). In general, synthetic pyrethroids produce much stronger irritant responses in *Anopheles* compared to repellency action (Table 1). For example, lambda-cyhalothrin and deltamethrin act as strong irritants on test populations of *An. minimus* complex mosquitoes while showing relatively weak repellency action (Chareonviriyaphap et al., 2001). Pothikasikorn et al. (2005) confirmed that Minimus complex species, *An. minimus* and *An. harrisoni* show a rapid irritancy to lambda-cyhalothrin and deltamethrin. Chareonviriyaphap et al., (2004) produced an extensive study to define the excito-repellency action of deltamethrin on 4 *Anopheles* species, all representing important malaria vectors in Thailand. Again, the findings demonstrated that deltamethrin produced a pronounced irritancy action compared to a much weaker repellency effect. Although repellency was less profound than contact excitation, the escape responses were statistically significant compared to the matched controls. A number of *Ae. aegypti* populations have been tested against a series of synthetic pyrethroids (deltamethrin, permethrin, alphacypermethrin, cyphenothrin, d-tetramethrin and tetramethrin) (Table 2). In general, all test populations of *Ae. aegypti* populations exhibit moderate to strong irritancy as compared with repellency (Grieco et al., 2005; Chareonviriyaphap et al., 2006; Paeporn et al., 2007; Thanispong et al., 2009, 2010). In addition, a few populations of *Culex quinqaefacsciatus* have been tested against the 3 principal classes of insecticides used in vector control; pyrethroids (deltamethrin), organophosphates (fenitrothion) and carbamates (propoxur) (Table 3). Striking differences in behavioral responses were seen between populations and active ingredients. Greater contact escape action was observed in a long-established colony exposed to deltamethrin and fenitrothion compared to two recent field populations (Sathantriphop et al., 2006).

To summarize, the behavioral responses to insecticides by mosquitoes are important components of a chemical's overall effectiveness in reducing human-vector contact and transmision of disease. To date, there is no convincing example of behavioral resistance in mosquito species to insecticides, rather all evidence indicates actions on the part of exposed mosquitoes are part of an innate behaviral repertoire. Behavioral response can be split into 2 distinct categories, stimulus-dependent and stimulus-independent actions (Georghiou, 1972). A stimulus-dependent response requires sensory stimulation of the insect in order for an avoidance action to proceed. In general, this form of avoidance enables the insect to detect a chemical on direct contact or spatially before acquiring a lethal dose (Muirhead- Thomson, 1960). On the other hand, a stimulus-independent response does not require direct sensory stimulation of insect for avoidance to occur but rather involves other natural behavioral components such as exophily (outside resting) or zoophily (non-human blood preference) in which an insect avoids exposure to a chemical by preferentially utilizing habitats without active ingredients present (Byford & Sparks, 1987). This type of response has also been included in so-called "phenotypic and genotypic behaviors" (WHO, 1986). Stimulus-dependent behavioral responses include the avoidance behaviors discussed in detail in this chapter. The term *avoidance behavior* is generally used to describe actions that are stimulated by some combination of excitation (irritancy) and repellency, the former taking place following physical contact while spatial repellency results without physical contact with an insecticide (Roberts et al., 1987).

Table 2. Degree of behavioral responses of female *Aedes aegypti* to synthetic pyrethroids and DDT at an operational field dose (mg/cm2).

Strains	Deltamethrin		Permethrin		α-cypermethrin		Cyphenothrin		d-tetramethrin		Tetramethrin		DDT		References
	I	R	I	R	I	R	I	R	I	R	I	R	I	R	
Field-R	++	-													Kongmee et al., 2004
Field-R	+	-													Kongmee et al., 2004
Field-R	+	-													Kongmee et al., 2004
Field-R	++++	-													Kongmee et al., 2004
Field-R	++	-													Kongmee et al., 2004
Field-S	+++	-													Kongmee et al., 2004
Lab-S	++	-													Kongmee et al., 2004
Lab-S	++	-													Kongmee et al., 2004
Lab-S	++	-													Kongmee et al., 2004
Field-RR			+	-											Paeporn et al., 2007
Field-RR			-	-											Paeporn et al., 2007
Field					++++	-							++	-	Thanispong et al., 2009
Field-S					+++	-							-	-	Thanispong et al., 2009
Lab-S					++++	++							++	++	Thanispong et al., 2009
Field-S	+++	+					++++	-	+++	+	++	+			Mongkalagoon et al., 2009
Field-S	++++	-					++++	+	+++	+	+++	-			Mongkalagoon et al., 2009
Field-S	++++	+					++++	+	+++	-	+++	+			Mongkalagoon et al., 2009

I: Irritancy (contact excitation), R: Repellency (noncontact/spatial), R: Resistant, S: Susceptible to test compound
++++: 81-100% escaped from treated chamber
+++: 61-79% escaped from treated chamber
++: 41-59% escaped from treated chamber
+: 21-40% escaped from treated chamber
-: ≤ 20% escaped from treated chamber but statistically significant ($P < 0.05$) from the matched control

Table 3. Degree of behavioral responses of *Culex quinquefasciatus* to deltamethrin, propoxur and fenitrothion at an operational field dose (mg/cm2).

Populations	Deltamethrin		Fenitrothion		Propoxur		References
	I	*R*	*I*	*R*	*I*	*R*	
Bangkok*	++	-	++	-	+	-	Sathantriphop et al., 2006
Nontaburi*	+	-	+	-	+	-	Sathantriphop et al., 2006
Mae Sot*	+	-	+	-	+	-	Sathantriphop et al., 2006

**Highly resistance to deltamethrin I: Irritancy, R: Repellency*

++++: 81-100% escaped from treated chamber

+++: 61-79% escaped from treated chamber

++: 41-59% escaped from treated chamber

+: 21-40% escaped from treated chamber

-: < 20% escaped from treated chamber but statistically significant (P < 0.05) from the matched control

4. CONCISE SUMMARY

Any compound used to control (eliminate, reduce or otherwise prevent harm) insect populations have been termed an "insecticide". An insecticide can also be defined as any compound that is used solely to kill insects. This single term of reference involving only one type of action (knockdown/death) is completely inadequate to describe the more meaningful and complete action of many insecticide compounds used against insect populations. Insects can respond to insecticides in at least 2 different ways; behavioral action, namely avoidance and toxicity. In the past, the prevailing practice has been to classify chemicals simply as toxicants for killing insects. As seen in this chapter, we introduce the term "chemical" in place of "insecticide" as it is more appropriate for recognizing the 2 other primary actions on mosquitoes vice toxicity alone.

Chemicals protect humans from the bites of mosquitoes through 3 different actions: excitation, repellency and toxcity (Roberts et al., 2000 & Grieco et al., 2007). Historically the vast majority of chemical studies have focused on the direct toxicological responses (susceptibility and resistance) of chemicals on mosquito populations whereas very little emphasis has been placed on the vector's behavior in response to sub-lethal exposure. Knowledge of the mosquito's behavioral responses to particular chemicals is highly significant in the prioritization and design of appropriate vector prevention and control strategies. Today, the development of insecticide resistance in insect pests and disease vectors occurs worldwide and on a increasing scale. However, resistance has still remained limited in many areas in spite of the long-term use of chemicals for control. This phenomenon suggests that behavioral responses likely play a significant role in how certain chemcials preform to interrupt human-vector contact while also reducing the selection pressure on a target insects for developing resistance (Roberts et al., 2000).

As discussed, at least 2 different types of mosqutio behavioral response outcomes to chemicals are recognized; excitation and repellency. Whether acting from direct contact or from a distance (spatially) both response activities are based on stimulus-response actions that result in clear movement away from an area with the chemical present. There have been numerous attempts to accurately measure the behavioral responses of mosquitoes to chemicals using various types of excito-repellency test systems (e.g., ERB and HITSS). However, no single system has yet been completely accepted as a standardized method of testing or analyzing behavioral responses. Currently, no test system as recommended by the WHO can discriminate between contact irritancy and noncontact repellency. The WHO tests are based on the concept that mosquitoes respond to chemicals only after physical contact and that spatial repellency plays little or no meaningful role in disease control, or may actually be a determental attribute of a chemical. The test systems outlined in this chapter possess the ability to distinguish essential behavioral responses and give relative significance to each in the potential prevention of disease transmission. The ERB and HITSS systems have been employed to examine the behavioral responses of significant mosquito vectors and pests (Anopheles, Aedes, and Culex) to various test compounds utilized in vector control programs, yielding valuable and highly reproducible results. This knowledge will facilitate improved decision-making about chemical selection, application methods, and future product development. Based on the aforementioned extensive findings, we believe that the behavioral reactions of mosquitoes to pesticides are essential components of disease control operations.

5. REFERENCES

Bangs, M.J. (1999). The Susceptibility and Behavior of *Anopheles albimanus* Weidemann and *Anopheles vestitipennis* Dyar & Knab (Diptera: Culicidae) to Insecticides in Northern Belize, Central America. Doctoral Dissertation, Uniformed Services University of the Health Sciences, Bethesda, Maryland, pp. 448.

Bondareva, N.L.; Artem'ev, M.M. & Gracheva, G.V. (1986). Susceptibility and Irritability Caused by Insecticides to Malaria Mosquitoes in the USSR. Part 1. *Anopheles pulcherrimus. Meditsinskaia Parazitologha I Parazitarnye Bolezni* (Moskva),Vol. 6, pp. 52-55.

Brown, A.W.A. (1983). Insecticide Resistance as a Factor in the Integrated Control of Culicidae. In Integrated Mosquito Control Methodologies. Volume 1. Laird M & Mile J. Eds. Academic Press: New York, pp. 161-235.

Byford, R.L & Sparks, T.C. (1987). Chemical Approaches to the Management of Resistant Horn Fly, *Haematobia irritans* (L.) Populations. *In* M. 0. Ford, D. W. Holloman, B. P. S. Khambay and R. M. Sawicki (eds.), Combating Resistance to Xenobiotics: Biological and Chemical Approaches, pp. 178-189. Ellis Horwood: Chichester UK.

Chareonviriyaphap, T.; Roberts, D.R., Andre, R. G., Harlan, H. & Bangs, M.J. (1997). Pesticide Avoidance Behavior in *Anopheles albimanus* Wiedemann. *Journal of the American Mosquito Control Association,* Vol. 13, No. 2, pp. 171-183, ISSN 1046-3607.

Chareonviriyaphap, T. & Aum-Aung, B. (2000). An Improved Excito-Repellency Escape Chamber for Behavioral Test in Mosquito Vectors. *Mekong Malaria Forum,* Vol. 5, pp.82-87.

Chareonviriyaphap, T.; Sungvornyothin, S., Ratanatham, S., & Prabaripai, A. (2001). Pesticide-Induced Behavioral Responses of *Anopheles minimus*, a Malaria Vector in Thailand. *Journal of the American Mosquito Control Association,* Vol. 17, No.1, (March 2001), pp, 13-22, ISSN 1046-3607.

Chareonviriyaphap, T.; Prabaripai, A., & Sungvornyothin, S. (2002). An Improved Excito- Repellency for Mosquito Behavioral Test. *Journal of Vector Ecology*, Vol. 27, No.2, (December 2002), pp, 250-252, ISSN 1081-1710.

Chareonviriyaphap, T.; Prabaripai, A. & Bangs, M.J. (2004). Excito-Repellency of Deltamethrin on the Malaria Vectors, *Anopheles minimus, Anopheles dirus, Anopheles sawadwongporni,* and *Anopheles maculatus*, in Thailand. *Journal of the American Mosquito Control Association,* Vol. 20, No.1, (March 2004), pp. 45-54, ISSN 1046-3607.

Chareonviriyaphap, T.; Parbaripai, A., Bangs, M.J., Kongmee, M., Sathantriphop, S., Muenvorn, V., Suwonkerd, W. & Akratanakul, P. (2006). Influence of Nutritional and Physiological Status on Behavioral Responses of *Aedes aegypti* (Diptera: Culicidae) to Deltamethrin and Cypermethrin. *Journal of Vector Ecology*, Vol. 31, No.1, (June 2006), pp. 89-101, ISSN 1081-1710.

Charlwood, J.D. & Paraluppi, N.D. (1978). The Use of Excito-Repellency Box with *Anopheles darlingi* Root, *An. nuneztovari* Gabaldon and *Culex pipiens quinquefaciatus* Say, Obtained from the Areas near Manaus, Amazonas. *Acta Amazonica*, Vol. 8, No. 4, pp. 605-611, ISSN 0044-5967.

de Zulueta, J. & Cullen, J.R. (1963). Deterrent Effect of Insecticides on Malaria Vectors. *Nature,* Vol.200, (November 1963), pp. 860-61, ISSN 0028-0836.

Evans, R. G. (1993). Laboratory Evaluation of the Irritancy of Bendiocarb, Lambdacyhalothrin, and DDT to *Anopheles gambiae. Journal of the American Mosquito Control Association,* Vol. 9, No.3, (September 1993), pp. 285-293, ISSN 1046-3607.

Elliott, R. & de Zulueta, J. (1975). Ethological Resistance in Malaria Vectors. Behavioral Response to Intradomestic residual insecticides. WHO/VBC75.569, pp. 1-15.

Gahan, J.B. & Lindquist, A.W. (1945). DDT Residual Sprays Applied in Buildings to Control *Anopheles quadrimaculatus*. *Journal of Economic Entomology,* Vol.38, No.2, pp. 223-230, ISSN 0022-0493.

Georghiou, G.P. (1972). The Evolution of Resistance to Pesticides. *Annual Review of Ecology and Systematics*, Vol.3, No.1 (November 1972), pp. 133-168, ISSN 0066-4162.

Giglioli, G. (1948). An Investigation of the House-Frequenting Habits of Mosquitoes of the British Guiana Coastland in Relation to the Use of DDT. *American Journal of Tropical Medicine and Hygiene,* Vol. 28, No.1, (January 1948), pp. 43-70, ISSN 0002-9637.

Grieco, J.P.; Achee, N.L., Andre, R.G. & Roberts, D.R. (2000). A Comparison Study of House Entering and Exiting Behavior of *Anopheles vestitipennis* (Diptera: Culicidae) using Experimental Huts Sprayed with DDT or Deltamethrin in the Southern District of Toledo, Belize, CA. *Journal of Vector Ecology,* Vol. 25 No.1, (June 2000), pp. 62-73, ISSN 1081-1710.

Grieco, J.P.; Achee, N.L., Sardelis, M.R., Chauhan, K .R. & Roberts, D.R. (2005). A Novel High Throughput Screening System to Evaluate the Behavioral Response of Adult Mosquitoes to Chemicals. *Journal of the American Mosquito Control Association*, Vol. 21, No.4, (December 2005), pp. 404-411, ISSN 1046-3607.

Grieco, J.P.; Achee, N.L., Chareonviriyaphap, T., Suwonkerd, W., Chauhan, K.R., Sardelis, M. & Roberts, D.R. (2007). A New Classification System for the Actions of IRS Chemicals Traditionally Used for Malaria Control. *PLoS ONE*, Vol.2, Issue. 8, (August 2008), pp. e716, ISSN 1932-6203.

Gubler, D.J. (1997). Dengue and Dengue Haemorrhagic Fever: Its History and Resurgence as a Global Public Health Problem. In: Gubler, DJ and Kuno, G (eds.). Dengue and Dengue Haemorrhagic Fever. New York: CAB International. 22 pp.

Kennedy, J.S. (1947). The Excitant and Repellent Effects on Mosquitoes of Sub-lethal Contacts with DDT. *Bulletin of Entomological Research,* Vol. 37, No. 4, (March 1947), pp. 593-607, ISSN 0007-4853.

Kongmee, M.; Prabaripai, A., Akaratanakul, P., Bangs, M.J. & Chareonviriyaphap, T. (2004). Behavioral Responses of *Aedes aegypti* (Diptera: Culicidae) Exposed to Deltamethrin and Possible Implications for Disease Control. *Journal of Medical Entomology*, Vol. 41, No. 6, (November 2004) pp.1055-1063, ISSN 0022-2585.

Lindsay, S.W.; Adiamah, J.H., Miller, J.E. & Armstrong, J.R.M. (1991). Pyrethroid-Treated Bednet Effects on Mosquitoes of *Anopheles gambiae* Complex in the Gambia. *Medical and Veterinary Entomology,* Vol. 5, No.4, (October 1991), pp. 477-483, ISSN 0269- 283X.

Lockwood, J.A.; Sparks, T.C. & Story, R.N. (1984). Evolution of Insect Resistance to Insecticide: a Reevaluation of the Roles of Physiology and Behavior. *Bulletin of the Entomological Society of America,* Vol. 30, No. 4, (November 1984), pp. 41-51, ISSN 0013-8754.

Malaithong, N.; Polsomboon, S., Poolprasert, P., Parbaripai, A., Bangs, M.J., Suwonkerd, W., Pothikasikorn, J., Akratanakul, P. & Chareonviriyaphap, T. (2010). Human-Landing Patterns of *Anopheles dirus* sensu lato (Diptera: Culicidae) in Experimental Huts Treated with DDT or Deltamethrin. *Journal of Medical Entomology,* Vol. 47, No. 4, (September 2010), pp. 823-32, ISSN 0022-2585.

Metcalf, R.L.; Hess, A.D., Smith, G.E., Jeffery, G.M. & Ludwig, G.W. 1945. Observations on the Use of DDT for the Control of *Anopheles quadrimaculatus*. Public Health Report, Vol. 60, No.27, pp. 753-774, ISSN 0033-3549.

Miller, J.E. (1990). Laboratory and Field Studies of Insecticide Impregnated Fibers for Mosquito Control. Ph.D. Dissertation, University of London. 336 pp.

Miller, J. R.; Siegert, P.Y., Amimo, F.A. & Walker, E.D. (2009). Designation of Chemicals in Terms of the Locomotor Responses They Elicit from Insects: an Update of Dethier et al. (1960). *Journal of Economic Entomology,* Vol. 102, No. 6, (December 2009), pp. 2056-2060, ISSN 0022-0493.

Mongkalangoon, P.; Grieco, J.P., Achee, N.L., Suwonkerd, W. & Chareonviriyaphap, T. (2009). Irritability and Repellency of Synthetic Pyrethroids on an *Aedes aegypti* Population from Thailand. *Journal of Vector Ecology*, Vol. 34, No. 1, (June 2009), pp. 217-24, ISSN 1081-1710.

Muenworn, V.; Akaratanakul, P., Bangs, M.J., Parbaripai, A. & Chareonviriyaphap, T. (2006). Insecticide Induced Behavioral Responses in two Populations of *Anopheles maculatus* and *Anopheles swadwongporni*, Malaria Vectors in Thailand. *Journal of the American Mosquito Control Association*, Vol. 22, No. 4, (December 2006), pp. 689-698, ISSN 1046-3607.

Muirhead-Thomson, R.C. (1951). Mosquito Behaviour in Relation to Malaria Transmission and Control in the Topics. pp. viii+219; 16 plates: 22 Text Figure. London: Arnold.

Muirhead-Thomson, R.C. (1960). The Significance of Irritability, Behaviouristic Avoidance and Allied Phenomena in Malaria Eradication. *Bulletin of World Health Organization*, Vol. 22, pp. 721-734, ISSN 0042-9686.

Peaporn, P.; Supaphathom, K., Sathantriphop, S., Chareonviriyaphap, T., Yaicharoen, R. (2007). Behavioral Responses of Deltamethrin and Permethrin Resistant Strains of *Aedes aegypti* When Exposed to Permethrin in an Excito-Repellency Test System. *Dengue Bulletin*, Vol. 31, pp. 153-159, ISSN 0250-8362.

Pates, H. & Curtis, C. (2005). Mosquito Behavior and Vector Control. *Annual Review of Entomology*, Vol. 50, pp.53–70, ISSN 0066-4170.

Polsomboon, S.; Poolprasert, P., Suwonkerd, W., Bangs, M.J., Tanasinchayakul, S., Akratanakul, P. & Chareonviriyaphap, T. (2008). Biting Patterns of *Anopheles minimus* Complex (Diptera: Culicidae) in Experimental Huts Treated with DDT and Deltamethrin. *Journal of Vector Ecology*, Vol. 33, No. 2, (December 2008), pp. 285-292, ISSN 1081-1710.

Pothikasikorn, J.; Chareonviriyaphap, T., Bangs, M.J. & Prabaripai, A. (2005). Behavioral Responses to DDT and Pyrethroids between *Anopheles minimus* Species A and C, Malaria Vectors in Thailand. *American Journal of Tropical Medicine and Hygiene*, Vol. 73, No. 2, (August 2005) pp. 343-349, ISSN 0002-9637.

Pothikasikorn, J.; Overgaard, H., Ketavan, C., Visetson, S., Bangs, M.J. & Chareonviriyaphap, T. (2007). Behavioral Responses of Malaria Vectors, *Anopheles minimus* Complex, to Three Classes of Agrochemicals in Thailand. *Journal of Medical Entomology*, Vol.44, No. 6, (November 2007), pp. 1032-1039, ISSN 0022-2585.

Quinones, M.L. & Suarez, M.F. (1989). Irritability to DDT of Natural Populations of the Primary Malaria Vectors in Colombia. *Journal of the American Mosquito Control Association*, Vol. 5, No. 1, (March 1989), pp. 56—59, ISSN 1046-3607.

Rachou, R.G. ; Moura- Lima, M., Duret, J.P.& Kerr, J.R. (1963). Experiences with the Excito- Repellency Test Box—Model OPS, pp. 442-447. *In* Proceedings, 50th Annual Meeting of the New Jersey Mosquito, the 19th American Mosquito Control Association, ISSN 1046-3607.

Ree, H.I. & Loong, K.P. (1989). Irritability of *Anopheles farauti, Anopheles maculatus,* and *Culex quinquefasciatus* to Permethrin. *Japanese Journal of Sanitary Zoology*, Vol. 40, No. 1, pp. 47—51, ISSN 0424-7086.

Reiter, P. & Gubler, D.J. (1997). Surveillance and Control of Urban Dengue Vectors. In: Gubler, DJ and Kuno, G (eds.). Dengue and Dengue Haemorrhagic Fever. New York: CAB International. pp. 425-462

Roberts, D.R. (1993). Insecticide Repellency in Malaria Vector Control: a Position Paper. VBC Report No. 81131. United States Agency for International Development, Arlington, VA. 72 pp, ISSN 0736-718X.

Roberts, D.R.; Alecrim, W.D., Tavares, A.M. & McNeil, K.M. (1984). Influence of Physiological Condition on the Behavioral Responses of *Anopheles darlingi* to DDT. *Mosquito News,* Vol.44, No.3, pp. 357-361, ISSN 0027-142X.

Roberts, D.R.; Alecrim, W.D., Tavares, A.M. & Radke, M.G. (1987). The House-Frequenting, Host Seeking and Resting Behavior of *Anopheles darlingi* in Southeastern Amazonas, Brazil. *Journal of the American Mosquito Control Association,* Vol.3, No.3, (September 1997), pp. 433-441, ISSN 1046-3607.

Roberts, D.R. & Alecrim, W.D. (1991). Behavioral response of *Anopheles darlingi* to DDT sprayed house walls in Amazonia. *The Bulletin of the Pan American Health Organization*, Vol.25, No. 3, pp. 210—217, ISSN 0042-9686.

Roberts DR, & Andre RG. (1994). Insecticide Resistance Issues in Vector Borne Disease Control. *American Journal of Tropical Medicine and Hygiene,* Vol.50, No.6, pp. 21—34, ISSN 0002-9637.

Roberts, D.R.; Chareonviriyaphap, T., Harlan, H.H. & Hshieh, P. (1997). Methods for Testing and Analyzing Excito-Repellency Responses of Malaria Vectors to Insecticides. *Journal of the American Mosquito Control Association*, Vol.13, No.1, (March 1997), pp. 13-17, ISSN 1046-3607.

Roberts, D.R.; Alecrim, W.D., Hshieh, P., Grieco, J., Bangs, M.J., Andre, R.G. & Chareonviriyaphap, T. (2000). A Probability Model of Vector Behavior: Effects of DDT Repellency, Irritability, and Toxicity in Malaria Control. *Journal of Vector Ecology,* Vol.25, No. 1, (June 200), pp.48-61, ISSN 1081-1710.

Roberts, D.R. & Tren, R. (2010). The Excellent Powder: DDT's Political and Scientific History. Dog Ear Publishing, pp. 432, ISBN: 978-160844-376-5.

Rozendaal, J.A.; Van Hoof, J.P.M., Voorham, J. & Oostburg, B.F.J. (1989). Behavioral Responses of *Anopheles darlingi* in Suriname to DDT Residues on Housewalls. *Journal of the American Mosquito Control Association*, Vol.5, No. 3, (September 1989), pp. 339-350, ISSN 1046-3607.

Sathantriphop, S.; Ketavan, C., Prabripai, A., Vietson, S., Bangs, M.J., Akatanakul, P. & Chareonviriyaphap, T. (2006). Susceptibility and Avoidance Behavior by *Culex quinquefasciatus* Say to three Classes of Residual Insecticides. *Journal of Vector Ecology,* Vol. 31, No. 2, (December 2006), pp. 266-274, ISSN 1081-1710.

Service, M.W. & Townson, H. (2002). The *Anopheles* Vector. pp. 59-84. In: Warrell DA, Gilles HM (eds). Essential malariology (4th ed.). Arnold, London UK, ISBN 9780340740644.

Smith, A. (1965). A Verandah-Trap Hut for Studying the House Frequenting Habits of Mosquitoes and for Assessing Insecticides. II. The Effect of Dichlorvos (DDVP) on Egress and Mortality of *Anopheles gambiae* Giles and *Mansonia uniformis*

(Theo.) entering naturally. *Bulletin of Entomological Research*, Vol. 56, pp. 275-286, ISSN 0007- 4853.

Smith, A. & Webley, D.J. (1969). A Varandah-Type Hut for Studying the House-Frequenting Habits of Mosquitoes and for Assessing Insecticide. III. The Effect of DDT on Behaviour and Mortality. *Bulletin of Entomological Research*, Vol. 59, No. 3, pp. 33-46, ISSN 0007-4853.

Sinka, M.E.; Bangs, M.J., Manguin, S., Chareonviriyaphap, T., Patil, A.P., Temperley, W.H., Gething, P.W., Elyazar, I.R., Kabaria, C.W., Harbach, R.E. & Hay, S.I. (2011). The Dominant *Anopheles* Vectors of Human Malaria in the Asia-Pacific Region: Occurrence Data, Distribution Maps and Bionomic Précis. *Parasites Vectors* , Vol.4, (May 2011), pp. 89, ISSN 1756-3305.

Sungvornyothin, S.; Chareonviriyaphap, T., Prabaripai, A., Trirakhupt, T., Ratanatham, S. & Bangs, M.J. (2001). Effects of Nutritional and Physiological Status on Behavioral Avoidance of *Anopheles minimus* (Diptera: Culicidae) to DDT, Deltamethrin and Lambdacyhalothrin. *Journal of Vector Ecology*, Vol. 26, No.2, (December 2001), pp. 202-215, ISSN 1081-1710.

Suwannachote, N.; Grieco, J.P., Achee, N.L., Suwonkerd, W., Wongtong, S. & Chareonviriyaphap, T. (2009). Effects of Environmental Conditions on the Movement Patterns of *Aedes aegypti* (Diptera: Culicidae) into and out of Experimental Huts in Thailand. *Journal of Vector Ecology*, Vol. 34, No. 2, (December 2009), pp. 267-275, ISSN 1081-1710.

Tanasinchayakul, S.; Polsomboon, S., Prabaripai, A. & Chareonviriyaphap, T. (2006). An Automated, Field Compatible Device for Excito-Repellency Assay in Mosquitoes. *Journal of Vector Ecology*, Vol.31, No. 1, (June 2006), pp. 210-212, ISSN 1081-1710.

Thanispong, K.; Achee, N.L., Bangs, M.J., Grieco, J.P., Suwonkerd, W., Prabaripai, A., & Chareonviriyaphap, T. (2009). Irritancy and Repellency Behavioral Responses of three strains of *Aedes aegypti* Exposed to DDT and Alpha-cypermethrin. *Journal of Medical Entomology*, Vol. 46, No. 6, (November 2009), pp. 1407-1414, ISSN 0022- 2585.

Thanispong, K.; Achee, N.L., Grieco, J.P., Bangs, M.J., Suwonkerd, W., Prabaripai, A., Chauhan, K.R. & Chareonviriyaphap, T. (2010). A High Throughput Screening System for Determining the three Actions of Insecticides against *Aedes aegypti* (Diptera: Culicidae) Populations in Thailand. *Journal of Medical Entomology*, Vol. 47, No. 5, (September 2010), 833-841, ISSN 0022-2585.

Tisgratog, R.; Thananchai, C., Bangs, M.J., Thainchum, K., Juntarajumnong, W., Prabaripai, A., Chauhan, K.R., Pothikasikorn, J.,& Chareonviriyaphap, T. (2011). Chemical Induced Behavioral Responses in *Anopheles minimus* and *Anopheles harrisoni* in Thailand. *Journal of Vector Ecology*, Vol. 36, No.2 (December 2011), pp. xxx-xxx, ISSN 1081-1710.

Trapido, H. (1954). Recent Experiments on Possible Resistance to DDT by *Anopheles albimanus* in Panama. *Bulletin of the World Health Organization*, 11: 885-889, ISSN 0042-9686.

World Health Organization. (1970). Insecticide Resistance and Vector Control. 17th Report of the World Health Organization Expert Committee on Insecticide. Instructions for Determining the Irritability of Adult Mosquitoes to Insecticide. *World Health Organization Technical Report Series,* pp. 433. ISSN 0512-3054.

World Health Organization. (1986). Resistance of Vectors and Reservoirs of Disease to Pesticides: *Tenth Report of the World Health Organization Offset Publication*, No 13, ISSN 0303-7878.

World Health Organization. (1992). Vector Resistance to Pesticides. Technical Report Series No. 818. *World Health Organization*, Geneva, pp. 62, ISSN 0042-9686.

World Health Organization. (1999). Prevention and Control and Dengue and Dengue Haemorrhagic Fever: Comprehensive Guidelines. *World Health Organization* Regional Publication, SEARO No. 29, New Delhi., pp. 134, ISSN 0378-2255.

CHAPTER

10

Crucial Plant Oils and Insecticidal Efficacy in Culex quinquefasciatus

1. INTRODUCTION: AN OVERVIEW

Plants are crucial for humanity as they serve as a food source, supply oxygen through photosynthesis, and are vital for maintaining ecological equilibrium (Corbino, 2000).

Essential oils are volatile, typically distillable liquid components that account for a plant's fragrance. The predominant majority emit a pleasant aroma, which holds metabolic and evolutionary significance by attracting pollinators, serving as a defense mechanism against parasites, herbivorous animals, and insects, facilitating plant adaptation during water scarcity, and contributing to reserve substances that provide H+ in electron transfer processes. The organoleptic features of essential oils may be attributed to predominant components, while in some instances, they are chemicals present in minimal quantities (traces) that determine the flavor, aroma, or therapeutic attributes (Scholes, 1995; Worwood, 1992).

These natural compounds are termed secondary metabolites, a designation that indicates they are not integral to the fundamental life processes of the plant but serve certain purposes. Corbino, 2000. They often amass in significant quantities without adverse effects, posing an issue within the cell or on the plant. These metabolites possess the ability to produce glycosides and are soluble within the plant. For many years, these metabolites were considered the ultimate results of metabolic processes, lacking defined function or being viewed only as waste from plant products (Lopez, 2008).

The investigation of these chemicals commenced with organic chemists in the 19th and early 20th centuries, motivated by their significance in the medical business and the production of flavorings, among other applications. The investigation of secondary metabolites prompted the advancement of separation techniques and spectroscopy for elucidating their structure and synthesis, becoming the foundation of modern organic chemistry (Lopez, 2008).They can be found in different parts of the plant: leaves (wormwood, basil, buchú, cidrón, eucalyptus, mint, lemongrass, marjoram, mint, patchouli, quenopod, rosemary, Sage, lemon balm, etc.), in the roots (angelic, asaro, saffron, calamus, turmeric, galanga, ginger, sandalwood, Sasafras, Valerian, vetiver, etc.), in the pericarp of the fruit (lemon, Tangerine, Orange, etc.), seeds (anise, cardamom, dill, fennel, cumin, etc.), in the stem (cinnamon, caparrapí, etc.), flowers (arnica, lavender, chamomile, pyrethrum, thyme, clove scent, rose, etc.) and in the fruit (caraway, coriander, bay leaf, nutmeg, parsley, pepper, etc.). (www.buscasalud.com)

2. ESSENTIAL OILS

Essential oils are complex mixtures of up to over 100 secondary metabolites, which can be classified in terpenoid compounds, phenolic and alkaloids based on their biosynthetic origins. Terpenes containing a single unit of isoprene are called monoterpenes, which contain three units named sesquiterpenes (Ikan, 1991; Silva, 2002).

Monoterpenes and sesquiterpenes are terpenes from 10 to 15 atoms of carbons biosynthetically derived from geranilpyrophosfates (SPG) and farnesilpyrophosfate (FPP) respectively. Monoterpenes and in general all the natural terpenoid compounds are synthetized by the route of the acetilCoA through a common intermediate which is the mevalonic acid. However, it has been proposed that some terpenoids are not originated by this route, and instead use an alternative route that may involve pyruvate, glyceraldehyde- 3-phosphate (Fig 1) (Adams et al, 1998; Sponsel, 1995).

Essential oils are widely distributed in some 60 families of plants including the Compositae, Labiadas, Lauraceae, Myrtaceae, Rosaceae, Rutaceae, Umbelliferae, etc. Monoterpenoides are mainly found in plants in the family of Ranunculales, Violales, and Primulales, while they are rare in Rutales, Cornales, Lamiales, and Asterales. Moreover, the sesquiterpenoides are commonly found in Rutales, Cornales, Magnoliales and Asterales. Although in both essential oils: mono - sesquiterpenes and phenylpropan are found in free-form, those which are linked to carbohydrates, have been more recently investigated, because it is considered to be the immediate precursor of the essential oil.

From the chemical point of view, despite its complex composition with different types of substances, essential oils can be classified according to the type of substances of which they are the major components. According to this fact, the essential oils rich in monoterpenes are called essential oils sesquiterpenoids, those rich in phenylpropan are the essential oils phenylpropanoids. Although this classification is very general there are more complex classifications that take into account other chemical aspects (Fig 2,3,4) (González, 1984; Ikan, 1991; Judd et al, 2002; Stachenko, 1996).

2.1 Methods of extraction

According to www.chemkeys.com 2000; Gascon et al. 2002; www.herbotecnia. com. ar, 2002; www.losandes.com.ar 2002. essential oils can be extracted from plant samples through various methods samples such as: expression, distillation with steam, extraction with volatile solvents, enfleurage, extraction with supercritical fluids and hydrodistillation.

Method of isolation

From such oils, it is possible to its isolation through the use of one or several chromatographic methods such as thin layer chromatography and HPLC.

For the chromatographic column and thin layer methods the silica gel are widely used as stationary phase. Polar pure or mixed solvents such as: Toluene-acetate of ethyl benzene, chloroform, dichloromethane, ethyl, chloroform-benzene, chloroform-ethanol - chloroform- benzene acetic acid benzene-acetate are used as mobile phase.

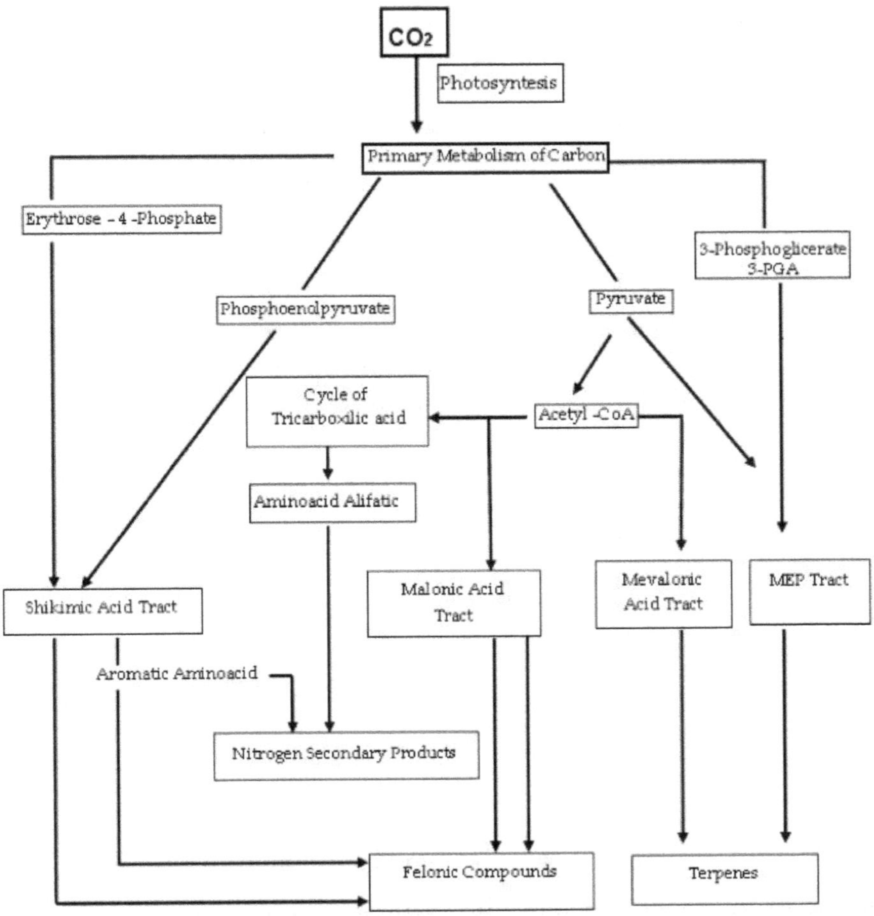

Fig. 1. General pathways of plant secondary metabolism.

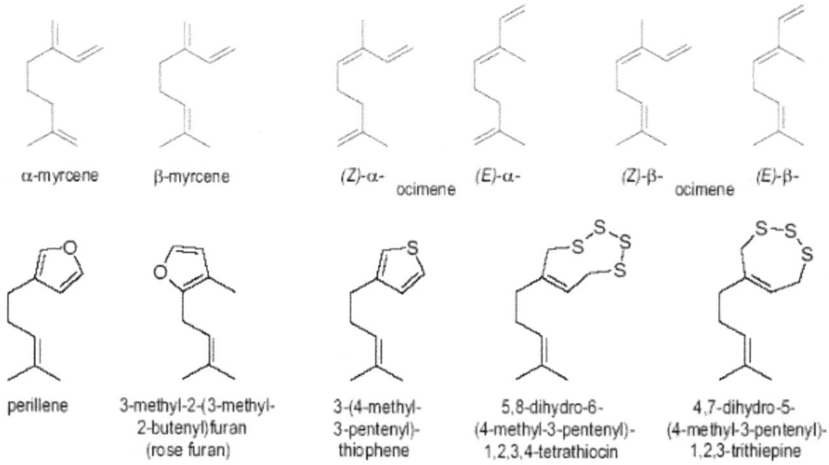

Fig. 2. Natural Monoterpenes.

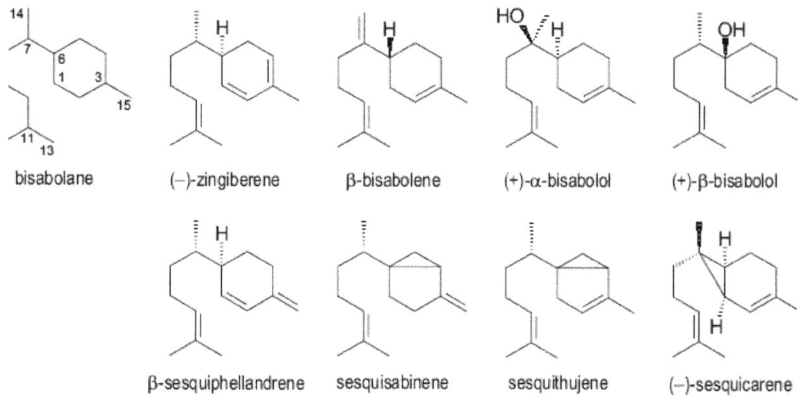

Fig. 3. Natural Sesquiterpenes.

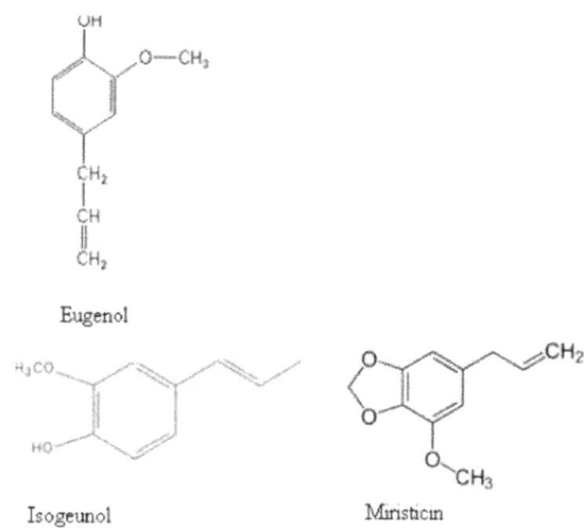

Fig. 4. Naturals phenylpropanoids.

However, currently more efficient and faster techniques of separation are used such as HPLC high efficiency liquid chromatography and gas (CG) chromatography, as well as also combinations "ON-LINE" HPLC- CG EM. These same methods are used for the analysis of the flower essences. The last technique, thanks to the recent development of capillary columns for high resolution, allows analyzing complex mixtures present in essential oils, and identify the components from the retention times through the so-called Kovats retention indexes (Ik). These values are characteristic for each component and there are databases with indexes of many components of essential oils. Ik values are determined in two chromatographic columns, one polar (e.g. CARBOWAX 20 M) and a non-polar (e.g. OV - 101 also called DB-1).

In addition, the docked technique chromatography of gases - mass spectrometry, (GC-MS) allows obtaining the spectrum of mass of each component which builds the molecular weight and structural information. Likewise there are databases with spectra of masses of many components, something similar goes for the index Kovats (determined in two columns of different polarity) and the spectra of mass are criteria for chemical allocation of many components of essential oils not just monoterpenes but also other types of substances characteristic of such oils. More recently the column chromatographic Chiral for separation of optically active components have been developed, and have developed methods for combined analysis HPLC HPLC-NMR of mixtures of sesquiterpenes and mass spectrometry.

2.3 Test of recognition of monoterpenes and sesquiterpenes (Pino, 1999; Stam, 1970)

Due to the diversity of functional groups that can be present in the components of mono - and sesquiterpenics of an essential oil, there is a specific test for their recognition. However there are few experimental procedures that allow to recognize some of them by their coloration with different reagents, its absorption of 254 nm UV light and its Rf in thin layer chromatography. For example, limonene is recognized in the plates of TLC because it does not absorb UV 254 nm light, adds bromine, does not form a derivative 2, 4- dinitrophenylhydrazone and produces Brown with sulfuric acid.

There is a procedure which combines the TLC with micro reactions (oxidation, reduction, dehydration, hydrolysis, etc.) described in the classic book already today of Ikan. Other reagents useful for revealing monoterpenes and sesquiterpenes are anisaldehid-sulfuric acid, sulfuric acid -vanillin and phosphomolibdic acid.

2.3.1 Spectral characterization

Monoterpenes and sesquiterpenes, in a number, can be characterized chemically from the gas chromatography and mass spectra data as noted above, but when there are doubts of such characterization some spectral methods such as infrared ultraviolet and magnetic resonance spectrometry can be used.

2.3.1.1 Infrared spectroscopy

Infrared Spectroscopy allows detection of the presence of groups hydroxyl, Carbonyl, aromatic rings, bond double C = C cis and trans, etc. To determine the spectrum just put a drop of the component in a cell of NaCl. For example, in the infrared spectrum of the 3-p- menten-7 present in the oil of cumin, the intense band in 1725 cm - 1 indicates a group not conjugated Carbonyl. The peak at 2710 cm - 1 is assigned to a proton aldehydic C-H. The bond around on 1375 cm - 1 indicates a group isopropyl, and the band of 817 cm- 1 average intensity indicates a double bond tri-substituted.

2.3.1.2 Ultraviolet spectroscopy

UV spectrum ultraviolet of monoterpenes and sesquiterpenes allows the recognition of functional groups and chromophoric. E.g. limonene presents an absorption maximum at 262 nm (ξ = 6400).

Nuclear magnetic resonance (1 H - 13 C); due to the developments of the NMR has database of Spectra, especially of 13C -NMR for the monoterpenes and more distributed sesquiterpenes. The NMR spectrometry-13 C has the additional advantage that the carbons of terpenoid chemical displacement (and other natural and synthetic substances) can be calculated by computer programs available in the market as Chemwind and ACD-Lab.

In addition, the recent development of two-dimensional methods homo - and heteronuclear, has allowed the fine structural determination of terpenoids and other natural substances, eliminating the ambiguity in the allocation of the observed signals.

2.3.1.3 Mass Spectrometry

There are databases that contain information and spectra of masses of the components of mono - and sesquiterpenoids of essential oils. These databases are now available in many commercial instruments of analysis as a chromatograph of gases coupled with mass spectrometers.

2.4 Phenylpropan

Phenylpropanes are secondary metabolites derived from phenylalanine, which is first converted to cinnamic acid and after a series of hidroxilations lead to acid cumaric, caffeic acid, pherulic acid etc. The conversion of these acids to their corresponding esters produces some of the volatile components responsible for the fragrance of herbs and flowers (Lopez, 2008).

Phenylpropanes of essential oils are extracted with the same methodology described above for mono - and sesquiterpenes. However, due to its aromatic ring they have advantages in detection by TLC and HPLC because absorb ultraviolet light (254 nm) and do not require to be disclosed with chemical agents, or need to be derivative, and therefore can be isolated and analyzed more easily.

2.4.1 Recognition of phenylpropan

There are test assay for recognition to the aromatic ring as a reaction with formaldehyde and sulfuric acid. Likewise in the case of phenylpropan with phenolic hydroxyl these can be recognized by the test assay of ferric chloride, which produces green and blue coloring with phenolic substances in general.

Substances with aromatic ring, their infrared spectra show signs characteristic of these compounds and give information about the type of replacement of the aromatic ring in addition to the functional groups present in the molecule. For example, the spectrum of eugenol shows, among other, bands in 3500 (wide) due to the hydroxyl group, 1510 characteristic of aromatic, and three bands in 990, 920 and 938 cm-1 characteristics of a vinyl mono substituted group. The IR from the cinamaldehyde spectrum displays bands in (weak) 3330, 3050, 2820, 2750, 1660 (intense, due to the carbonyl group), 975, 740 and 695 cm^{-1} among others.

Unlike the majority of mono - and sesquiterpenes, the phenylpropanes absorb UV light with a maximum around 254 nm depending on the groups present in the molecule chromophors. For example, the isoeugenol shows maxima at 260 (15850) and 305 (7000), safrole in 286 nm, the myristicine at 276 nm, the isosafrole in 264 nm, trans- cinnamic acid in 273 nm and cis- cinnamic acid in 264 nm.

Phenylpropan 1H-NMR spectra show signs of aromatic protons around 6-8 ppm whose multiplicities and coupling constants allow a clear structural assignment even with low resolution spectra. Trans-anethole there is a double signal around 1.9 ppm due to the protons from the methyl group, a singlet in 3.9 due to the protons of the methoxy group, a complex signal around 6.1 ppm due to the two protons olefinics provision trans each other, and a double around 6.9 ppm characteristic of 4 protons of an aromatic ring p-based.

2.5 Natural insecticides way-of-action (Silva 2002)

As growth regulators, this effect can manifest itself in several ways. The first is related to the molecules that inhibit the metamorphosis. Other compounds make the insect to have an earlier metamorphosis, occurring at a time which is not favorable. It has been observed that certain molecules can alter the function of the hormones that regulate these mechanisms so that there are insects with malformations, dead or sterile (Gunderson, 1985; Sangykurn, 1999).

Studies carried out from different concentrations of extract of Meliaceas show that this extract inhibits the feeding and negatively affects the development and survival of different species of insects. The anti-alimentary activity of this compound shows that at doses ranging from 5.5 to 27. 6 μg/cm2 caused an inhibitory activity of more than 75% the insects. The way of action of these compounds extracted from various species of Meliaceas are taken from a combination between an anti feeding effect and post-digestive toxicity. (www.cannabiscultura.com)

The use of plants as repellents is an ancient knowledge. This practice is basically done with compounds that have odor or irritant effects such as chili and garlic. There are homemade recipes that describe the use of fennel (*Foniculum vulgare*), rude (*Ruta graveolens*) and eucalyptus (*Eucalyptus globules*) among other aromatic plants to repel moths of clothing.

2.5.1 Mode of action of the natural insecticide

Treatments with natural compounds such as essential oils or pure compounds (Awde & Ryan, 1992; Keane & Ryan, 1999;Ryan & Byrne, 1988) may cause symptoms that indicate neurotoxic activity including hyperactivity, seizures and tremors followed by paralysis (knock down), which are very similar to those produced by the insecticides pyrethroids (Kostyukovsky et al., 2002).

Acetyl cholinesterase enzyme catalyzes the hydrolysis of the neurotransmitter excess acetylcholine in the synaptic space between choline and acetic acid. It has been recognized that essential oils could their effect through ACE inhibition.

Many hydrophobic compounds incite the deactivation of protein and enzyme inhibition; the acetilcolinesterase (AchE) is an enzyme that is particularly susceptible to the hydrophobic inhibition (Ryan et al. 1992).

Is described in this context essential oil interfering with AchE are acting as potent of the central nervous system where all synapses cholinergic are virtually located (Bloomquist, 1999) of the cholinesterase inhibitors are known as anticolinesterasic, the chemical products that interfere with the action of this enzyme are potent neurotoxins (López, 2008).

The majority of plant insecticide are extracts made by a group of active compounds of diverse chemical nature, which is hardly found in the same concentrations so the pressure of selection on the plague will not be always the same, i.e. in general, insects take more to develop resistance to a blend of compounds than any of its components separately (Silva, 2003).

2.6 Future prospects

With the beginning of the new millennium, assessment of a large number of essential oils from native or typical flora has increased in many countries of the Americas and Africa due to its potential use as an alternative method of control.

There have been studies with oils in laboratory conditions reporting larvicide action in species of the genus *Culex, Anopheles, Aedes* (Ansari, 2005; Assarn, 2003;

Albuquerque et al, 2004; Amer et al, 2006a ; Bassole et al., 2003; Carvalho et to 2003; Cavalcanti et al, 2004; Cetin et al, 2006; Cheng et to 2004; Chung et 2010 b Dharmagda et al, 2005; of Mendonca et al., 2005; Faley et al., 2005; Hafeez et 2011; Khandagle et al, 2001; Mathew & Thoppil 2011;Morais et al, 2006; Prajapati et al., 2005; Pushpanathan et al, 2006; Phasomkusolsil & Soowera 2010; Pitarokili et al, 2011; Raví et al, 2006; Tare et al, 2004; Tomas et al, 2004).

In the same way it has been evaluated the repellent capacity of some oils in the same species of mosquitoes; Amer et al, 2006b;Barnard et al, 2004; Byeoung-Soo, 2005; Caballero-Gallardo, et-2011; Chang et al, 2006; Jaenson et al, 2006; Das et al., 2003; Kim et al, 2004; Oyedale et al, 2002; Odalo et al., 2005; Pohilt et 2011; Phasomkusolsil & Soowera 2010; Tawatsin et al, 2006; Trongtokit et al. 2005a; 2005b; Yang et al., 2005; Zhu et al, 2006; Less has been evaluated adulticide effect found only two reports described by Chaiyasit et al 2006 and Miot et al 2004.

3. STUDIES DEVELOPING IN CUBA WITH ESSENTIAL OILS FOR CONTROLLING OF PUBLIC HEALTH PESTS

In Cuba there is a large plant biodiversity with a lot of endemic species. Myrtaceas, Piperaceas, Zingiberaceae, Pinaceae families possess insecticidal action potential based to their essential oils. To this day, there have been verified in public health pests such as; *Blatella germanica, Musca domestica* and *Aedes aegypti* (Aguilera et al 2003; 2004;Leyva et al, 2007 a, b, 2008a, b; 2009a, b; 2010) and more recently in *Culex quinquefasciatus.*

Evaluated essential oils are essentially from *Pimenta racemosa, Eugenia melanadenia, Psidium rotundatum* and *Melaleuca leucadendron* of the botanical family Myrtaceae, *Piper auritum, Piper aduncum* of the family Piperacea, *Curcuma longa* in the family Zingiberaceae, *Artemisia abrotamum* in the family Asteraceae, *Pinus tropicalis* and *Pinus caribbaea* of the Pinacea family. A derivative of essential oil of pinaceas has also been evaluated; turpentine oil obtained by hidrodestylation of the resin of endemic pine trees, mostly composed of terpenes (α-and β- pinene and a modified turpentine where a part of the original turpentine oil was modified by photoizomerization of the α and β-pinene to compound of the type verbenone and pulegone.) In all tested oils, there has been found necessary dosages that elicited from 5% to 95% of mortality. After testing CHI2 they have a probability greater than 0.05 by which can raise all the mortalities occurring in each are associated with the used dose.

Studies with these oils in larvae of *Aedes aegypti* show high insecticidal action by the low lethal concentrations obtained and the high slopes of the regression lines; however, in *Blatella germanica* and *Musca domestica* the bioassays were conducted in adults, showing a very low activity given the high lethal concentrations obtained.

Natural essences of plants owe its insecticide and repellent action to the presence in its composition of derivative monoterpenic as d-limonene, α -terpineol, β - myrceno, linolool, 1.8-cineole, 4-terpineol, thymol, carvacrol, α-pinene and β-pinene.) Studied oils possess structures lactones as the 1.8 cineole, 4-terpinol, α and β - pinene, cineole, safrole, turmenon, chavicol, eugenol, beta-phellandrene, longifoleno among others, which account for a 18-68% of its total membership, and are that they can be attributed the high insecticidal action in *Aedes aegypti.*

One aspect assessed for reasons of eco-sustainability was ovicidal and inhibitory action of the development by the oil of turpentine in larvae of *Ae aegypti.*

The largest ovicidal effect was shown with the dose diagnosis of oil of turpentine with photochemical treatment. In a test of hypothesis of different proportions for each dose of oil, the diagnostic doses were compared with every CL_{95} finding differences between the doses tested with modified oil of turpentine for p=0.05 while for oil not modified there was only significant difference comparing the CL_{95} and the diagnostic doses. (p=0.05). By comparing the CL_{95} and the diagnostic dose, he found significant difference for a p = 0.05. If we compare both types of essences we may conclude that although modified turpentine oil presents a CL_{95} dose higher than oil without changing this it has a greater effect ovicidal and therefore more protector.

The percentage of hatching inhibition was 36.47% of the total number of larvae. The larvae completing their development to adults corresponded to 60.54% of surviving larvae. As a figure of interest the greatest number of semi emergency in adults occurred at 72 hours after exposure to the oil of turpentine (with a predominance of males). At 96 hour, the largest proportion was in favor of females, who became non-existent for this sex at 120 and 168 hours (5-7 d).

More recently, oil of turpentine and the bicyclic α-pinene obtained by fractional distillation of oil of turpentine have been evaluated in *Culex quinquefasciatus*, important vector of West Nile virus and encephalitis.

The purification of the α-pinene was made using column chromatography in system of variable stationary phase (SiO_2, Al_2O_3, and their mixtures in dependence of the polarities of the mobile phases for elution and separation).) Molecular characterization was developed, using techniques of FTIR and NMR, as well as methodologies for its extension to field conditions type TLC (chromatography on thin layer in the presence of plates of glass with silica gel 200-G254 activated with AgNO3 / acetonitril via wet impregnation with methanol). The plates are eluted with mixtures of solvents of media polarity (ethyl-acetone - alcohol isopropyl acetate). Figure 5 shows the spectroscopic characterization of the α pinene. In table 1 and 2 is shown the concentrations finally used, the obtained mortalities and lethal concentrations that causes 50% to 95% percent of mortality, the slope of the regression line and diagnostic dose.

The lowest lethal concentrations were obtained with oils of turpentine and turpentine modified followed by *Curcuma longa, Psidium rotundatum* and *Chenopodium ambrosoide*. By applying a $\chi 2$-square test ($p < 0.05$) to the results of the bioassays, the results showed that the mortalities are associated with the used dose. The higher slopes of the lines were equal with the oil of turpentine, *Eugenia melanadenia, Psidium rotundatum* and *Chenopodium ambrosiode*, corresponding to the lower value of the slope to the terpene α-pinene. To the slope values obtained were applied a $\chi 2$ test finding significant difference between them for $p < 0.0001$, indicating that these oils have a different response for the same strain.

In our assessed results we find the insecticidal action of oil in *Cx quinquefasciatus* and is important to note that lethal concentrations obtained are higher than concentrations found in *Aedes aegypti* to these same oils (Leyva,2008a; 2009a,b). This response could be that this species has developed physiological resistance and some tolerance to insecticides with which it has active defense before any insecticide action.

Oil of *Curcuma longa* presents in its composition more than 50% of sesquiterpenes, apart from the monoterpenes and α -pinene to which can be attributed the insecticide action. If we compare studies conducted with *Curcuma aromatica* (Choochate et al., 2005) and *Curcuma zedoaria* in *Aedes aegypti* these oils show lower values of LC_{50} and CL_{99} (Champakaew et al, 2007).

Within the family Piperaceae, notably members of the genus Piper have noted insecticidal action, molluscicide (Chansang et al., 2005; Parmar et 1997) this is due to the presence of the secondary metabolites as alkaloids, phenylpropanoids, lignans, neolignanos, terpenes, flavononas, among others (Pino et al, 2004; Smith & Kassim 1979). *Piper aduncum* specifically shows in its composition a high percent of dilapiol, α-pinene and 1.8- cineole (Bottia et 2007). In Mexico studies were not mortalities obtained with a kind of Piper in *Culex quinquefasciatus* in aqueous extracts to 5 and 15% where used in the aerial part of plants (Perez-Pacheco et al 2004).

The larvicidal activity of *Chenopodium ambrosioide* was evaluated in *Ae aegypti*, with extracts metanolics to different concentrations resulting most important inhibition of pupal development and an increase in the time of larval development (Suparvarn et al 1974) has also been evaluated in mosquito repellent activity (Gillij et al 2008). *Ch ambrosiodes* has as one of its components majority the carvacrol reported in the literature as insecticide terpene (Silva et al 2008; Kordali et al 2008). *Pimenta racemosa, Eugenia melanadenia* and *Psidium rotundatum* presented in its composition more than 50% of monoterpenes and sesquiterpenes standing out in his majority composition 1,8 cineole, 4-terpineole and α- pinene terpenes with high insecticidal action (Pino et al, 2004,2005).

Oil of turpentine in *Aedes aegypti* obtained strong insecticide and inhibitory action of development in previous studies (Lucia et al, 2007; Leyva et al, 2009b 2010). In this study isolated terpene α -pinene (one of the major components of turpentine) was higher than the own turpentine lethal concentrations, this may be due to the synergistic effect with several monoterpens (α and β -pinene) at the same time and not isolated and evaluated by themselves in insects.

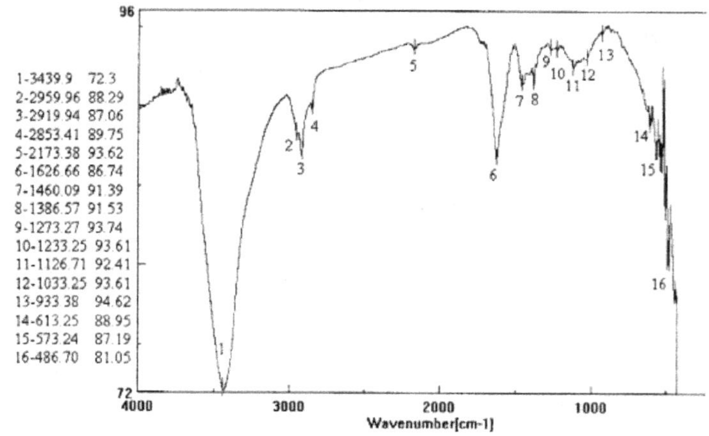

Fig. 5. Spectroscopic characterization of the α pinene.

Table 1. Concentrations used and rates obtained with derivatives of Pinaceae in *Cx quinquefasciatus* larvae.

Conc (%)	Mortality (%)	Lethal Conc		Slope of the Line
Turpentine				
0.0035	5	CL_{50}	0.0042	20.9 Diagnostic Dose 0.0108 (108 mg/L)
0.0040	31			
0.0050	94	CL_{95}	0.0055	
0.0060	100			
Modified Turpentine				
0.0030	5	CL_{50}	0.0038	14.4 Diagnostic Dose 0.0112 (112 mg/L)
0.0040	60			
0.0050	93	CL_{95}	0.0050	
0.0060	100			

Conc (%)	Mortality (%)	Lethal Conc		Slope of the Line
α-pinene				
0.0040	35	CL_{50}	0.0052	3.7 Diagnostic Dose 0.044 (440 mg/L)
0.0050	46			
0.0060	56	CL_{95}	0.014	
0.0070	67			
0.0080	78			

Table 2. Concentrations used and rates obtained with essential oils tested in *Cx quinquefasciatus* larvae.

Conc (%)	Mortality (%)	Lethal Conc		Slope of the Line
Chenopodium ambrosoide				
0.0040	10			11.8 Diagnostic Dose : 0.0162 (162 mg/L)
0.0050	42	CL50	0.0051	
0.0060	80			
0.0070	95	CL95	0.0071	
0.0080	98			
Curcuma longa				
0.0040	46			5.22 Diagnostic Dose : 0.0232 (232 mg/L)
0.0050	66	CL50	0.0041	
0.0060	80			
0.0070	86	CL95	0.0086	
0.0080	94			
Psidium rotundatum				
0.0040	28			8.16 Diagnostic Dose 0.018 (180 mg/L)
0.0050	58	CL50	0.0047	
0.0060	80			
0.0070	92	CL95	0.007	
0.0080	97			
Piper aduncum				

Conc (%)	Mortality (%)	Lethal Conc		Slope of the Line
0.0100	36			6.72 Diagnostic Dose 0.050 (500 mg/L)
0.0120	53	CL50	0.0110	
0.0140	73			
0.0160	85	CL95	0.0200	
0.0180	90			
0.0200	95			
		Pimenta racemosa		
0.0040	10			5.23 Diagnostic Dose 0.0380 (380 mg/L)
0.0050	20	CL50	0.0070	
0.0060	44			
0.0070	50	CL95	0.0140	
0.0080	62			
0.0100	79			
		Eugenia melanadenia		
0.0240	14			18.5 Diagnostic Dose 0.0718 (718 mg/L)
0.0260	41	CL50	0.026	
0.0280	70			
0.0300	78	CL95	0.033	
0.0320	90			

4. CONCISE SUMMARY

The interest in plant-derived natural products has intensified due to the emergence of resistance to synthetic insecticides used to diminish insect populations.

The biological activity of natural substances depends on their structure and the dosages employed. Vegetable pesticides has the significant advantage of compatibility with other ecologically acceptable low-risk alternatives for insect control. Plants' secondary metabolites, which defend against predators and insects, render them natural candidates for controlling insect species that are both disease vectors and agricultural pests. It is illogical to assume that they will entirely supplant synthetic insecticides. Logical thinking has a complementary role in optimizing and enhancing the sustainability of existing integrated pest management techniques.

Numerous publications exist that enumerate plants possessing insecticidal qualities. Utilizing such plants requires more than only being deemed promising or have established insecticidal qualities. An assessment of environmental and health issues should also be conducted. The usage of endangered plants, those with little biomass, or those whose utilization necessitates significant modifications to their natural density is not advisable. An optimal insecticide plant should be perennial, extensively distributed in nature or amenable to cultivation, utilizing renewable plant parts such as leaves, flowers, or fruits, without necessitating the destruction of the plant for material collection (thus avoiding the use of roots and bark). It must meet minimum agro-technician requirements and adhere to principles of ecological sustainability, possess additional applications (such as medicinal uses), maintain a low economic value, exhibit efficacy at minimal dosages, and demonstrate potential for biotechnological scaling. Silva (2002).

Insecticide plants possess the advantage of additional medical applications, quick decomposition that mitigates the risk of food residues, and specificity for pest insects while being less harmful to natural foes. They also exhibit a slower development of resistance relative to synthetic pesticides. Conversely, the drawbacks include a more rapid degradation by ultraviolet radiation, resulting in a diminished residual effect; however, not all plant-derived insecticides exhibit lower toxicity than synthetic alternatives, and their residual properties are not well-defined. Research in our country indicates that essential oils and their derivatives from various evaluated plant families exhibit significant insecticidal activity against the larvae of Aedes aegypti and Culex quinquefasciatus. Specifically, those derived from Pinaceae demonstrate ovicidal properties and inhibit the development of Aedes aegypti, positioning them as promising alternatives for controlling these insect species..

5. REFERENCES

Adam, KP; Thiel, R; Zapp, J; Becker, H. (1998). Involment of the mevalonic acid pathway and glyceraldehyde-pyruvate pathway in the terpenoid biosynthesis of the liverworts *Ricciocarpos natansand Conocephalum conicum. Archives of Biochemistry and Biophysics.* 354: 181-187.

Aguilera, L; Navarro, A; Tacoronte, JE; Leyva, M ;Marquetti MC. (2003) Efecto letal de Myrtaceas cubanas sobre *Aedes aegypti* (Díptera: Culicidae) *Rev Cubana Med Trop*;55(2):100-4.

Aguilera L, Tacoronte JE, Navarro A, Leyva M, Bello A, Cabreras MT. (2004) Composición química y actividad biológica del aceite esencial de *Eugenia melanadenia* (Myrtales: Myrtaceae) sobre *Blattella germanica* (Dictyoptera: Blattellidae). *Revista CENIC Ciencias Químicas.* : 35(3).

Alburquerque, MR; Silveira,ER; De Ochoa, DE; Lemus, TL; Souza, EB; Santiago, GM; Pessoa, OD. (2004) Chemical composition and larvicidal activity of the essential oils from *Eupatorium bentonicaea* Baker (Astaraceae) *J Agric Food Chem.*;52 (22): 6708-11.

Amer, A & Mehlhorn, H. (2006) Repellency effect of forty-one essential oils against *Aedes, Anopheles* and *Culex* mosquitoes. *Parassitol Res.*99 (4):478-90.

Amer, A & Methlhorn, H. (2006).Larvicidal effects of various essentials oils against, Aedes, Anopheles and Culex larvae (Diptera: Culicidae). *Parasitol Res.*99(4):466-72.

Ansari, MA; Mittal, PK; Razdan, RK; Sreehari, U. (2005) Larvicidal and mosquito repellent activities of Pine (*Pinus longifolia*, Familia: Pinaceae) oil. *J Vect Borne Dis.* 95-99.

Assarm, AA; El-Sobky, MM. (2003) Biological and histopathological studies of some plant extracts on larvae of *Culex pipiens* (Diptera:Culicidae) *J. Egypt Soc Parasitol l.*33(1):189-200.

Awde, JA; Ryan, MF (1992) Piperazine as a reversible competitive inhibitor of acetylcholinesterase. *Cienc. Biol.Ecol.Syst.* 12: 37-42.

Barnard, DR & Xue, RD. (2004) Laboratory evaluation of mosquito repellent against *Ae albopictus, Culex nigripalpus* (Diptera: Culicidae). *J Med Entomol.* 41 (4): 726-30.

Bassole, IH; Guelbeogo, WM; Nebie, R; Constantini, C; Sagnon, N; Kabore, Z; Traore, SA. (2003) Ovicidal and larvicidal activity against *Ae aegypti* and *An gambiae* complex mosquitoes of essential oils extracted from three spontaneus plants of Burkina Faso. *Parassitologia* 45 (1):23-6.

Bottia, E; Díaz, O; Mendivelso, D; Martínez, J; Stashenko, E. (2007) Comparación de la composición química de los metabolitos secundarios volátiles de cuatro plantas de la familia Piperaceae obtenidos por destilación–extracción simultánea. Scientia et Technica;13(33). ISSN 0122-1701

Bloomquist, J. (1999) Insecticides: Chemistries and Characteristics. Departmentof entomology. Virginia Polytechnic institute and state university. Blacksslomg Virginia. In http://ipmworld.umn.edu/chapters/bloomq.htm.

Byeoung-Soo, P; Won-Sik, C; Jeong-Han, K; Kap-Ho, K; SPNG-Eun, L. (2005). Monoterpenes from Thyme (Thymus vulgaris) as potencial mosquito repellents. *J Am Mosq. Control Association* .21 (1);80-83.

Caballero-Gallardo, K; Olivero-Verbel, J; Stashenko, EE.(2011) Repellent Activity of Essential Oils and Some of Their Individual Constituents against *Tribolium castaneum* Herbst.*J Agric Food Chem.* Feb 3. [Epub ahead of print]

Carvalho, AF; Melo, VM; Carveiro, AA; Machoo, MI; Beantim, MB; Rabelo, EF. (2003) Larvicidal activity of the essential oil fro *Lippia sidioide* against *Aedes aegypti. Mem Oswaldo Cruz.*98 (4): 569-71.

Cavalcanti, ES; Morais, SM; Lima, MA; Santana, EW. (2004) Larvicidal activity of essentials oils from Brazilian plants against *Ae aegypti. Mem Inst Oswaldo Cruz*.99(5):541-4.

Cetin H, Yani A. Study of the larvicidal activity of Origanum (Labiatae) species from Southwest Turkey. *J. Vector Ecol.* 2006; 31 (1): 118-22.

Chaiyasit, D; Choochate, W; Rattanachanpichai, E; Chaithong, U; Chaiwong P et al. (2006) Essential oils as potencial adulticides against two populations of *Aedes aegypti*, the laboratory and natural field strain in Chiang Mai province northern Thailand. *Parasitol Res* 1

Chang, KS; Talk, JH; Kim, SI; Lee, WJ; Ahn, YJ. (2006) Repellency of *Cinnamomum cassia* bark compounds and cream countain cassia oil to de *Ae aegypti* under laboaratory and indoor conditions. *Pest Manag Sci.* 62 (11):1032-1038.

Choochate, W; Chaiyasit, D; Kanjarapotti, D; Rottana, E; Champiochi, E et al (2005). Chemical composition and antimosquito potencial of rhizome extract and volatile oil derivade from *Curcuma aromatica* against *Aedes aegypti* (Diptera: Culicidae). *J Vector Ecol.*30 (2):302-9.

Champakaew, D; Choochote, W; Pongpaibul, Y; Chaithong, U; Jitpakdi, A; Tuetun, B; Pitasawat, B. (2007) Larvicidal efficacy and biological stability of a botanical natural product, zedoary oil-impregnated sand granules, against *Aedes aegypti* (Diptera, Culicidae). *Parasitol Res.*100(4):729-37.

Chansang, U ; Zahiri ,NS ; Bansiddhi, J ; Boonruad, T ; Thongsrirak, P ; Mingmuang, J ; Benjapong , N ; Mulla, MS. (2005) Mosquito larvicidal activity of aqueous extracts of long pepper (*Piper retrofractum* vahl) from Thailand. *J Vector Ecol.*30(2):195-200.

Cheng, SS; Liu, JY; Tasai, KH ; Chen, WJ; Chang, ST. (2004) Chemical composition and mosquito larvicidal activity of essential oils from leaves of different *Cinnamomum osmophloem* provenanaces. *J Agric Food Chem.*52 (14):4395-400.

Chung, IM; Song, HK; Yeo, MA; Moon, HI. (2010) Composition and immunotoxicity activity of major essential oils from stems of *Allium victorialis* L. var. platyphyllum Makino against *Aedes aegypti L. Immunopharmacol Immunotoxicol.* Dec 16. [Epub ahead of print]

Chung, IM; Ro, HM; Moon, HI (2010) Major essential oils composition and immunotoxicity activity from leaves of *Foeniculum vulgare* against *Aedes aegypti* L. *Immunopharmacol Immunotoxicol.* Nov 16. [Epub ahead of print]

Corbino, G . (2000) Productos naturales vegetales de interés económico Instituto Nacional de Tecnología Agropecuaria. Estación Experimental San Pedro. Buenos Aires, Argentina.Lab.Prod.InVitro.(11-2-2005)In : http://www.inta .gov.ar/ sanpedro /1_sala_de_lectura/difusion/novedades.../corbinocondim.ht.

Das, NG; Barauah, I; Talukdar, PK; Das, SC. Evaluation of botanicals as repellent against mosquitoes. *J Vect Borne Dis* ;40 (1-2): 49-53.

De Mendoca, FA; Da Silva, KF; Dos Santos, KK; Ribeneiro Junior, KA; Sant'Ana, AE. (2005) Activities of some brazilian plants against larvae of the mosquito *Ae aegypti.Fitoterapia*;76 (7-8):629-36.

Dharmagada, VS; Naik, SN; Mittal, PK; Vasudevan, P. (2005) Larvicidal activity of *Tagetes patula* essential oil against three mosquito species. *Bioresour Technol* 96 (11);1235-40.

Faley, DH & Frances, SP. (2005) Laboratory evaluation of coconut oil as a larvicide for *Anopheles faraoti* and *Culex annulirostris*. *J Am Mosq.Control Assoc.*;21(4): 477-9.

Gascón, AD; Pelayes, MA. (2002). *Generalidades sobre los procesos extractivos en la obtención de aceites esenciales* Departamento de Tecnología Agroindustrial Facultad de Ciencias Agrarias Chacras de Coria – Mendoza.

Gillij, YG ; Gleiser, RM ; Zygadlo, JA. (2008) Mosquito repellent activity of essential oils of aromatic plants growing in Argentina . *Bioresour Technol.*99(7): 2501-15.

González, P.(1984) Capítulo VI In "*Utilización Terapeútica de Nuestras Plantas Medicinales*", Universidad de La Salle, Bogotá.

Gunderson, CA; Samuelian, JH; Evans ,CK. (1985) Effects of the mint monoterpene pulegone on *Spodoptera eridinan* (Lepidoptera: Noctuidae). *Environ Entomol.*14:859-63.

Hafeez, F; Akram, W; Shaalan, EA. (2011).Mosquito larvicidal activity of citrus limonoids against *Aedes albopictus*. *Parasitol Res.* Jan 7. [Epub ahead of print]

http://www.buscasalud.com/boletín/analisis/2003_01_1117_09_32.html (12-1-2005)

http://www.chemkeys.com/esp/md/tee_6/ecoqui_2/metqui_5/metqui_5.htm (8-1-2007)

http://www.herbotecnia.com.ar/poscosecha-esencias.html (8-1-2007)

http://www.losandes.com.ar/imprimir.asp?nrc=163985 (8-1-2007)

http://www.cannabiscultura.com/f1/showthread.php?p=84457 (4-3-2011)

Ikan, R.(1991) Natural products. A laboratory guide. Second Edition. Academia Press (Eds).360 pp

Jaenson, TG; Palsson, K; Borg-Karlson, AK. (2006) Evaluation of extracts and oils of mosquito (Diptera: Culicidae) repellent plants from Sweden and Guinea Bissau. *J Med. Entomol.*43 (1):113-9.

Judd, WS; Campell, CS; Kellogg, EA; Stevens, PF; Donoghue, MJ. (2002) Secondary plants compounds. Plant systematic: a phylogenetic approach. Chapter 4, Structural and Biochemical Characters. Second Edition. Sinaeur Axxoc, USA.

Keane, S;Ryan, MF. (1999). Purification, characterization and inhabitation by monoterpenesof acetylcholinesterase from the waxmoth *Galleria mellonella*. *Insect Biochemistry and Molecular Biology* 29:1097-1104.

Khandagle, AJ; Tare, VS; Raut, KD; Morey, RA. (2011).Bioactivity of essential oils of *Zingiber officinalis* and Achyranthes aspera against mosquitoes. *Parasitol Res.* Feb 11. [Epub ahead of print]

Kim, SI; Chang, KS; Yang, YC; Kim, BS; Ahn, YJ. (2004) Repellency of aereosol and cream products containg fennel oil to mosquitoes under laboratory and field conditions. *Pest Manang Sci.*;60 (11):1125-30.

Kordali, S; Cakir, A; Ozer, H; Cakmakci, R; Kesdek, M; Mete, E. (2008)Antifungal, phytotoxic and insecticidal properties of essential oil isolated from Turkish *Origanum acutidens* and its three components, carvacrol, thymol and p-cymene. *Bioresour Technol.* May 28.

Kostyukovsky, M; Rafaeli, A; Gileadi, C; Demchenko, N; Shaaya, E. (2002). Activation of octopaminergic receptors by essential oil constituent isolated from aromatic plants: posible mode of action against insect pest. *Pest Management Science*. 58: 1101-06.

Leyva M , Aguilera L, Tacoronte J, Montada D, Bello A, Marquetti MC. (2007) Estudio de laboratorio del aceite esencial de *Pimenta racemosa* (Myrtales: Myrtaceae) y su posible utilización para el control de *Musca domestica* (Diptera: Muscidae) *Rev CNIC Ciencias Químicas*

Leyva M, Tacoronte J, Marquetti MC.(2007) Composición química y efecto letal de *Pimenta racemosa* (Myrtales: Myrtaceae) en *Blatella germanica* (Dictyoptera: Blattellidae) Rev. Cub. Med. Trop . 59 (2)

Leyva M, Tacoronte JE, Marquetti MC, Scull R, Montada D, Rodríguez Y, Bruzón R.(2008)Actividad insecticida de aceites esenciales de plantas en larvas de *Aedes aegypti* (Diptera: Culicidae) *Rev Cubana Med Trop* 60(1):78-82.

Leyva M, Tacoronte JE,Marquetti MC, Montada D.(2008) Actividad insecticida de tres aceites esenciales en *Musca domestica* (Diptera: Muscidae) Rev Cubana Med Trop 60(3).

Leyva M, Tacoronte JE, Marquetti MC ,Scull R , Tiomno O , Mesa A, Montada D. (2009) Utilización de aceites esenciales de pinaceas endémicas como una alternativa de control en *Aedes Aegypti Rev .Cub . Med.Trop*;61(3):239-43.

Leyva M, Marquetti MC, Tacoronte JE, Scull R , Tiomno O , Mesa A, Montada D. (2009)Actividad larvicida de aceites esenciales de plantas contra *Aedes aegypti* (L) (Díptera: Culicidae) *Biomed* 20: 5-13.

Leyva M, Marquetti MC, Tacoronte JÁ, Tiomno O, Montada D (2010) Efecto inhibidor del aceite de trementina sobre el desarrollo de larvas de *Aedes aegypti* (Diptera: Culicidae) *Rev.Cub.Med.Trop.*;62(3):

López, MD.(2008) Toxicidad volatil de monoterpenoides y mecanismos bioquímicos en insectos plagas del arroz almacenado. Tesis Doctoral. Universidad de Murcia. Facultad de Quimica. España 245 pp

Lucia, A; Gonzalez Audino, P; Saccacini, E; Licastro, S; Zerba, E; Masuh, H (2007). Larvicidal effect of *Eucalyptus grandis* essential oil and turpentine and their major components on *Ae aegypti* larvae *J.Am.Mosq Control Assoc.*23:293-303

Mathew, J; Thoppil, JE.(2011) Chemical composition and mosquito larvicidal activities of Salvia essential oils. *Pharm Biol.* Feb 2. [Epub ahead of print]

Miot, HA; Batistilla, RF; Batista, K; de A Volpato, DE, Augusto, LS; Madeira, NG; Haddad, V; Miot, LD. (2004) Comparative study of the topical effectiveness of the Andiroba oil (*Coparais guianensis*) and DEET 50 % as repellent for Ae sp. *Rev Inst Med. Trop Sao Paaulo.*46 (5): 253-6.

Morais, SM; Cavalvanti, ES; Bertini, LM; Oliveira, CL; Rodríguez, JR; Cardoso, JH. (2006) Larvicidal activity of essential oils from Brazilian Croton sepecies against *Ae aegypti. J Am Mosq. Control Assoc.* 22(1):161-4.

Odalo, JO; Omolo, MO; Malebo, H; Angira, J; Njeru, PM; Ndiege, IO; Hassanali ,A. (2005) Repellency of esential oils of some plants from Kenya coast against *An gambiae. Acta Tropical.*95 (3):210-8.

Oyedale, AO; Gbolade, AA; Sosan, MB; Adewoyin, FB; Sayelu, OL; Orafidaya, OO. (2002) Formulation of an effective mosquito-repellent topical product from lemon gross. *Phytomedicine.*9(3):259-62.

Parmar, V.S;Jain, S.C;Bisht, K.S;Jain, R;Taneja, P;Jha, A ;et al. (1997) Phytochemistry of the genus *Piper*. *Phytochemistry*. 46:597-673

Pérez-Pacheco, R; Rodríguez –Hernández, C ; Lara-Reyna, J; Montes –Belmont, R; Ramírez – Valverde, G. (2004) Toxicidad de aceites, esencias y extractos vegetales en larvas de mosquito *Culex quinquefasciatus* Say (Diptera: Culicidae) *Acta Zoológica Mexicana*. 20(1): 141-152.

Phasomkusolsil, S; Soonwera, M. (2010) Potential larvicidal and pupacidal activities of herbal essential oils against *Culex quinquefasciatus* say and *Anopheles minimus* (theobald).*Southeast Asian J Trop Med Public Health.* Nov;41(6):1342-51.

Phasomkusolsil, S; Soonwera, M. (2010) Insect repellent activity of medicinal plant oils against *Aedes aegypti* (Linn.), *Anopheles minimus* (Theobald) and *Culex quinquefasciatus* Say based on protection time and biting rate.*Southeast Asian J Trop Med Public Health.* Jul;41(4):831-40.

Pino, J. Borges P. (1999) Los componentes volátiles de las especias. I. Métodos de obtención y de análisis. *Alimentaria.*30 (1) 39-45.

Pino, JA; Marbot, R; Bello, A; Urquiola, A (2004) Essential Oils of *Piper peltata* (L.) Miq. and *Piper aduncum* L. from *Cuba Journal of Essential Oil Research: JEOR*, Mar/Apr 2004.

Pino, J A; Marbor, R; Bello, A;Urquiola, A. (2005) Essential oil of *Eugenia melanadenia* Krug et Urb. from Cuba". *Journal of Essential Oil Research* JEOR. FindArticles.com. (09-2- 2011) In: . http://findarticles.com/p/articles/mi_qa4091/is_200307/ai_n9287900/

Pitarokili, D; Michaelakis, A; Koliopoulos, G; Giatropoulos, A; Tzakou, O. (2011). Chemical composition, larvicidal evaluation, and adult repellency of endemic Greek Thymus essential oils against the mosquito vector of West Nile virus. *Parasitol Res.* Feb 8.

Pohlit, AM; Lopes, NP; Gama, RA; Tadei, WP; de Andrade-Neto, VF. (2011) Patent Literature on Mosquito Repellent Inventions which Contain Plant Essential Oils - A Review.*Planta Med* Feb 15. [Epub ahead of print]

Prajapati, V; Tripathi, AK; Aggarwall, KK; Kanuja, SP (2005) Insecticidal repellent and oviposition-deterrente activity of selected essential oils against *An stephensi, Ae aegypti* and *Cx quinquefasciatus. Bioresour Technol.* 96 (16): 1749-57.

Pushpanathan, T; Jebanesan, A; Govindarajan, M. (2006) Larvicidal, ovicidal and repellent activities of *Cymbopogan citratus* Stapf (Graminae) essential oil against the filarial mosquito *Culex quinquefasciatus* (Say) (Diptera : Culicidae). *Trop Biomed.* 23(2):208- 12.

Ravi, K; Bhavani, K; Sita Devi, P; Rajaswara Rao, BR; Janarahan, RK. (2006) Composition and larvicidal activity of leave and stem essential oil of *Chloroxylon swetenia* against *Ae aegypti* and *An stephensi. Bioresour Technol.* 97(18):24-31.

Ryan, MF; Byrne, O (1988) Plant-insect coevolution and inhibition of acetilcholinesterase. *Journal of Chemical Ecology* 14:1965-75.

Ryan, MF; Awde, J; Moran, S. (1992) Insect pheromones as reversible competitive inhibitors of acetilcholinesterase. *Invertebrate reproduction and development.* 22:31-38.

Scholes, M. (1995). *La Terapia de los Aromas.* Sterling Press, New York.

Stamm, M. D. (1970).Modernos Conocimientos sobre la Química de los Aceites Esenciales. Métodos de Análisis. *C.S.I.C. Manuales de Ciencia Actual 3*. Patronato "Alonso de Herrera", Madrid.

Sangkyun L; Tsao, R; Coats, J. (1999) Influence of dietary applied monoterpenoids and derivates on survival and growth of european corn borer (Lepidoptera: Pyralidae). *J Econ Entomol*.92(1):56-67.

Smith, RM & Kassim, H. (1979) The essential oil of *Piper aduncum* from Fiji. New Zealand. *J Sci* . 22:127-128.

Sponsel VM. (1995). The biosyntesis and metabolism of gibberellins in higher plants. In Plant hormones. Davies PJ (Ed). Kuwer Academic Publishers Boston. 66-97 pp.

Suparvarn, P; Knapp, F; Sigafus, R. (1974) Biologically active plant extracts for control of mosquito larvae. *Mosquito News*.34(4):398-402

Silva, WJ; Dória, GA; Maia, RT; Nunes, RS; Carvalho, GA; Blank, AF; Alves, PB; Marçal, RM; Cavalcanti, SC. (2008) Effects of essential oils on *Aedes aegypti* larvae: Alternatives to environmentally safe insecticides. *Bioresour Technol*.;99(8):3251-5

Silva, G. (2002) *Insecticidas vegetales*. (11-12-2006) In: http://ipmworld.umn.edu/cancelado /Spchapters/GSilvaSp.htm

Silva, G.(2003) Resistencia a los insecticidas In: Silva, G y Hepp, R (Eds). Bases para el manejo racional de insecticidas. Universidad de Concepción, Fundación para la Innovación Agraria: 237-60pp

Stashenko, E. (1996) In: *Memorias del IV Congreso Nacional de Fitoquímica*, Universidad Industrial de Santander, Escuela de Química, Bucaramanga, febrero de 1996, pp. 29-53.

Tare, V; Deshpade, S; Sharma, RN. (2004) Susceptibility of two different strains of *Ae aegypti* (Diptera: Culicidae) to plants oils. *J Econ. Entomol*. 97 (5):1734-6.

Tawatsin, A; Asavadachanukorn, P; Tahavara, U; Wongsenkongman, P; Bansidhi, J; Boonroad, J. (2006) Repellency of essential oils extracted from plantas in Thailand against four mosquito vectors (Diptera: Culicidae) *Southeast Asian J Med Trop. Med Public Heath*. 37 (5):915-31.

Thomas, TG. (2004) Mosquito larvicidal properties of essential oil of indigenus plant *Ipomoea airica* L. *Jpn J Infect Dis*.57 (4).

Trongotokit, Y; Curtis, CF; Rangsriyum, Y. (2005) Efficacy of repellent products against caged and free flying *An stephensi* mosquitoes Southeast Asian. *J Trop Med. Public Health*. 36 (6): 1423-31.

Trongotokit, Y; Rangsriyum, Y; Komalamisra, N; Apiwathnarson, C. (2005) Comparative repellency of 38 essential oils against mosquito bites. *Phytother Res.*; 19(4):303-9.

Worwood, V. (1992) *The Complete Book of Essential Oils and Aromatherapy*. New World Library, New York.

Yang, P& Ma, Y. (2005) Repellent effects of plants essential oils against *Ae albopictus. J Vector Ecol.* 30 (2): 231-4.

Zhu, J; Zeng, X; Yanma, J; Liu, J; Qian, K; Tucker, B; Schultz, G; Coats, J; Zang, A. (2006) Adults repellency and larvicidal activity of five planta essential oil against mosquitoes. *J Am Mosq Control Assoc* 22(3):515-22.

CHAPTER 11

Biological Control of Mosquito Larvae Utilizing Bacillus thuringiensis subsp. israelensis

1. INTRODUCTION: AN OVERVIEW

Chemical pesticides offer numerous advantages for food production and human health, demonstrating significant efficacy in enhancing agricultural and forestry productivity. Nonetheless, they present some risks, including contamination of water and food supplies, poisoning of non-target organisms, accumulation in the food chain, and the emergence of insect pest populations resistant to chemical insecticides (Wojciech & Korsten 2002). Chemical pesticides are known to diminish populations of natural enemies, disrupt biological control, and potentially lead to outbreaks of secondary pests that were previously suppressed by these natural enemies (Bartlett, 1964). Additionally, pest species can develop resistance to pesticides, while natural enemies do not (Johnson & Tabashnick, 1999).

The application of synthetic organic pesticides has resulted in significant economic, societal, and environmental consequences. The escalating expenses associated with the development and production of petrochemically produced pesticides, coupled with their diminishing efficacy due to pervasive insect resistance. Consequently, the chemical pesticide industry persists in formulating novel, costlier chemicals, thereby escalating pesticide prices. They have inflicted mortality and illness on humans and harmed the environment both socially and ecologically. It is believed that only a minuscule proportion of the administered insecticides is necessary for the suppression of the target pest. Over 99.9% of the remainder is released into the environment via soil, water, and food cycles (Metcalf, 1986).

Alternative insect management techniques provide sufficient pest control while presenting reduced risks. An option is the utilization of microbiological insecticides, which comprise bacteria or their by-products. Microbial pesticides are particularly advantageous due to their minimal toxicity to non-target organisms and humans. In comparison to other widely utilized insecticides, they are safe for both pesticide applicators and consumers of treated crops. Microbial insecticides are often referred to as biological pathogens and biological control agents. Chemical pesticides are predominantly utilized globally compared to microbial control; nonetheless, certain microbial control agents can partially substitute some dangerous chemical pest control substances. Various biological control agents composed of bacteria, fungi, viruses, pheromones, and plant extracts have been utilized primarily for the management of insect populations that threaten forests and agricultural crops (McDonald & Linde, 2002).

The microbial insecticides most widely used in the world are preparations of *Bacillus thuringiensis* (*Bt*). The insecticidal activity of *Bt* is due to the proteic parasporal inclusions that are produced during sporulation. Insecticides based on the proteinaceous δ-endotoxin of *Bt* constitute part of a more ecologically rational pest control strategy. *Bt* strains have been isolated world wide from many habitats, including soil, insects, stored-product dust, and deciduous and coniferous leaves, all of which have a limited host range, however together span a wide range of insects orders which include: Lepidoptera, Diptera, Coleoptera, Hymenoptera, Homoptera, Phthiraptera, Orthoptera, Acari, and Mallophaga and other organisms such as nematodes, mites, and protozoa (Federici, 1999). *Bacillus thuringiensis* subsp. *israelensis* (*Bti*) or serotype H-14, exhibit acute toxicity towards dipteran insects such as larval mosquitoes and black flies (de Barjac 1978) and is currently used in mosquito control programs world wide (Priest, 1992). The World Health Organization's (WHO) Onchocerciasis Control Program (OCP) in West Africa using *Bti* toxins has been one of the success stories of international co-operation in the control of infectious diseases program (Webb 1992; Drobniewski 1993). This chapter aims to furnish non-Bt professionals with a fundamental understanding of Bti, given its significance in controlling many tropical diseases such as malaria and dengue..

2. BACILLUS THURINGIENSIS STRAINS

2.1 Origin of some strains of B. thuringiensis

A large number of strain of *Bt* have been isolated from which to date. *Bt* as currently recognized is actually a complex of subspecies. They have grouped in 79 serotypes (Zeigler, 1999). The first *Bt* strain was isolated from diseased larvae of the silkworm, *Bombi mori*, in Japan by Ishiwata (1901). Iwabushi (1908) describe the bacillus as *Bacillus sotto.* Aoki & Chigasaki (1915) and Mitani & Wataral (1916) purified a highly toxic substance from sporulated *Bt* cultures. It was not officially described, however, until it was reisolated by Berliner in 1915 from diseased larvae of the Mediterranean flour moth, *Anagasta kuehniella*, in Thuringia, Germany, hence the derivation of species name thuringiensis (Federici, 1999). A *Bt* strain was isolated again by Mattes (1927) and described briefly the inclusion rhomboidal body. The activity of the *Bt* strains against lepidopteran larvae was described by Metalnikov (1930) and by Husz (1931). The association of the inclusion bodies of *B. thuringiensis* with toxicity against insects was established by Steinhaus. In 1951 published a paper which described the morphology of *Bt* and its possible use in the biological control against the alfalfa caterpillar. In France, a product named "Sporeine" was developed and used against *Ephestia kuhniella* (Lepidoptera) (Jacobs, 1950). Hannay, (1953) described a parasporal body in bipyramidal shape produced by the bacterium during sporulation and suggested that the crystal was involved in the toxic activity. The protein nature of the crystals was determined by Hanna and Fitz-James (1955).

2.2 Features of B. thuringiensis strains

Bt is a facultative anaerobic, gram-positive bacterium that forms characteristic protein inclusions adjacent to the endospore (Fig.1). The crystalline inclusion are composed of proteins known as ICPs crystal proteins. Cry proteins, or δ-endotoxins is the basis for commercial insecticidal formulations of *Bt*. Insecticides containing *Bt* in pest control programs is now considered as a viable strategy, which has proven to be both safe and reliable over the last 45 years (Chungjatupornchai et al. 1988).

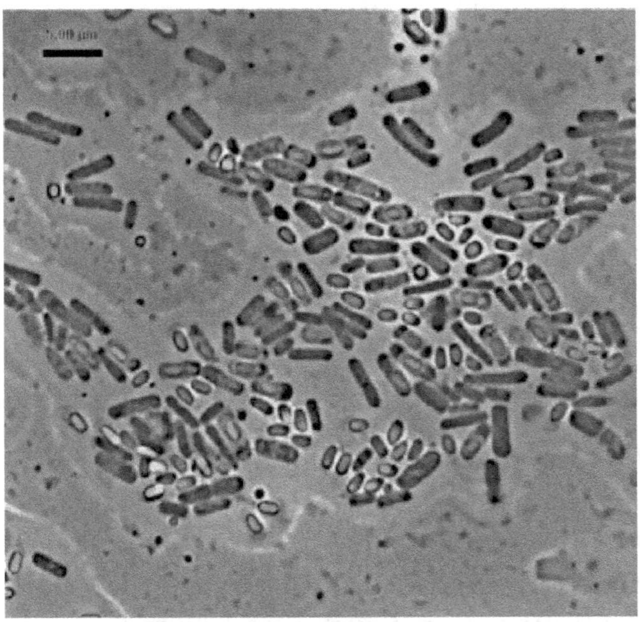

Fig. 1. Bacillus thuringiensis subsp. israelensis.

2.3 Advantages and disadvantages of Bt

According to Federici (1999) the main reasons for the success of *Bt* include (i) the high efficacy of its insecticidal proteins (ii) the existence of a diversity of proteins that are effective against a range of important pests (iii) its relatively safety to nontarget insect predators and parasites (iv) its easy to mass production at a relatively low cost, and (v) its adaptability to conventional formulations and application technology.

Advantages and disadvantages of *Bt* with chemical insecticides were summarized by Rowe,and Margaritis (1987).

Advantages of *Bt:*

- High specificity, i.e. no mammalian or nontarget effects; use permitted up to date of harvest.
- No broad insect resistance observed or expected to develop.
- Adaptable to many types of formulations; potential to incorporate feeding stimulants or baits to increase the attractiveness of the formulations to the insects and thereby increase their efficacy.
- Probability of producing more potent formulations and reducing production costs through improved-fermentation technology.
- High probability that strain selection and/or genetic engineering will lead to better control of pest insects by newly found or created strains of Bt having novel host spectra or increased activity

Disadvantages of *Bt:*

- Narrow host spectrum
- Lack of patent protection on new strains
- Proper timing of application require due to slower effect than chemical insecticides
- Activity dependent of ingestion, and feeding activity depends on environmental conditions
- Relatively higher cost compared to chemical insecticides

2.4 Clasification of strains of Bacillus thuringiensis

Bt is a crystalliferous spore-forming bacterium close genetic relationship with *B. cereus, B. anthracis* and *B. mycoides* (Höffe & Whiteley, 1989). The classification of *Bt* is difficult because DNA sequencing studies of conserved gene regions of theses species have suggested that they belong to a single specie. *Bt* strains are distinguished from *B. cereus, B. anthracis* and *B. Mycoides* by the ability to produce parasporal crystalline inclusions during sporulation (de Barjac, 1978). Crystal formation is the criterion for distinguishing between *B. cereus* and *Bt*, otherwise they could be considered as the same specie. Research based on a comparative study of 16s rRNA sequences, *Bt* and *B. cereus* var. *mycoides* differed from each other and from B. *anthracis* and *cereus* by less than nine nucleotides (Ash and Collins 1992). Chen & Tsen (2002) amplified 16S rDNA and *gyrB* gene by PCR and they found that the discrimination between *B. cereus* and *Bt* strains, when a large number of Bacillus strains were tested was difficult. They proposed, to distinguish *Bt* from *B. cereus*, a single feature, such as the presence of a parasporal crystal protein or cry gene is reliable.

Several attemps were made to classify *Bt* strains. de Barjac and Frachon (1990) didn´t find correlation between biochemical reactions and 27 H serotypes using 1,600 *B. thuringiensis* isolates. They demonstrated that the current biochemical tests have no value as the sole criteria for differentiating *Bt* strains. Another approach ot these authors for classification was the use of susceptibility to certain bacterial viruses called phages. There are 14 bacteriophages that have been used for *Bt* but phage typing is inconsistent with serotyping and does not permit classification. They found frequent cross-reactions.

One widely clasification system for *Bt* strains is based in the determination of the H-flagellar antigen technique described by Barjac & Bonnefoi (1962). This technique needs very motile bacterial cultures to prepare flagellar suspensions. These suspensions are titrated against antisera directed against *B. thuringiensis* strains of each serotype. Presently *B. thuringiensis* strains are classified within 79 serotypes. Table 1 shows the classification of *Bt* strains by serotypes.

Bt strains that had discovered previously to 1977 were pathogens towards lepidopteran larvae. However, in 1975 was discovered *B. thuringiensis* serovar *israelensis* toxic to mosquito larvae (Goldberg & Margarit 1977) and in 1983 a strain from *Bt* serovar *morrisoni* was found to be pathogenic to Coleoptera larvae (Krieg et al. 1983). According to these findings, serological clasification, although still in use as a basic method to clasify *Bt* strains could not be related with pathogenicity. Subsequently studies showed that, within a serotype, different activity, spectra can be found. For example, some strains of *B. thuringiensis* serotype *morrisoni* in their parasporal inclusión bodies contain different proteins and have activity against Diptera, Coleoptera or Lepidoptera.

With the knowledge of sequence of the genes that encode the proteins cry it was proposed a classification based on the cry toxin genes. Höfte and Whiteley (1989) proposed a nomenclature clasification scheme for *Bt* crystal proteins based in their structural aminoacid sequence, deduced from the DNA and host range. They named *cry* (crystal protein) genes and their related proteins, "Cry proteins". They clasified 42 *Bt* crystal protein genes into 14 distinct genes grouped into 4 major clases. The classes were CryI (Lepidoptera-specific), CryII (Lepidoptera-and Diptera-specific), CryIII (Coleoptera-specific), and CryIV (Dipter-specific).

Table 1. Classification of *B.thuringiensis* strains by serotype (modified after Zeigler, 1999).

Serotype	Serovar	Serotype	Serovar
1	thuringiensis	28a, 28c	jeghatensan
2	finitimus	29	amagiensis
3a, 3b, 3c	kurstaki	31	toguchini
3a, 3c	alesti	32	cameroun
3a, 3d	sumiyoshiensis	33	leesis
3a, 3d, 3e	fukuokaensis	34	konkukian
4a, 4b	soto/dendrolimus	35	seoulensis
4a, 4c	kenyae	36	malayensis
5a, 5b	galleriae	37	andralousiensis
5a,5c	canadensis	38	owaldocruzi
6	entomocidus/subtoxicus	39	brasilensis
7	aizawai/pacificus	40	huazhongensis
8a.8b	morrisoni	41	sooncheon
8a, 8c	ostriniae	42	jinghongiensis
8b, 8d	nigeriensis	43	guiyangiensis
9	tolworthi	44	higo
10a, 10b	darmstadiensis	45	roskildensis
10a, 10c	londrina	46	chanpaisis
11a, 11b	toumanoffi	47	wratislaviensis
11a, 11c	kyushuensis	48	balearica
12	thompsoni	49	muju
13	pakistani	50	navarrensis
14	israelensis	51	xiaguangiensis
15	dakota	52	kim
16	indiana	53	asturiensis
17	tohokuensis	54	poloniensis
18a, 18b	kumamtoensis	55	palmanyolensis
18a, 18c	yosoo	56	rongseni
19	tochigiensis	57	pirenaica
20a, 20b	yunnanensis	58	argentinensis

Serotype	Serovar	Serotype	Serovar
20a, 20c	pondicheriensis	59	iberica
21	colmeri	60	pingluonsis
22	shanongiensis	61	sylvestriensis
23	japonensis	62	zhaodongensis
24a, 24b	neolonensis	64	azorensis
24a, 24c	novosibirsk	65	pulsiensis
25	coreanensis	66	graciosensis
26	silo	67	vazensis
27	mexicanensis	none	wuhanensis
28a, 28b	monterrey		

The *cryI* genes can be distinguished from the other *cry* genes simply by sequence homology (>50% aminoacid identy) and encode 130 to 140 kDa proteins which accumulate in bypiramidal crystalline inclusions during the sporulation of *Bt. cryII* genes encode 65-kDa proteins which form cuboidal inclusions in strains of several species. *cryIII* genes produces romboidal crystals containing one major protein, a 72kDa protein. cryIV class genes (*cryIVA*, cryIVB, cryIVC, and cryIVD) as well as *cytA* were all isolated from the same 72-Mda plasmid present in *Bt* subsp. *israelensis.*

3. BACILLUS THURINGIENSIS SUBSP. ISRAELENSIS AS AN IMPORTANT PART OF MOSQUITO CONTROL

3.1 Properties of Bti

In 1975-76 under a World Health Organization sponsored project, a new *Bt* strain was discovered in the Negev desert in Israel by Goldberg and Margalit (1977). The strain was isolated from *Culex* sp. dead larvae mosquito. Later was identified as *Bt israelensis*, serotype H14 according to its fagellar antigenicity by de Barjac (1978).

Bti has all the features taxonomic, morphological, growth, sporulation, of isolation, cultivation of other varieties of *B. thuringiensis* (Fig.1). The larvicidal activity of *Bt israelensis* on mosquito transmitted diseases was the most important feature of the strain. The insecticidal properties of this bacteria are due primarily to insecticidal proteins produced during sporulation. The key proteins are Cyt1A(27.3·kDa), Cry4A (128·kDa), Cry4B (134·kDa) and Cry11A (72·kDa) and in three different inclusion types assembled into a spherical parasporal body held together by lamellar envelope (Ibarra & Federici, 1986). The inclusions are realtively small (0.1 to 0.5µm) and there are usually two to four inclusions per cell which vary in shape from cuboidal to bipyramidal, ovoid or anamorph (Charles & de Barjac 1982; Mikola *et al.* 1982; Yamamoto *et al.* 1983)

3.2 Mosquitoes, important vectors of tropical diseases

Mosquitoes are important vectors of several tropical diseases that suck blood from human and animals. They are vectors of multiple of diseases of man through transmision of pathogenic viruses, bacteria, protozoa and nematodes (Priest, 1992). From the medical point of view, mosquitoes are among the most important insects due their capacity to transmit human diseases such as malaria and dengue.

Table 2. Some diseases transmitted by mosquito (Rawlins, 1989; Walsh 1986).

Vector	Disease
Anopheles	Malaria, lymphatic filariasis
Culex	Lymphatic filariasis, Japanese encephalitis, other viral diseases
Aedes	Yellow fever, dengue, dengue hemorrhagic fever, other viral diseases, lymphatic filariasis
Mansonia	Lymphatic filariasis

There are about 3000 species of mosquito, of which about 100 are vectors of human diseases. Some of the more important of these diseases are listed in Table 2. It is estimated two billion people worldwide living in areas where these are endemic (World Health Organization, 1999). Thus, there is an urgent need for new agents and strategies to control these diseases.

3.3 Susceptibility of mosquito species to B. thuringiensis serotype israelensis

Bti is highly pathogenic against *Culicidae* (mosquitoes) and *Simuliidae* (blackflies), and has some virulence against certain others Diptera, especially Chironomidae (midges). Mosquito have four distinct stages in their life cycle: egg, larva, pupa and adult. Depending on the specie a female lays between 30 and 300 eggs at a time on the surface of the water, singly (*Anopheles*), in floating rafts (*Culex*) or just above the water line or on wet mud (*Aedes*). Once hatched the larvae grow in four different stages (instars). The first instar measures 1.5 mm in length, the fourth instar about 8-10 mm. The fully grown larvae then changes into a comma shaped pupa. When mature, the pupal skin splits at one end and a fully developed adult emerges. The entire period from egg to adult takes about 7-13 days under good conditions (Wada, 1989). First instar is more suceptible to *Bti* than fourth instar (Mulla *et al.* 1990). Pupa does not feed and therefore is not affected by *Bti*. For almost all species tested, increasing age of the larvae resulted in reduced susceptibility in mosquito (Chen et al. 1984: Mulla *et al.* 1985). *Bti* was found to be specific toxic to larvae of 109 mosquito species (Table 3).

Table 3. Larvicidal activity of *Bti* on mosquito species (Glare & O´Callaghan 1998).

Mosquito genus	Species
Aedes	40
Anopheles	27
Culex	19
Culiseta	5
Mansonia	5
Psorophora	3
Armigeres	3
Toxorhynchites	2
Limatus	2
Trichophospon	1
Uranotaenia	1
Tripteroides	1
	Total 109

Among mosquitoes, different preparations of *Bti* have shown different levels of toxicity to host species. Others factors influencing the susceptibility of mosquito larva to *Bti*. For example, the effect of a given dosage of toxin could produce different results depending on weather the lethal dose is administered all at once or in same doses over a long period (Aly et al. 1988).

Bti has an LC50 in the range of 10–13 ng/ml against the fourth instar of many mosquito species (Federici *et al.* 2003). Generally, *Culex* and *Aedes* are highly susceptible while *Anopheles* are less suceptible, but can be killed with *Bti* (Balaraman *et al.* 1983). Much higher concentrations of *Bti* are required to induce mortality in anopheline larvae than in *Aedes aegypti* larvae (Goldberg & Margalit 1977; Nugud & White 1982).

The exception is *A. franciscanus* that is as susceptible as other genera (García *et al.* 1980, Sun *et al.* 1980). Furhtermore, even within one genus, some species are more suceptible than others (Chui et al. 1993). Sun et al. (1980) suggested that a difference in feeding behavior might account for differences in susceptibilities. Filtering rates vary between genera and species. For example Aly (1988) showed that in the absence of *Bti*, larvae cleared the suspensions with constant relative filtration rates of 632 (*Ae. aegypti*), 515 (*Cx. quinquefasciatus*) or 83.9 μL/Larvae/h (*An. albimanus*). Another factor to be considered for the suceptibility to *Bti* is the behavior of the larvae. *Anopheles* larvae filter-feed on food particles present at the surface of water or a few centimeters below it, where as *Culex* and *Aedes* larvae not only feed faster but are capable of filter-feeding at much deeper water dephts (Aly et al. 1988).

Mosquito species which are not filter feeders do not seem suceptible. For example, against *Culicoides occidentalis* (Colwell 1982) and *Coquillettidia perturbans*, *Bti* larvicides had no effect (Walker et al 1985). Solubilization of the protein reduced the toxicity dramatically (50- to - 100 fold) this is attributed to reduced level of toxin ingestion by larvae owing to their filter feeding behaviour (Chungjatupornchai et al. 1988).

3.4 Susceptibility of mosquito species to purified Bt proteic crystals

Several efforts have been made to purify the proteins Cry of *Bti* and other *Bt* subps. with the purpose of studying the chemistry of Cry proteins, the synergism between them, and the effect of each crystal on larvae of different mosquito species. It is somewhat difficult to separate the *Bt* spores and crystals because they are of similar size and surface characteristics. For that reason several methods have been used to purify *Bt* crystals proteins. Using NaBr gradients (Chang et al. 1992), sucrose gradients (Debro et al. 1986), renografin gradients (Aronson et al. 1991), and in a separatory funnels (Delafield et al. 1968). Bioassays of purified Cry proteins have been allowed to know that not only *Bti* Cry proteins have activity on mosquito larva. Other purified Cry proteins of other *Bt* strains have also activity on mosquito larvae. Table 4 shows the mosquito toxicity of purified crystals of some *Bt* strains.

Table 4. Mosquitocidal activity of Cry and Cyt proteins (Modified after "Zeigler, 1999").

Name	Source Strain	Mosquito toxicity
Cry4Aa1	*B.t. israelensis* 4Q2-72	*Aedes agypti, Anopheles stephensi, Culex pipiens* (Diptera: Cuclidae)
Cry4Ba1	*B.t. israelensis* 4Q2-72	*Aedes agypti* (Diptera: Cuclidae)
Cry10Aa1	*B.t. israelensis* ONR60A	*Aedes agypti,* (Diptera: Cuclidae)
Cry11Aa1	*B.t. israelensis* HD-567	*Aedes agypti, Anopheles stephensi, Culex pipiens* (Diptera: Cuclidae)
Cry11Ba1	*B.t. jegathensan* 367	*Aedes agypti, Anopheles stephensi, Culex pipiens* (Diptera: Cuclidae)
Cry11Bb1	*B.t. medellin*	*Aedes agypti, Anopheles albimanus, Culex quinquefasciatus* (Diptera: Cuclidae)
Cry16Aa1	*Clostridium bifermentans malasya* CH18	*Aedes agypti, Anopheles stephensi, Culex pipiens* (Diptera: Cuclidae)
Cry19Aa1	*B.t. jegethesan*	*Anopheles stephensi, Culex pipiens* (Diptera: Cuclidae)

Name	Source Strain	Mosquito toxicity
Cry20Aa1	*B.t. fukuokaensis*	*Aedes agypti,* (Diptera: Cuclidae)
Cry21Aa1	*B.t. higo*	*Culex pipiens molestus* (Diptera: Cuclidae)
Cyt1Aa1	*B.t. israelensis* IPS82	*Aedes agypti, Anopheles stephensi, Culex pipiens* (Diptera: Cuclidae)
Cyt1Ab1	*B.t. medellin* 163-131	*Aedes agypti, Anopheles stephensi, Culex pipiens* (Diptera: Cuclidae)
Cyt2Aa1	*B.t. kyushuensis*	*Aedes agypti, Anopheles stephensi, Culex pipiens* (Diptera: Cuclidae)

3.5 Resistance

One major problem with insects control via chemical insecticides is the evolution in insects of resistance to those insecticides. The use of *Bti* on biological control of mosquitoes has no resulted in the development of resistence in host populations. Laboratory attemps to induce resistance by continual exposure to *Bti* have generally failed to detect resistence (Lee & Chong 1985; Georghiou & Wirth 1997).

The lack of resistance development to *Bti* could be due to its complex mode of action, involving synergistic interactions between up to four proteins (Becker & Maragrit 1993). Use of a single protein from *Bti* for mosquito control resulted in resistance after only a few generations in the laboratory (Becker and Margalit 1993). Georghiou & Wirth (1997) also showed that resistance could be raised in only a few generations when single *Bti* toxin was used (i.e. Cry 4Aa, 4Ba, 10Aa or 11Aa), and was progressively more difficult to raise in mosquitoes with combinations of two, and three toxins. When all four *Bti* toxins were used, resistance incidence was remarkably low. On the other hand Wirth et al. (2005) have shown that the lack of resistance in *Bti* is due to the presence of the Cyt1Aa protein in the crystal. For example, *Culex quinquefascitus* populations resistant to CryIVA, Cry4B and Cry11A have been obtained in the laboratory but not mosquito larvae resistant to Cry and Cyt1Aa proteins (Georghiou & Wirth 1997).

3.6 Synergism

Bti produces four crystal proteins Cry (4Aa, 4Ba, 10Aa, and 11Aa) and two Cyt (1Aa and 2Ba) (Guerchicoff *et al.* 1997). No single crystal component is as toxic as the intact crystal complex (Chan *et al.* 1993; Wu *et al.* 1994; Chilcott & Ellar, 1998). One possible explanation for this is that two or more proteins act synergisticall, yielding a higher activity than would be expected on the basis of the specific toxicity of the individual protein (Finney, 1971). For example, the toxicity against mosquito larvae of Cyt1Aa is lower compared to each of the four Cry proteins (Crickmore et al. 1995). However, cytA can potentiate the activity of the toxins and synergistic interactions that seems to account for the high toxicity of the *Bti* strains (Delecluse et al. 1993).

Tabashnick (1992) proposed a method to measure synergistic effect. Using the proposed method discuss the data reported by two authors: in bioassays with *A. aegypti* larvae, Wu and Chang (1985) found that mixtures of the 27- and 65-kDa proteins from *B. thuringiensis* subsp. *israelensis* were more toxic than expected on the basis of their individual toxicities, however, Chilcott and Ellar (1988) concluded from their own data that no synergism between these two proteins occurred. With this new interpretation of Tabashnick method, both studies support the same conclusions: positive synergism between the 27-kDa protein (CytA) and either of the CryIV proteins (65 and 130 kDa) and no such synergism between CryIV proteins (65 and 130 kDa). Other studies have been carried out with the aim of increasing the synergistic activity of *Bti.* Ramirez-Suero et al. (2011) evaluated the synergistic effect of *S. griseus* and *Bt aizawai* chitinases with *Bt israelensis* spore-toxin complex against *Aedes aegypti* larvae. The synergistic factor values according to Tabashnik (1992) method were 2 and 1.4, respectively.

3.7 Effect of Bti on no-target organisms

Bti has no direct effect on aquatic organisms other than mosquitoes, blackflies and chironomids. Other aquatic organisms, such as shrimps, mites and oysters are generally unaffected (Glare & O'Callaghan, 1998). This large safety margin of preparations of *Bti* for non-target organisms indicate their suitability for mosquito control programs in areas where protection of the natural ecosystem is important (Sinegre et al. 1980).

Several authors have reviewed the non target effects of *Bti* (Becker & Margalit 1993; Lacey & Mulla 1990). Field applications have often been monitored for effects on non-target organisms but no significant non-target effects have been reported (Ali, 1981; Jackson et al. 1994; Hershey et al. 1995).

4. PRODUCTION OF BTI BY FERMENTATION

4.1 Culture medium for Bti production

Commercial production of *Bti* is performed using culture media based on complex nutrients sources. The main purpose of the fermentation is to obtain high quantities of *Bti* crystals. The *Bti* parasporal crystal can account for up to 25% of the sporulated cell dry weight. To optimize the cry production it is necessary to have a suitable culture medium because the toxicity obtained at the end of the fermentation depend on the culture medium and operating conditions. The culture media that have been reported in the literature for high growth and sporulation can be used for any variety of *Bt*. Not always a high cell growth ensures an elevated Cry protein production or an increased insecticidal activity. Various culture mediums have been used for high growth and sporulation of *Bt* in the laboratory: 2XSG medium (Leighton & Doi, 1971), PA medium (Thorne, 1968), G-Tris medium (Aronson et al. 1971), CDGS medium (Nakata, 1964). Other media with inexpensive substrates have been reported by Pearson & Ward (1988), Smith (1982), Foda et al. (1985), Dulmage et al. (1970), Salama et al. (1983), Goldberg et al. (1980).

4.2 Factors affecting Cry production

There are several factors that influence the production of crystals: (1) Carbon source. Glucose is the most appropiate carbon source either for high *Bt* growth and sporulation (Smith, 1982). When glucose has been exhausted in the fermentation, the abscence of this can trigger sporulation. The use of one or other carbon source affects the biological activity and the morphology of the crystals (Dulmage, 1970). (2) Nitrogen source. An appropiate source of aminoacids provides high growth rates and high sporulation of *Bt* strains. Its absence delays sporulation and low yield in Cry proteins (Goldberg et al. 1980) (3) Carbon:Nitrogen ratio. Higher C:N rates glucose do not deplete at the end of fermentation and biomass yield decrease. Several authors have recomended a carbon nitrogen ratio of 7.5:1. (Salama et al. 1983; Foda et al. 1985) (4) Oxygen. High aeration rates are important for high spore and toxin formation. As k_La, increase biomass and Cry protein formation are increased (Rowe & Margaritis, 1987)(5) pH. Optimum pH for *Bt* growth is 6.8-7.2. If pH rises to 9.0 Cry protein can be dissolved (6) Temperature. Optimum temperature of *Bt* is 28-32°C. Higher temperatures favours plasmid losses or *Bt* mutants (Rowe & Margaritis, 1987).

5. MOLECULAR BIOLOGY OF BTI

5.1 Bti cry and cytA genes

All *Bt* strains contain extrachromosomal DNA. *Bt* strains are well known for its numerous plasmids ranging in size from 1.5 MDa to 130 MDa. Plasmids have been found in each variety examined and the plasmid profiles appear to be strain specific. However these plasmid profiles depend on the media type and growth rate of the strain and can be readily gained or lost (Federici, 1999). The mere presence of plasmids in *Bt* does not prove that they are involved in crystal formation, many non-crystalliferous bacteria also contain plasmids The *cry* genes are located on large plasmids although some Cry genes have been reported on the chromosome (Baume & Malvar, 1995). As mentioned earlier, *Bti* produces four different Cry proteins: CryIVA, CryIVB, Cry11A, and the cytolytic CytA protein (Hoffe and Whiteley, 1989). The Cry proteins are codified by *cryIVA*, *cryIVB*, *cry11A*, and *cytA* genes, respectively. These genes responsible for the toxicity of *Bti* have been sequenced by various researchers (Table 5).

Table 5. *cry* **and** *cytA* **genes DNA sequences of** *Bti.*

Gene name	GenBank Accesion No.	Coding Region	Reference
cryIVAa1	Y00423	1-3540	Ward &Ellar, 1987
cry4Aa2	D00248, E01676	393-3935	Sen *et al.* 1988
cry4Ba1	X07423, X05692	157-3564	Chungjatupornchai *et al.* 1988
cry4Ba2	X07082	151-3558	Tungpradubkul *et al.* 1988
cry4Ba3	M20242	526-3930	Yamamoto *et al.* 1988
cry4Ba4	D00247, E01905	461-3865	Sen *et al.* 1988
cry11Aa1	M31737	41-1969	Donovan *et al.* 1989
cry11Aa2	M22860	1-235	Adams *et al.* 1989
cyt1Aa1	X03182	140-886	Waalwijk *et al.* 1985
cyt1Aa2	X04338	509-1255	Ward & Ellar, 1986
cyt1Aa3	Y00135	36-782	Earp a& Ellar, 1987
cyt1Aa4	M35968	67-813	Galjart *et al.* 1987

In *Bti* the elements responsible of the toxicity against mosquito larvae are located in a large plasmid of 72 MDa (125 kb) and contribute to the formation of a complex parasporal body (Aronson 1993). Figure 2 shows the partial map of the *Bti* 125 kb plasmid.

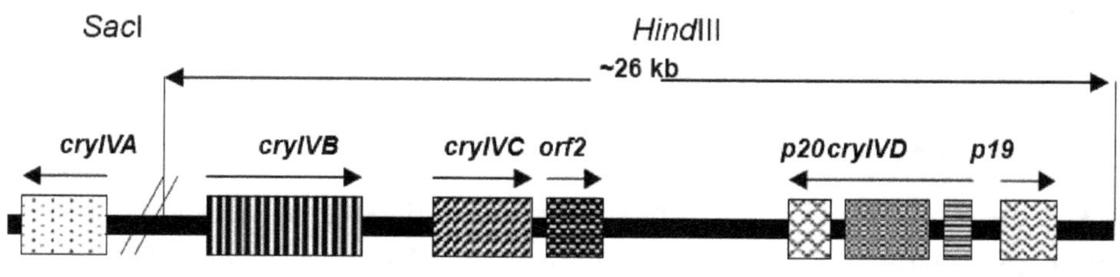

Fig. 2. Map of 26kb DNA fragment carrying the DNA genes responsibles of the toxicity of *Bti* **(Modified after Ben-Dov** *et al.* **1996).**

5.2 Bti operon of cryIVD gene

The DNA sequence indicated that *cryIVD* gene is the second gene of an operon which includes three genes. A gene that encodes a 19 kDa polypeptide, *cryIVD* gene and a gene that encodes a 20 kDa polypeptide (Dervyn et al. 1995). Transcription of *cryIVD* gene in *Bti* is induced 9 h after the beginning of the sporulation. DNA sequence analysis and potential promoters are recognized by the RNA polimerase associated with the σ35 and σ28 , specific sigma subunit of the RNA polymerase genes related with control of sporulation of *B. thuringiensis*. *cryIVA* and *cryIVB* promotors are activated in the mid-sporulation phase (Ben- Dov et al. 1996), *cryIVA* is regulated by the σ35 RNA polymerase gene and *cryIVB* is under control of σ35 (Yoshisue et al. 1994). *cytA* gene is transcribed by two promotors, pBtI and pBtII, regulated by the RNA polymerase σ35 and σ28, respectively (Brown & Whiteley, 1988, 1990). These results have demostrated that *cryIVD* transcription is subjected to σ35 regulation.

5.3 Identication of Bacillus thuringiensis pesticidal crystal genes by PCR

The polymerase chain reaction (PCR) is a molecular tool widely used to amplify a single copy or a few copies of a piece of DNA across several orders of magnitude, generating thousands to millions of copies of a particular DNA sequence. The identification of *Bt* toxin genes by PCR can partially predict the insecticidal activity of a given strain. Several studies have reported that the type of *cry* and *cyt* genes present in a *Bt* strain correlates to some extent with its insecticidal activity (Porcar & Juarez-Perez). Thus, the identification of the gene content in a *Bt* strain can be used to predict its insecticidal potential. The PCR-based identication of *B. thuringiensis* cry genes was first developed by Carozzi et al. (1991), who introduced this technique as a tool to predict insecticidal activity. They found correspondence with the toxicity predicted on the basis of the amplification product profiles. Carozzi et al. proposed PCR as an accurate, fast methodology for the identification of novel strains and the prediction of insecticidal activity of new isolates, and they also forecast the possible use of PCR for the discovery of previously unknown cry genes. highly conserved regions and recognizing entire cry gene subfamilies are often used in a preliminary screening prior to performing a second PCR with specific primers. The primers used to amplify *cry4A, cry4B, cry11A*, and *cytA* genes have been designed by various researchers and are shown in Table 6.

Table 6. PCR primers pairs and the cry or cyt genes they amplify of ***Bti.*** **Sources: Ben-Dov** ***et al.*** **(1997)*, Carozzi et al. (1991)**, Bravo** ***et al.*** **(1998)*****

Direct	Sequences (5′→3)	Reverse Sequences(5′→3)	Amplifies/Product (bp)
EE-4A(d)	GGGTATGGCACTC AACCCCACTT*	Un4(r) GCGTGACATACCCATTTCCAGGTCC*	cry4A/1529
Dip2A(d)	GGTGCTTCCTATTC TTTGGC**	Dip2B TGACCAGGTCCCTTGATTAC**	*cry4A*/1290
EE-4B(d)	GAGAACACACCTA ATCAACCAACT*	Un4(r) GCGTGACATACCCATTTCCAGGTCC*	*cry4B*/1951
EE-11A(d)	CCGAACCTACTAT TGCGCCA*	EE-11A(r)CTCCCTGCTAGGA TTCCGTC*	*cry11A*/445
gral cyt1(d)	AACCCCTCAATCA ACAGCAAGG***	gral cyt(r) GGTACACAATACATAACGCCACC***	cyt1/522-525

5.4 Expression of Bti genes in other strains

Expression of *Bti* genes either individually or in combination in crystal-negative *Bt* or other strains have being carried on by several researchers. The genes encoding these proteins have been expressed in *Caulobacter* (Thanabalu et al. 1992), cyanobacteria (Manasherob et al. 2002; Murphy & Stevens 1992), *Escherichia coli* (McLean & Whiteley 1987; Tanapongpipat et al. 2003), *Bacillus subtilis* (Ward et al. 1986), and *Bt* (Crickmore et al. 1990). However, *Bt* toxins have been expressed as active or inactive toxins, especially when expressed in *E. coli* (Ogunjimi et al. 2002).

Quintano et al. (2005) reported the expression of *cry11A* from *Bti* in *S. cerevisiae*. The *cry11A* gene was expressed as fusion proteins with gluthathione *S*-transferase under the control of the *S. cerevisiae HXK1* promoter. The protein was purified by affinity chromatography using gluthathione *S*-transferase–Sepharose beads. Insecticidal activity against third-instar *Aedes aegypti* larvae of the recombinant *S. cerevisiae* cell extracts (LC_{50} = 4.10μg protein/mL) and purified GST-cry11A fusion protein (LC_{50} = 4.10μg protein/mL) was detected in cells grown in ethanol.

Servant et al. (1999) constructed a recombinant *B. sphaericus* strain containing the *cry11A* gene from *Bti*. They found an LC_{50} for the cry11A protein of 1.175 μg/mL against fourth-instar *A. aegypti* larvae. Poncet *et al.* (1997) constructed a recombinant *B. sphaericus* strain with *cry11A* and *p20* genes integrated into the chromosome. In this case, the LC_{50} value reported against third- instar *A. aegypti* larvae was 0.023 μg/mL. Xu *et al.* (2001) studied the expression of cry11A and *cry11A+p20* gene cluster in recombinant *E. coli* and *Pseudomonas putida*. They found that both recombinant bacteria contained higher levels of Cry11A protein when the adjacent *p20* gene was present on the same DNA fragment. Yamagiwa et al. (2004) reported that the solubilized cry11A protein, obtained from a nonrecombinant *Bti*, was less toxic against *Culex pipiens* larvae than the crystal itself (LC50 of 0.267 and 0.008 μg/mL, respectively). These authors obtained 2 GST fusion proteins of 36 and 32 kDa from cry11A. The LC50 against *C. pipiens* larvae obtained using both proteins were 0.818 μg/mL. In other study with the purpose to preserve the toxicity of sunlight-sensitive Cry proteins, Manasherob et al. (2002) constructed a transgenic cyanobacterium *Anabaena* PCC 7120 to express the genes *cry4Aa*, *cry11Aa* and an accesory protein (*p20*) under control of two tandem strong promotors. Cyanobacterium *Anabaena* can multiply in breeding sites of mosquito larvae and serve as their food source. Higher toxicity against *Aedes aegypti* larvae was obtained in this study.

6. MODE OF ACTION OF BTI CRY PROTEINS AGAINST MOSQUITO LARVAE

6.1 Pore-forming-toxins

Bt Cry and Cyt toxins belong to a class of bacterial toxins known as pore-forming toxins (PFT) that are secreted as water-soluble proteins undergoing conformational changes in order to insert into, or to translocate across, cell membranes of their host (Bravo et al. 2007). In most cases, PFT are activated by host proteases after receptor binding inducing the formation of an oligomeric structure that is insertion competent. Finally membrane insertion is triggered, in most cases, by a decrease in pH that induces a molten globule state of the protein (Parker and Feil, 2005). Cry and Cyt proteins are PFT proteins. Both proteins are solubilized in the gut of suceptible diperans insects and proteolytic activated by midgut proteases. For the Cry 11Aa protoxin, proteolytic activation involves amino-terminal processing and intramolecular cleavage leading to two fragments of 36 and 32 kDa that remain associated and retain insect toxicity.

6.2 Mechanism of action of Bti toxins

An acepted model for Cry toxin action against mosquito larvae is that it is a multistage process. (i) Ingestion of Cry protein by the larvae (ii) Solubilization of the crystals in the alkaline midgut (iii) Proteolytic activation of the insecticidal solubilised protein (iv) Toxin binds to receptors located on the apical microvellus membrana of ephitelial midgut cell walls (v) Alter the toxin binds the receptor, it is though that there is a change in the toxin conformation allowing toxin insertion into the membrane (vi) Electrophysiological and biochemical evidence suggest that the toxins generate pores in the cell membrane, thus disturbing the osmotic balance, consecuently the cells swell and lyse (vii) The gut becomes paralyzed and the insect stops feeding. Most mosquito larvae die whitin few hours of ingestion, generally cease feeding within 1 hour, show reduced activity by two hours and die six hours after ingestion (Chicott et al. 1990; Marrone & Macintosh, 1993).

Several authors have studied the mechanisms of action of the δ-endotoxin of *Bti* on mosquito larvae. Thomas and Ellar (1983) found that δ-endotoxin active against mosquito larvae was inactivated by prior incubation with lipids extracted from ephitelial midgut *Aedes albopictus* cells. They reported that toxin binds to membrane lipids (phosphatidyl choline, sphingomyelin and phosphatidyl ethanolamine). According to their results, they proposed a mechanism in which the interaction of toxin with membrane lipids causes a detergent-like rearrangement of the lipids and as a consequence cytolisis. Others authors have corroborated these results: Cyt protein, unlike Cry toxins, do not recognize specific binding sites and do not bind to protein receptors, directly interact with membrane lipids inserting into the membrane and forming pores (Knowles *et al.* 1989; Promdonkoy & Ellar, 2000) or destroying the membrane by a detergent-like interaction (Butko, 2003).

The high efficacy of *Bti* is because of the production of multiple toxins with different modes of action. Perez *et al.* (2005) reported that Cyt1Aa protein functions as a receptor for the Cry11Aa toxin and suggest that this interaction explains the synergism between the Cyt1A and Cry11A proteins. Further, the Cyt proteins in *Bti* synergize the toxic effect of Cry11A and Cry4 toxins and, even more, suppresses the resistance to these Cry toxins (Wirth *et al.* 1997).

7. FORMULATION OF BTI TOXINS

7.1 Potency in Bti formulations

Formulation is a preparation of an insecticide for a particular application method. Formulation plays an important rol in determining final virulence. The vast majority of the formulations of *Bt* have been developed to control agricultural and forest pests, mainly *Lepidoptera*. However, the feeding habits of *Lepidoptera* are different to feeding habits of *Diptera*. Mosquito larvae feed by filtering water and concentrate organic particles. Product formulations based on *Bti* should consider the mosquito larvae habits and the environmental conditions, promote that Cry proteins retain their toxic activity and promote that the larva have access to them. Products based on strains of *Bti* are given a potency based on bioassays on third or fourth instar mosquito larvae. Bioassays are conducted using 6 to 7 dilutions of the toxin by duplicate in 100 mL cups containing 20 third instar *Aedes aegypti* larvae. Duplicate cups with 20 mosquito larvae in 100 mL of deionized water without test material serves as a control (McLaughlin *et al.* 2004). Concentration-mortality data are obtained, transformed to a log-probit scale, and potency is obtained by comparing the estimated LC_{50} of a test substance with that of a standard with a known potency (de Barjac 1985). The international standard recognized for *Bti* is IPS82. LC_{50} and LC_{90} are the dose require to kill 50 and 90 percent of the mosquito larvae of a tested population after 24h tested duration. Each sample is bioassayed at least 3 times on various days and the results are average values. LC_{50} and LC_{90} are measured in micrograms or milligrams of material per liter, or parts per million (ppm).

For potency calculations, it is used the international recognized standard for mosquito assay, IPS82 (15 000 ITU/mg) provided by Institute Pasteur, Paris, France. Standard vials are kept at -18°C.

Product potency is calculated by the Abbott (1925) formula:

$$\text{Potency (A)} = \text{Potency (std)}\ LC_{50}\ \text{(std)}/LC_{50}\ \text{(A)}$$

where (a) is the product and (std) is the standard.

The size of the particle could be a factor that influences the potency of the toxin. A product with small particles is more homogenously distributed in the water than a product with larger particles and small particles which sink slower. Changes in LC_{50} are not necesary regarded to reflect changes in amount of toxin, but could be a function of particle size/distribution (Skovmand et al. 1997). Change of particle size also change LC_{50} . Decreasing particle increased LC_{50} and thus decrease the calculated potency. The slope measured between LC_{50} and LC_{90} values should have high value. Higher value will require a smaller quantity to kill a greater number of larvae. The slope of the dosage-mortality curve is in function of the heterogeneity of the product effect. If product availability is in function of particle sizes, particles with broad ranges of particle size distribution will also have low slopes.

7.2 Bti formulations

A variety of *Bti* formulations have been studied for mosquito control under laboratory and field conditions. *Bti* fluid formulations are not stable in heat and high humidity and cannot be stored for months under tropical conditions (McGuirre et al. 1996) . In many cases, and especially in areas exposed to the sun, the residual effect is very short (Leong et al. 1980) and the product has to be reapplied. Photoinactivation seems to be one of the major environment factors affecting the stability of *Bti* delta-endotoxin (Morris, 1983). Yu-Tien et al. (1993) reported that *Bti* completely lost its toxicity to mosquito larvae when exposed to irradiation at 253nm. Poszgay et al. (1987) showed that exposure of *B. thuringiensis* toxin to 40 h of ultraviolet light irradiation resulted in lost activity. Cry proteins inactivation by the solar radiation is the result of the destruction of the tryphtophan (Pusztai et al. 1991). Research and development efforts are focusing on formulations to avoid the ultraviolet light effect. Ramirez-Lepe et al. (2003) encapsulated *Bti* spore-toxin within aluminum/ carboxymethylcellulose using green malachite, congo red or ponceau red as photoprotective agents against ultraviolet light in lab conditions. The encapsulated form of the *Bti* spore- toxin complex with photoprotectors avoided the limitation in controlling mosquito larvae caused by ultraviolet light. Yu-Tien et al. (1993) achieved photoprotection of the spore-toxin complex by addition of melanin.

Other *Bti* formulations have been developed. For example, Ramirez-Suero et al. (2005) evaluated maltodextrin, nixtamalized corn flour and corn starch for entraping active materials in *Bti* spore-toxin complexes dried by aspersion. Dried products had water activity values below 0.7 suggesting that the formulations are long shelf-life because keep the products without microorganisms for longer periods and increases the larval feeding and as a consequence have higher activity against mosquito larvae. Fig. 3 shows the scanning electron microscope of spray/ dried formulations with grits and nixtamalized corn flour. It shows smoth spherical grits and corn particles entrapping the spore/toxin ingredient. Manasherob et al. (1996) encapsulated Bti in the protozoan Tetrahymena pyriformis and the activity against A. stephensi was enhanced 8 times when exposed to protozoan cells filled with Bti compared to exposed to the same concentration of Bti alone. Elcin (1995) encapsulated Bti in alginate microcapsules and increase its stability and its effect against Culex sp larvae. Another alternative to protect Bti crystals from ultraviolet light is obtaining mutants that protectrs Bti crystal. Hoti & Balaraman (1993) obtained a mutant of Bti that produced a brown pigment during sporulation, the pigment was identified as melanin. Other Bti formulations have been developed. Culigel superabsorbent polymer controlled-release system for the slow release of Bti to mosquito larvae (Levy et al. 1990). Combination of chemical and biological agents susch as insect growth regulator s- methoprene + Bti (Bassi et al. 1989). Sprayed-dried Bti powder as a fizzy tablet (Skovmand & Eriksen 1993). Floating bait formulations designed to improve the effect of bacterial toxins, especially against Anopheles spp (Aly et al. 1987).

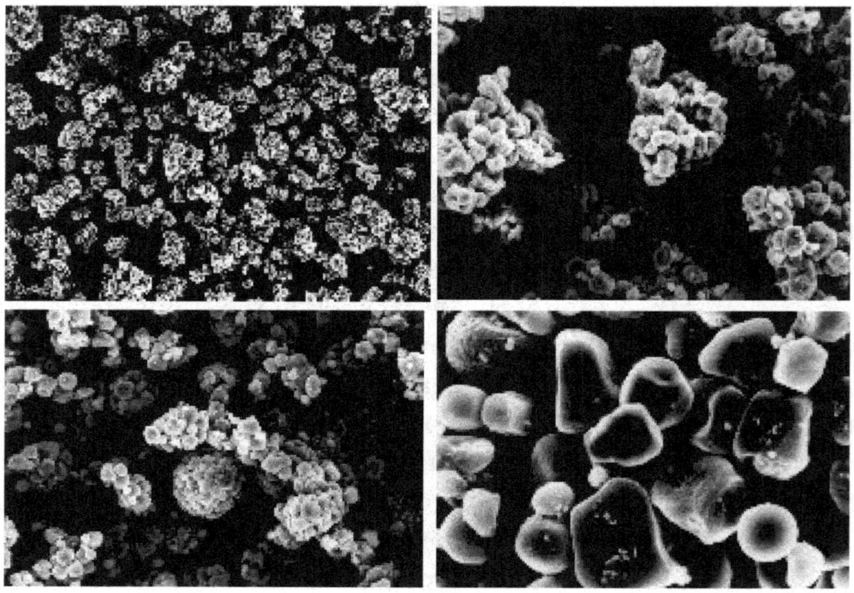

Fig. 3. Scanning electron micrograph of spray/dried formulation of *Bti* spore/ toxin complex with Grits a) 379X and b)1500X and Nixtamalized corn flour c) 370X and d) 1500X.

Highly concentrated liquid formulations are available for control of floodwater mosquitoes while formulations which float for as long possible have been developed for use in fast- flowing or turbulent waters. Formulations which settle and persist at the bottom are required for buttom feeders. Granules which float on the surface are the most effective against surface feeders such as *Anopheles* spp. (Mulla et al. 2004).

Briquettes for mosquitoes with continual successive generations like *Culex* spp (Becker & Margalit 1993; Kase & Branton 1986). Briquetts or pellets, in particular seem to be useful for overcoming lack of persistence, which is one of main limitations of *Bti.* Granules using plant, such corn (maiz) grits or clay carriers are particular useful in aerial application to breeding sites with dense foliage as salt marshes or rice fields. Sustained release formulations such as floating briquettes or semi submersible pellets are designed to provide long-lasting larvicidal activity in containers or small ponds. Ingestion of the toxin depends on the rate of feeding, the rate at which the toxin falls to the bottom of the pool and becomes inaccesible, and competition to ingestion from other suspended organic materials. In turbid and polluted waters the rate of application needs to be at least two-fold greater than in clear water (Mulla et al. 1985). The feeding habits of mosquito larvae influence formulation design. *Anopheles* larvae are surface feeders and ingest particulate material from water surface such as yeast or flour and filter feed poorly. This has led to the development of formulations that present the toxin at, or just below, the water surface and such prepartions are particulary effective against certain *Anopheles* larvae (Aly et al. 1987)

8. FIELD APPLICATION

One of the major drawbacks in the use of *Bti* is its rapid inactivation (24-48 h) in the environment (Mulla et al. 1993). Thus larvae populations of stagnant water mosquitoes recover within 5-7 days following treatment. Therefore the use of *Bti* is limited by the low efficacies of current preparations under field conditions (Tyanum & Mulla 1999). Since there is a little persistence of the toxin further applications are necessary to effect continuous control. Formulation and applications techniques can extend the persistence of activity for over one month in some situations, but activity remains sensitive to factors like UV degradation. In areas exposed to the sun, the residual effect is very short (Leong et al. 1980) and the product has to be reapplied. Other factors that affect the toxicity of *Bti* are particle sedimentation (Rushed & Mulla, 1989), protein adsorption onto silt particles, organic matter, elevated temperatures (Ohana et al. 1987), consumption by other organisms to which the toxin is not lethal (Blaustein and Margalit, 1991), dissolved tannins (Lord & Undeen, 1990) and inactivation by sunlight. Other factors that have been observed are that *Bti* does not recycle, under simulated field conditions, is unable to germinate and multiply in mud at the botton of pools although it did remain viable viable for up to 22 days, and higher water depth where applied decrease its activity (Ohana et al. 1987).

Early reports showed that a primary powder formulations of *Bti* had virtually no residual effect against mosquito larvae beyond application (Aguilar-Meza et al. 2010), although the delta-endotoxin remained chemically stable in neutral and acid waters (Sinegre et al 1980). Extend persistence with *Bti* is possible through use of improved formulations. Gunasekaran *et al.* (2002) tested a floating sustained release formulation of *Bti* in polluted water habitats against *Culex quinquefasciatus* larvae. Briquetts may result in more prolonged control than liquid formulations as these products have greater persistence through slow release (Kase & Branton 1986).

On the other hand, Aguilar-Meza et al. (2010) tested the residual insecticidal activity after field exposure of an aluminum-carboxymethylcellulose microencapsulated formulation of *Bti* spore-toxin complex with malachite green as photoprotective agent. The formulation improved the activity against *Aedes aegypti* larvae for 30 days and was comparable to that of the chemical insecticide temephos.

9. CONCISE SUMMARY

Bti is a bacterium that has been effectively utilized in biological control initiatives targeting mosquito and fly larvae *globally. This review examines several aspects of the study, which will provide fresh insights to enhance effectiveness in biological control.*

10. REFERENCES

Abbott, W.S. (1925). A method of computing the effectiveness of an insecticide. *Journal of Economic Entomology*, Vol. 18, pp. 265-267.

Adams, L.F.; Visick, J.E. & Whiteley H.R. (1989). A 20-kilodalton protein is required for efficient production of the *Bacillus thuringiensis* subsp. *israelensis* 27-kilodalton crystal protein in *Escherichia coli*. *Journal of Bacteriology*, Vol 171, pp. 521-530

Aguilar-Meza, O.; Ram rez-Suero, M.; Bernal, J.S. & Ram rez-Lepe, M. (2010). Field evaluation against *Aedes aegypti* larvae of aluminum-carboxymethylcellulose-encapsulated formulation of *Bacillus thuringiensis* serovar *israelensis*. *Journal of Economic Entomology*, Vol.103, pp. 570–576

Aronson, A.I. Angelo, N. & Holt, S.C. (1971). Organic Nutrients Required For Growth and Sporulation of *Bacillus Cereus*. *Journal of Bacteriology*, Vol.106, p.p.1016-1025.

Ali, A. (1981). *Bacillus thuringiensis* serovar. *israelensis* (ABG-6108) against chironomid midges and some nontarget invertebrates. *Journal of Invertebrate Pathology* Vol 38, pp. 264–272.

Aly, C.; Mulla, M.S. & Schnetter, W. (1987). Floating formulations increase activity of *Bacillus thuringiensis* var. *israelensis* against *Anopheles* larvae. *Journal of the American Mosquito Control Association* Vol 3, pp. 583–588.

Aly, C.; Mulla, M.S.; Xu, B.Z. & Schnetter, W. (1988). Rate of ingestion by mosquito larvae (Diptera:*Culicidae*) as a factor in the effectiveness of a bacterial stomach toxin. *Journal of Medical Entomology*, Vol 25, pp.191–196.

Aoki, K & Chigasaki, Y. (1915) Uber des toxin von sog. Sotto-bacillen Mitt. Med. Fak. Kais. Univ. Tokyo. Vol. 14, pp. 59-80

Aronson, A.I.; Han, E.S; McGaughey, W & Johnson, D. (1991). The solubility of inclusion proteins from *Bacillus thuringiensis* is dependent upon protoxin composition and is a factor in toxicity to insects. *Applied and Environmental Microbiology*, Vol 57, pp.981-986

Ash, C. & Collins, M.D. (1992). Comparative analysis of 23s ribosomal RNA gene sequences of *Bacillus anthracis* and emetic *Bacillus cereus* determined by PCR-direct sequencing. *FEMS Microbiology Letters,* Vol 73, pp. 75-80

Bassi, D.G.; Weathersbee, A.A.; Meisch, M.V. & Inman, A. (1989). Efficacy of Duplex and Vectobac against *Psorophora columbiae* and *Anopheles quadrimaculatus* larvae in Arkansas ricefields. *Journal of the American Mosquito Control Association,* Vol 5, pp. 264-266.

Bauer, L.S. (1995). A threat of the insecticidal crystal protein of *Bacillus thuringiensis. Florida Entomologist*, Vol 78, pp. 414-443

Baum, J. A. & T. Malvar, T. (1995). Regulation of insecticidal crystal protein production in *Bacillus thuringiensis. Molecular Microbiology,* Vol 18, pp. 1-12.

Bartlet, B.R. (1964). Integration of Chemical and Biological Control. In: *Biological Control of insect pests and weeds*, P. DeBach, (Ed.), 489-511, Chapman & Hall, London.

Becker, N. (2000). Bacterial control of vector-mosquitoes and blackies. In *Entomopathogenic Bacteria: From Laboratory to Field Application,* (Ed.) J.F. Charles, A. Delécluse and C. Nielsen-LaRoux), pp. 383-398. Dordrecht, The Netherlands: Kluwer.

Becker, N. & Margalit, J. (1993). Use of *Bacillus thuringiensis israelensis* against mosquitoes and blackflies. In *Bacillus thuringiensis, an EnvironmentalBiopesticide: Theory and Practice,* P. F. Entwistle, J. S. Cory, M. J. Bailey and S. Higgs (Ed.)), pp. 147-170. John Wiley & Sons, New York.

Ben-Dov, E.; Zaritsky, A.; Dahan, E.; Barak, Z.; Sinai, R.; Manasherob, R.; Khamraev, A.; Troitskaya, E.; Dubitsky, A.; Berezina, N. & Margalith, Y. (1997). Extended screening by PCR for seven *cry*-group genes from field-collected strains of *Bacillus thuringiensis. Applied and Environmental Microbiology*, Vol 63, pp. 4883-4890

Bravo, A.; Sarabia, S.; Lopez, L.; Oniveros, H.; Abarca, C.; Ortiz, A.; Ortiz, M.; Lina, L.; Villalobos F.J.; Peña, G.; Nuñez-Valdez, M.E.; Soberón, M. & Quintero. R. (1998). Characterization of *cry* genes in a Mexican *Bacillus thuringiensis* strain collection. *Applied and Environmental Microbiology,* Vol 64, pp. 4965-4972

Bravo, A.; Gill, S.S. & Soberon, M. (2007). Mode of Action of *Bacillus thuringiensis* Cry and Cyt toxins and their potential for insect control. *Toxicon,* Vol.49, No.4 (March 2007). pp.423-435

Brown, K. L. & Whiteley, H.R. (1988). Isolation of a *Bacillus thuringiensis* RNA polymerase capable of transcribing crystal protein genes. *Proceedings of the National Academy of Sciences,* Vol 85, pp. 4166–4170.

Brown, K. L., & H. R. Whiteley. (1990). Isolation of the second *Bacillus thuringiensis* RNA polymerase that transcribes from a crystal protein gene promoter. *Journal of Bacteriology,* Vol 172, pp. 6682–6688.

Butko, P. (2003). Cytolytic toxin Cyt11A and its Mechanism of Membrane Damage: Data and Hypotesis. *Applied and Environmental Microbiology.* Vol69, No.5, (May 2003) pp 2415-2422

Carozzi, N. B.; Kramer, V.C.; Warren, G.W.; Evola, S. & Koziel, M.G. (1991). Prediction of insecticidal activity of *Bacillus thuringiensis* strains by polymerase chain reaction product profiles. *Applied and Environmental Microbiology,* Vol 57, pp. 3057-3061

Crickmore, N.; Zeigler, D.R.; Feitelson, J.; Schnepf, E.; Vanrie, J.; Lereclus, D.; Baum, J. & Dean, D.H. (1998). Revision of the nomenclature for the *Bacillus thuringiensis* pesticidal crystal proteins. *Microbiology and Molecular Biology Reviews,* Vol 62, pp. 807-813

Crickmore, N.; Nicholls, C.; Earp, D.J.; Hodgman, T.C.; & Ellar, D.J. (1990). The construction of *Bacillus thuringiensis* strains expressing novel entomocidal ä-endotoxin combinations. *Journal of Biochemistry,* Vol.270, pp. 133–136.

Chang, C.; Dai, S.M Frutos, R.; Federici, B.A. & Gill, S.S. (1992). Properties of a 72-kilodalton mosquitocidal protein from *Bacillus thuringiensis* subsp. morrisoni PG-14 expressed in *Bacillus thuringiensis* subsp. *kurstaki* by using the shuttle vector pHT3101. *Applied and Environmental Microbiology,* Vol 58, pp. 507-512

Charles, J. F. & deBarjac, H. (1982). Sporulation et crystal- logenese de *Bacillus thuringiensis* var *israelensis* en microscopie electronique.*Annals of Microbiology* (Inst.Pasteur)133A:425-442.

Chen, M.L.& Tsen H.Y. (2002). Discrimination of Bacillus cereus and Bacillus thuringiensis with 16S rRNA and gyrB gene based PCR primers and sequencing of their annealing sites. *Journal of Applied Microbiology.* Vol.92, pp. 912–919

Chilcott, C. N. & Ellar, D.J. (1988). Comparative toxicity of *Bacillus thuringiensis* var. *israelensis* crystal proteins *in vivo* and *in vitro. Journal of General Microbiology,* Vol.134, pp.2552-2558

Chilcott, C.N.; Knowles, B.H.; Ellar, D.J. & Drobniewski, F.A. (1990). Mechanism of Action of *Bacillus thuringiensis israelensis* Parasporal Body. In *Bacterial Control of Mosquitoes and Blackfles,* eds de Barjac, H. & Sutherland, D.J. pp. 45-65. New Brunswick : Rutgers University Press.

Chungjatupornchai, W.; Höfte, H.; Seurinck, J.; Angsuthanasombat, C. & Vaeck, M. (1988). Common features of *Bacillus thuringiensis* toxins specific for *Diptera* and *Lepidoptera. European Journal of Biochemistry,* Vol.173, pp.9-16 de Barjac, H. & Bonnefoi, A. (1973). Mise au point sur la classification des *Bacillus thuringiensis. Entomophaga.* Vol 18, pp. 5-17

de Barjac, H. 1978. A new subspecies of *Bacillus thuringiensis* very toxic for mosquitoes; *Bacillus thuringiensis* serotype H-14. *Compte Rendu Académie Sciences Paris Series D.* Vol. 286 pp.797-800.

de Barjac, H. & Frachon, E. (1990). Classification of *Bacillus thuringiensis* Strains Entomophaga. Vol. 35, pp. 233-240.

de Barjac, H. & Larget-Thiery, I. (1985). Characteristics of IPS82 standard for bioassay of *B. thuringiensis* H-14 preparations. WHO Rep/VBC/84.892.

Dervyn, E.; Poncet, S.; Klier, A. & Rapoport, G. (1995). Transcriptional regulation of the *cryIVD* gene operon from *Bacillus thuringiensis* subsp. *israelensis. Journal of Bacteriology,* Vol.177, pp.2283-2291

Donovan, W. P.; Dankocsik, C. & Gilbert, M.P. (1988). Molecular characterization of a gene encoding a 72-kilodalton mosquito-toxic crystal protein from *Bacillus thuringiensis* subsp. *israelensis. Journal of Bacteriology,* Vol.170, pp.4732-4738

Dulmage, H.T. (1970). Production of the spore- endotoxin complex by variants of *Bacillus thuringiensis* in two fermentation media. *Journal of Invertebrate Pathology,* Vol.16, pp.385-389.

Earp, D.J. & Ellar, D.J. (1987). *Bacillus thuringiensis* var. *morrisoni* strain PG14: nucleotide sequence of a gene encoding a 27 kDa crystal protein. *Nucleic Acids Research,* Vol 15, pp. 3619-3624

Dervyn, E.; Poncet, S.; Klier, A,; Rapoport, A.; Yoshisue, H.; Nishimoto, T.; Sakai, H. & Komano, T. (1994). Identification of a promoter for the crystal protein-encoding gene *cryIVB* from *Bacillus thuringiensis* subsp. *israelensis. Gene* Vol.137, pp.247–251

Debro, L.; Fitz-James, P.C. &Aronson, A. (1986). Two different parasporal inclusions are produced by *Bacillus thuringiensis subsp. finitimus. Journal of Bacteriology,* Vol.58, pp. 507-512

Delafield, F.P.; Somerville, H.J. & Rittenberg S.C. (1968). Immunological homology between crystal and spore protein of *Bacillus thuringiensis. Journal of Bacteriology,* Vol.96 pp.713-720

Dunkle, R.L., & B.S. Shasha. 1989. Response of starch encapsulated *Bacillus thuringiensis* containing ultraviolet screens to sunlight. Environ. Entomol. 18: 1035-1041.

Drobniewski, F.A. (1993) Onchocerciasis and the Onchocerciasis Control Programme. *Microbiology Europe* Vol.1, pp. 26-28,

Elcin, Y.M.; Cokmus, C. & Sacilik, C.S. (1995). Aluminum carboximethylcellulose encapsulation of *Bacillus sphaericus* 2362 for control of *Culex* spp. (Diptera: *Culicidae*) larvae. *Journal of Economic Entomology* Vol.88 pp. 830-834

Federici, B.A. (1999). *Bacillus thuringiensis* in Biological Control. . In: *Handbook of Biological Control.* T. Fisher (Ed.)Academic Press (Ed.) 575-593, ISBN 10: 0-12-257305-6

Federici, B.A.; Park, H.W.; Bideshi, D.K.; Wirth M.C. & Johnson, J.J. (2003). Review. Recombinant Bacteria for Mosquito Control. *The Journal of Experimental Biology*, Vol.206, pp.3877-3885

Finney, D.J. (1971). Probit Analysis. Cambridge Univ. Press, Cambridge, United Kingdom.

Foda M.S.; Salama, H.S. & Selim, M. (1985). Factors afecting growth physiology of *Bacillus thuringiensis. Applied Microbiology and Biotechnology*, Vol.22, pp.50-52.

Galjart, N. J.; Sivasubramanian, N. & Federici, B.A. (1987). Plasmid location, cloning and sequence analysis of the gene encoding a 23-kilodalton cytolytic protein from *Bacillus thuringiensis* subsp. *morrisoni* (PG-14). Current Microbiology, Vol.16, pp. 171-177

Glare T.R. & O'Callaghan M. (1998). Environmental and Health Impacts of *Bacillus thuringiensis israelensis. Report for the Ministry of Health.* Biocontrol & Biodiversity, Grasslands Division, Ag. Research. Lincoln, New Zealand.

Goldberg, L. J. & Margalit, J. (1977). A bacterial spore demonstrating rapid larvicidal activity against *Anopheles sergentii, Uranotaenia unguiculata, Culex univitattus, Aedes aegypti* and *Culex univitattus. Mosquto News,* Vol.37, pp. 355–358

Georghiou, G.P. & Wirth, M.C. (1997). Influence of Exposure to Single versus Multiple Toxins of *Bacillus thuringiensis* subsp. *israelensis* on Development of Resistance in the Mosquito *Culex quinquefasciatus* (Diptera: *Culicidae*). *Applied and Environmental Microbiology,* Vol.63, No. 3, pp.1095-1101

Goldberg, I.; Sneh, B.; Battae, E. & Klein, D. (1980). Opt i mi zat i on o f a medium for a high product i on o f spore-crystal preparat i on o f *Bacillus thuringiensis* effective against Egyptian cotton leaf worm *Spodoptera littorallis. Biotechnology Letters* Vol.2, pp. 419- 426.

Hannay, L.C. (1953). Cristalline inclusion in aerobic spore forming bacteria. Nature. Vol. 172, pp. 1004.

Hannay, C.L. & Fitz-James P.C. (1955). The protein crystals of *Bacillus thuringiensis Berl. Canadian Journal of Microbiology.* Vol 1, pp. 694-710.

Hershey, A. E., Shannon, L., Axler, R., Ernst, C. & Mickelson, P. (1995). Effects of methoprene and *Bti* (*Bacillus thuringiensis* var. *israelensis*) on non-target insects. *Hydrobiologia,* Vol. 308, pp.219-227

Höfte, H. & Whiteley, H.R. (1989). Insecticidal crystal proteins of *Bacillus thuringiensis*. *Microbiology Reviews* Vol.53 pp. 242-255.

Hoti, S.L, & Balaraman, K. (1993). Formation of melanin-pigment by a mutant of *Bacillus thuringiensis* H-14. *Journal of General Microbiology* Vol.139, pp. 2365-2369.

Husz, B. (1931). Experiments during 1931 on the use of *Bacillus thuringiensis Berl.* in controlling the corn borer. *Science Reports International of Corn Borer Investigation*, Vol. 4, pp. 22-23.

Ibarra, J. E. & Federici, B. A. (1986). Isolation of a relatively nontoxic 65-kilodalton protein inclusion from the parasporal body of *Bacillus thuringiensis*subsp. *israelensis*. *Journal of Bacteriology*, Vol.165, pp.527-531

Ignofo, C.M., Shasha, B.S. & Shapiro, M (1991). Sunlight ultraviolet protection of the *Heliothis* nuclear polyhedrosis virus through starch encapsulation technology. *Journal of Invertebrate Pathology*, Vol 57: 134-136.

Ishiwata, S. (1901). On a kind of severe flacherie (sotto disease). *Dainihon Sanshi Kaiho* Vol.114, pp.1-5

Iwabushi, H. (1908) Popular silkworm Pathology A.A. Meibundo (Ed.) Tokyo, pp. 428.

Kase, L.E. & Branton, L. (1986). Floating particle for improved control of aquatic insects. U.S. Pat. 4631857

Knowles, B. H.; Blatt, M.R.; Tester, M.; Horsnell, J.M.; Carroll, J.; Menestrina G. & Ellar, D.J. (1989). A cytolytic delta-endotoxin from *Bacillus thuringiensis* var. *israelensis* forms cation-selective channels in planar lipid bilayers. *FEBS Letters*, Vol.244, pp.259–262

Krieg V.A.; Huger A.M.; Langenbruch G.A. & Schnetter W. (1983). *Bacillus thuringiensis* var. *tenebrionis*: Ein neuer, gegenüber lair ven von coleopteren wirksamer pathotyp. *Z. Äng. Ent.* Vol.96, pp. 500-508

Jacobs, S.E. (1950).Bacteriological control of the flour moth *Ephestia kuhniella*. *Proceedings of the Society of Apply Bacteriology*, Vol.13, pp. 83-91.

Jackson, J. K.; Sweeney, B. W.; Bott, T. L.; Newbold, J. D. & Kaplan, L. A. (1994). Transport of *Bacillus thuringiensis* var. *israelensis* and its effect on drift and benthic densities of nontarget macroinvertebrates in the Susquehanna River, northern Pennsylvania. *Canadian Journal of Fisheries and Aquatic Sciences* Vol.51, pp.295-314

Johnson, M.W. & Tabashnik B.E. (1999). Enhanced Biolgical Control Through Pesticide Selectivity. In: *Handbook of Biological Control.* T. Fisher (Ed.)Academic Press (Ed.) 297-317, ISBN 10: 0-12-257305-6

Lacey L.A.; Urbina, M.J. & Heitzzman, C.M.(1989). Sustained release formulations of *Bacillus sphaericus* and *Bacillus thuringiensis* (H-14) for control of container-breeding *Culex quinquefasciatus. Mosquito News* Vol.44, pp.26-32.

Lacey, L. A. & Mulla, M. S. (1990). Safety of *Bacillus thuringiensis* var. *israelensis* and *Bacillus sphaericus* to nontarget organisms in the aquatic environment. In *Safety of Microbial Insecticides.* Laird-M, Lacey-LA and Davidson-EWSO (Ed.), pp. 259 pp. CRC Press Inc.; Boca Raton; USA.

Leighton, T.J. & Doi, R.H.(1971). The Stability of Messenger Kibonucleic Acid during Sporulation in *Bacillus subtilis. The Journal of Biological Chemistry.* Vol.246, pp.3189- 3195.

Leong, K.L.; Cano, R.J. & Kubinski, A.M (1980). Factors affecting *Bacillus thuringiensis* total field persistence. *Environmental Entomology,* Vol.9 pp.593-599.

Levy, R.; Nichols, M.A. & Miller, T. W. (1990). Culigel superabsorbent polymer controlled- release system: application to mosquito larvicidal bacilli. *Proceedings and Abstracts, Vth International Colloquium on Invertebrate Pathology and Microbial Control, Adelaide, Australia* , 107

Lord, J.C. & Undeen A.H. (1990). Inhibition of the *Bacillus thuringiensis* var. *israelensis* toxin by dissolved tannins. Environmental Entomology, Vol.19 pp.1547-1551

Manasherob, R., Ben-Dov, E., Margalit, J., Zaritsky, A. & Barak, Z. (1996). Raising activity of *Bacillus thuringiensis* var. *israelensis* against *Anopheles stephensi* larvae by encapsulation in *Tetrahymena pyriformis* (Hymenostomatidia: Tetrahymenidae). *Journal of the American Mosquito Control Association,* Vol.12, pp. 627-631

Manasherob, R.; Ben-Dov, E.; Xiaoqiang, W.; Boussiba, S., & Zaritsky, A. (2002). Protection from UV-B damage of mosquito larvicidal toxins from *Bacillus thuringiensis* subsp. *israelensis* expressed in *Anabaena* PCC 7120. *Current Microbiology,* Vol 45, pp. 217- 220.

Marrone, P.G. & MacIntosh, S.C. (1993). Resistance to *Bacillus thuringiensis* and Resistance Management. *In: Bacillus thuringiensis*, an Environmental Biopesticide: Theory and

Practice, Entwistle, P. F., Cory, J. S., Bailey, M. J., and Higgs, S., Eds., John Wiley and Sons, Chichester, UK, pp.221-235.

McDonald B.A. & Linde C. (2002). Pathogen Population Genetics, Evolutionary Potential, and Durable Resistance. *Annual Review of Phytopathology,* Vol. 40, pp. 349-379

McGuirre, M.R.; Shasha, B.S.; Eastman, C.E. & Sagedchi H.O. (1996). Starch and fluor-based sprayable formulations: Effect on rainfastness and solar stability of *Bacillus thuringiensis. Journal of Economic Entomology,* Vol.89, pp. 863-869

McLaughlin, R.E.; Dulmage, H.T.; Alls, R.; Couch, T.L.; Dame, D.A.; Hall, I.M.; Rose, R.I. &.Versol, P.L. (1984). U.S. standard bioassay for the potency assessment of *Bacillus thuringiensis* var. *israelensis* serotype H-14 against mosquito larvae. *Bulletin of the Entomological Society of America*, Vol.30, pp. 26-29

McLean, K.M. & Whiteley, H.R. (1987). Expression in *Escherichia coli* of a cloned crystal protein gene of *Bacillus thuringiensis* subsp. *israelensis*. *Journal of Bacteriology*, Vol.169, pp. 1017–1023

Mattes, O. (1927) Parasitare kronkheiten der mehlmotten larven und versuche uber ihre Verwendborkeit als biologische Bekampfungsmittel. *Sisher Ges. Beforder. Ges Naturw. Marburg.* Vol.62., pp. 381-417

Metalnikov, S.; Hengula, B. & Strail, D.M. (1930). Experiments on the application of bacteria against the corn borer. *Science Reports International Corn Borer Investigation,* Vol3, pp. 148-151.

Mikkola, A. R.; Carlberg, G.A.; Vaara, T. & Gyllenberg, H.G. (1982). Comparison of inclusions in different *Bacillus thuringiensis* strains. An electron microscope study. FEMS Microbiology Letters, Vol.13, pp.401-408

Mitani, K & Watari, J. (1916). A new method to isolate the toxin of *Bacillus sotto* Ishiwata by passing through a bacerial filter and a preliminary report on the toxic action of this toxin to the silkworm larvae. *Archi Gensanshu Serzojo Hokoku.* Vol 3, pp. 33-42.

Metcalf, R.L. (1986). The Ecology of Insecticides and the Chemical Control of Insects. *In Ecological Theory and Integrated Pest Managment in Practice*, M. Kogan (Ed.) 251 297 John Wiley & Sons, New York.

Morris, O.N. (1983). Protection of *Bacillus thuringiensis* from inactivation from sunlight. *Canadian Entomology*, Vol 115, pp.1215-1227.

Mulla, M. S. & Darwazeh, H.A. (1985). Efficacy of formulations of *Bacillus thuringiensis* H-14 against mosquito larvae. *Bulletin fo the Society of Vector Ecology.* Vol.10, pp.14–19

Mulla, M. S.; Darwazeh, H.A.; Ede, L.; Kennedy, B. & Dulmage H.T. (1985). Efficacy and field evaluation of *Bacillus thuringiensis* (H-14) and *B. sphaericus* against floodwater mosquitoes in California. *Journal of American Mosquito Control Association* Vol1, pp.310–315.

Mulla, M. S.; Darwazeh, H.A. & Zgomba, M. (1990). Effect of some environmental factors on the efficacy of *Bacillus sphaericus* 2362 and *Bacillus thuringiensis* (H-14) against mosquitoes. *Bulletin of the Society of Vector Ecology,* Vol.15, pp.166–175.

Mulla, M. S.; Chaney, J.D. & Rodchareon, J. (1993). Elevated dosages of *Bacillus thuringiensis* var. *israelensis* fail to extend control of *Culex* larvae. *Bulletin of the Society of Vector Ecology*, Vol.18, pp.125–132.

Mulla, M. S.; Thavara, U.; Tawatsin, A. & Chompoosri, J. (2004). Procedures for the evaluation of field efficacy of slow-release formulations of larvicides against *Aedes aegypti* in water-storage containers. *Journal of the American Mosquito Control Association* Vol.20, pp.64–73

Nakata, H.M. (1964). Organic Nutrients Required For Growth And Sporulation Of *Bacillus Cereus*, *Journal of Bacteriology*, Vol.88, pp.1522-1524;

Ohana, B.; Margalit, J. & Barak, Z. (1987). Fate of *Bacillus thuringiensis* subsp *israelensis* under simulated field conditions. *Applied and Environmental Microbiology*, Vol.53, pp.828- 831.

Ogunjimi, A.A.; Chandler, J.M.; Gbenle, G.O.; Olukoya, D.K., & Akinrimisi, E.O. (2002). Heterologous expresión of *cry2* gene from a local strain of *Bacillus thuringiensis* isolated in Nigeria. *Biotechnology Applied Biochemistry*, Vol.36, pp.241–246

Pearson, D. & Ward, O.P. (1988). Effect of culture conditions on growth and sporulation of *Bacillus thuringiensis* susp. *israelensis* and development media for production of the protein crystal endotoxin. *Biotechnology Letters*, Vol.10, pp.451-456.

Perez, C.; Fernandez, L.E.; Sun, J.; Folch, J.L.; Gill, S.S. & Soberon M. (2005*). Bacillus thuringiensis* subsp. *israelensis* Cyt1Aa synergizes Cry11Aa Toxin by Functioning as a Membrane-Bound Receptor. *Proceedings of the National Academy of Sciences.* Vol 102, No. 51. pp. 18303-18308

Poncet, S.; Bernard, C.; Dervyn, E.; Cayley, J.; Klier, A.; & Raport, G. (1997). Improvement of *Bacillus sphaericus* toxicity against dipteran larvae by integration, via homologous recombination, of the *cry11A* toxin gene from *Bacillus thuringiensis* subsp. *israelensis*. *Applied and Environmental Microbiology,* Vol.63, pp 4413–4420

Porcar, M. & Juarez-Perez, V. (2003). PCR-based identification of *Bacillus thuringiensis* pesticidal crystal genes. *FEMS Microbiology Reviews*. Vol. 26, pp. 419-432.

Pozsgay, M.; Fast. P.; Kaplan, H. & Carey, P.R. (1987). The effect of sunlight on the protein crystals from *Bacillus thuringiensis* var. *kurstaki* HD1 and NRD12: A raman spectroscopy study. *Journal of Invertebrate Pathology,* Vol.50 pp.620-622

Priest, F.G. (1992). A review. Biological control of mosquitoes and other biting flies by *Bacillus sphaericus* and *Bacillus thuringiensis*. *Journal of Applied Bacteriology,* Vol.72, pp.357-369.

Promdonkoy, B. & Ellar, D.J. (2000). Membrane pore architecture of a cytolytic toxin from *Bacillus thuringiensis*. *Journal of Biochemistry*.Vol. 350, pp.275–282.

Pusztai, M.; Fast, P.; Gringorten, L., Kaplan, H.; Lessard, T. & and Carey, P.R. (1991). The mechanism of sunlight-mediated inactivation of *Bacillus thuringiensis* crystals. *Journal of Biochemistry*, Vol.273, pp.43-47

Quintana-Castro, R.; Ramírez-Suero, M.; Moreno-Sanz, F. & Ramírez-Lepe, M. (2005). Expression of the *cry11A* gene of *Bacillus thuringiensis* ssp. *israelensis* in *Saccharomyces cerevisiae,.Canadian Journal of Microbiology, Vol.*51, pp.165-170

Ram rez-Lepe, M.; Aguilar, O.; Ram rez-Suero, M. & Escudero, B. (2003). Protection of the spore-toxin complex of *Bacillus thuringiensis* serovar *israelensis* from ultraviolet irradiation with aluminum-CMC encapsulation and photoprotectors. *Southwestern. Entomology,* Vol.28, pp.137-143.

Ramírez-Suero, M.; Robles-Olvera, V. & Ramírez-Lepe, M. (2005). Spray-dried *Bacillus thuringiensis* serovar *israelensis* formulations for control of *Aedes aegypti* larvae. *Journal of Economic Entomology*, Vol.98, pp.1494-1498

Ram rez-Suero, M.; Valerio-Alfaro, G.; Bernal, J.S. & Ramirez-Lepe M. (2011). Synergisitic effect of chitinases and *Bacillus thuringiensis israelensis* spore-toxin complex against *Aedes aegypti* larvae. *The Canadian Entomologist* Vol,143: pp. 157–164.

Rawlins, S.C. (1989) Biological control of insect pests affecting man and animals in the tropics. *CRC Critical Reviews in Microbiology* Vol.16, pp.235–252

Rowe, G.E. & Margaritis, A.M. (1987). Bioprocess Developments in the Production of Bioinsecticides by *Bacillus thuringiensis. CRC Critical Reviews in Biotechnology.* Vol.6, No.4, pp. 87-127.

Rushed, S.S. & Mulla, M.S. (1989). Factors influencing ingestion of particulate materials by mosquito larvae (Diptera: Culicidae). *Journal of Medical Entomology,* Vol.26, pp. 210- 216

Salama, H.S.; Foda, M.S. & Dulmage H.T. (1983). Novel Media for Production of δ-endotoxins form *Bacillus thuringiensis. Journal of Invertebrate Pathology,* Vol.41, pp.8- 19.

Sen, K.; Honda, G.; Koyama, N.; Nishida, M.; Neki, A.; Sakai, H.; Himeno, M. & Komano, T. (1988). Cloning and nucleotide sequences of the two 130 kDa insecticidal protein genes of *Bacillus thuringiensis* var. *israelensis. Agriculture Biology Chemistry*, Vol.52, pp.873-878

Servant, P.; Rosso, M.L.; Hamon, S.; Poncet, S.; Delécluse, A. & Rapport, G. (1999). Production of cry11A and cry11B toxins in *Bacillus sphaericus* confers toxicity towards *Aedes aegypti* and resistant *Culex* populations. *Applied and Environmental Microbiology*, Vol.65, pp.3021–3026.

Sinegre, G.; Gaven, B.; Jullien, J. L. & Crespo, O. (1980). Effectiveness of the serotype H-14 of *Bacillus thuringiensis* against the principal species of man-biting mosquitoes on the Mediterranean coast of France. *Parassitologia* Vol.22, pp.223-231.

Skovmand, O. & Eriksen, A. G. (1993). Field trials of a fizzy tablet with *Bacillus thuringiensis* subsp. *israelensis* in forest spring ponds in Denmark. *Bulletin of the Society for Vector Ecology, Vol.* 18, pp.160-163.

Skovmand, O.; Hoeg, D.; Pedersen, H.S., & Rasmussen T. (1997). Parameters Influencing Potency of *Bacillus thuringiensis* var. *israelensis* Products. *Journal of Economic Entomology* Vol 90, No.2. pp. 361-369.

Smith, A.R. (1982). Effect of strain and medium variation on mosquit production by *Bacillus thuringiensis* var. *israelensis. Canadian Journal of Microbiology,* Vol.28, pp.1090-1092.

Tabashnik B.E. (1992). Evaluation of Synergism among *Bacillus thuringiensis* Toxins. *Applied and Environmental Microbiology,* Vol.58, pp.3343-3346

Tanapongpipat, S.; Luxananil, P.; Boonhiang, P.; Chewawiwat, N.; Audtho, M. & Panyim, S. (2003). A plasmid encoding a combination of mosquito-larvicidal genes from *Bacillus thuringiensis* subsp. *israelensis* and *Bacillus sphaericus* confers toxicity against broad range of mosquito larvae when expressed in Gram-negative bacteria. FEMS Microbiology Letters, Vol.228, pp.259–263

Thanabalu, T.; Hindley, J.; Brenner, S.; Oei, C. & Berry, C. (1992). Expression of the mosquitocidal toxins of *Bacillus sphaericus* and *Bacillus thuringiensis* subsp. *israelensis* by recombinant *Caulobacter crescentus*, a vehicle for biological control of aquatic insect larvae. *Applied and Environmental Microbiology,* Vol.58, pp. 905–910

Thomas, W.E. & Ellar, D.J. (1983) *Bacillus thuringiensis* var. *israelensis* crystal delta-endotoxin: effects on insect and mammalian cells *in vitro* and *in vivo. Journal of Cell Science* Vol.60, pp.181-197.

Thorne, C.B. (1968). Transducing Bacteriophage for *Bacillus cereus. Journal Of Virology,* Vol.2, pp. 657-662.

Tungpradubkul, S.; C. Settasatien, & Panyim, S. (1988). The complete nucleotide sequence of a 130 kDa mosquito-larvicidal delta-endotoxin gene of *Bacillus thuringiensis* var. *israelensis. Nucleic Acids Research* Vol.16, pp.1637-1638.

Tyanyun, S. & Mulla, M.S. (1999). Field evaluation of new water dispersible granular formulations of *Bacillus thuringiensis* subp. *israelensis* and *Bacillus sphaericus* against Culex mosquito in microcosms. *Journal of the American Mosquito Control Association* Vol 15 pp. 356-361.

Waalwijk, C.; A.; Dullemans, M.; VanWorkum, M.E.S. & Visser, B. (1985). Molecular cloning and the nucleotide sequence of the Mr28,000 crystal protein gene of *Bacillus thuringiensis* subsp. *israelensis. Nucleic Acids Research,* Vol.13, pp. 8207-8217

Walsh, J. (1986) River blindness, a gamble pays off. *Science*, Vol.232, pp.922–925.

Ward, E.S.; Ridley, A.R.; Ellar, D.J. & Todd, J.A. (1986). *Bacillus thuringiensis* var. *israelensis* endotoxin. Cloning and expression of the toxin in sporogenic and asporogenic strains of *Bacillus subtilis*. *Journal of Molecular Biology*, Vol.191, pp. 13–22

Ward, E.S., & Ellar, D.J. (1986). *Bacillus thuringiensis* var. *israelensis* delta-endotoxin: nucleotide sequence and characterization of the transcripts in *Bacillus thuringiensis* and *Escherichia coli*. *Journal of Molecular Biology*, Vol.191, pp.1-11

Ward, E. S., & Ellar, D.J. (1987). Nucleotide sequence of a *Bacillus thuringiensis* var. *israelensis* gene encoding a 130 kDa delta-endotoxin. *Nucleic Acids Research*, Vol 15, pp. 7195 Webb, G. (1992) The Onchocerciasis Control Programme. *Transactions of the Royal Society of Tropical Medicine and Hygiene,* Vol. 86, pp.113-114

Wirth, M. C.; Georghiou, G. P.; Federici, B. A. (1997). CytA enables CryIV endotoxins of *Bacillus thuringiensis* to overcome high levels of CryIV resistance in the mosquito, Culex quinquefasciatus. *Proceedings of the National Academy of Sciences,* Vol.94, pp.10536-10540

Wojciech J. J. & Korsten L. (2002). Biological Control of Postharvest Diseases of Fruits. *Annual Review of Phytopathology* Vol. 40, pp. 411-441.

World Health Organization (1999). *World Health Report.* Geneva: World Health Organization. Wu, D. & Chang, F.N (1985). Synergism in mosquitocidal activity of 26 and 65 kDa proteins from *Bacillus thuringiensis* subsp. *israelensis* crystal. FEBS Letters, Vol.190, pp.232-236

Xu,Y.; Nagai, M.; Bagdasarian, M.; Smith,T.W. & Walker, E.D. (2001). Expression of the *P20* Gene from *Bacillus thuringiensis* H-14 Increases Cry11a Toxin Production and Enhances Mosquito- Larvicidal Activity in Recombinant Gram-Negative Bacteria. *Applied And Environmental Microbiology*, Vol.67, p.p.3010–3015

Wada Y. (1989). Control of Japanese encephalitis vectors. *Southeast Asian journal of tropical medicine and public health,* Vol.20, pp. 623–626

Yamagiwa, M., Sakagawa, K., & Sakai, H. (2004). Functional Analysis of two processed fragments of *Bacillus thuringiensis* cry11A toxin. *Bioscience Biotechnology and Biochemistry*, Vol.68, pp.523–528

Yamamoto, T.; IWatkinson, I.A.; Kim, L.; Sage, M.V.; Stratton, R.; Akande, N.; Li, D.; D.P. & Roe, B.A. (1988). Nucleotide sequence of the gene coding for a 130-kDa mosquitocidal protein of *Bacillus thuringiensis israelensis*. *Gene*, Vol 66, pp.107-120.

Yamamoto, T.; Iizuku, T. & Aronson, J.N. (1983). Mosquitocidal protein of *Bacillus thuringiensis* subsp. *israelensis*: identification and partial isolation of the protein. *Current Microbiology* Vol.9, pp.279-284.

Yu-Tien, L.; Men-Ju, S.; Dar-Der, J.; Wang, W.I.; Chin-Chi, C. & Cheng-Chen, C. (1993). Protection from ultraviolet irradiation by melanin mosquitocidal activity of *Bacillus thuringiensis* var. *israelensis*. *Journal of Invertebrate Pathology,* Vol 62:131-136.

Zeigler, D.R. (1999). *Bacillus thuringiensis Bacillus cereus. Bacillus Genetic Stock Center.* Catalog of strains 7th. ed. Vol.2

Index

www.ingramcontent.com/pod-product-compliance
Lightning Source LLC
LaVergne TN
LVHW081939200926
848484LV00009B/2468